Handbook of Civil Engineering

Handbook of Civil Engineering

Edited by **Sarah Crowe**

CLANRYE
INTERNATIONAL

New Jersey

Published by Clanrye International,
55 Van Reypen Street,
Jersey City, NJ 07306, USA
www.clanryeinternational.com

Handbook of Civil Engineering
Edited by Sarah Crowe

International Standard Book Number: 978-1-63240-258-5 (Hardback)

Printed in the United States of America.

Contents

Preface

Civil engineering is a field of engineering sciences that deals with the study of designing, constructions, and maintenance procedures of all man-made and naturally built structures such as roads, bridges, dams, irrigation canals, building structures and monuments. This engineering discipline is considered as the second oldest engineering science, the oldest being military engineering. In fact, civil engineering is known as the gap between non military engineering and military engineering. This area of study stands as a pillar at the centre of many other engineering disciplines such as architecture science, environment science, structural engineering, forensic engineering, material engineering, construction engineering, etc. Civil engineering plays a crucial role in many sectors such as human development, government, private home owners to multinational companies among others.

The principles of Civil Engineering touch almost every aspect of our everyday lives. From the water we use to brush our teeth to the road we drive on, and the power that charges our cell phone, all involve the applications of Civil Engineering. The origins of Civil Engineering can be traced to Ancient Egypt between the duration 4000 and 2000 BC. This was the period when humans began abandoning their nomadic existence and started looking for alternate sources of shelter.

I would like to acknowledge the significant efforts made by the entire editorial team in compiling this work. Their significant effort is greatly appreciated and the high quality of this publication is a testament to their dedication. I especially wish to acknowledge the contributing authors, without whom a work of this magnitude would clearly not have been realizable. Not only do I appreciate their participation, but also their adherence as a group to the time parameters set for this publication.

Editor

Smoothed Particle Hydrodynamics Simulation of Wave Overtopping Characteristics for Different Coastal Structures

Jaan Hui Pu[1] and Songdong Shao[2]

[1] *School of Engineering, Nazarbayev University, 53 Kabanbay Batyr Avenue, Astana 010000, Kazakhstan*
[2] *School of Engineering, Design and Technology, University of Bradford, West Yorkshire BD7 1DP, UK*

Correspondence should be addressed to Jaan Hui Pu, jhpu@nu.edu.kz

Academic Editors: S. Chen and A. Li

This research paper presents an incompressible smoothed particle hydrodynamics (ISPH) technique to investigate a regular wave overtopping on the coastal structure of different types. The SPH method is a mesh-free particle modeling approach that can efficiently treat the large deformation of free surface. The incompressible SPH approach employs a true hydrodynamic formulation to solve the fluid pressure that has less pressure fluctuations. The generation of flow turbulence during the wave breaking and overtopping is modeled by a subparticle scale (SPS) turbulence model. Here the ISPH model is used to investigate the wave overtopping over a coastal structure with and without the porous material. The computations disclosed the features of flow velocity, turbulence, and pressure distributions for different structure types and indicated that the existence of a layer of porous material can effectively reduce the wave impact pressure and overtopping rate. The proposed numerical model is expected to provide a promising practical tool to investigate the complicated wave-structure interactions.

1. Introduction

Many types of breakwaters have been developed for the purpose of shore and harbor protections. The common goal of such structures is to reduce the wave height and energy to an acceptable level in the coastal areas. When the structure is made of porous materials, additional wave energy is dissipated inside the structure due to the flow friction within the porous media. The wave overtopping of coastal structure has always been of great interest and many studies have been carried out to evaluate the flow overtopping discharge for different breakwater designs. For example, the European Overtopping Manual (EurOtop 2007) [1] provides a very comprehensive and practical tool for evaluating the wave overtopping for different sea defenses and has been widely used in the engineering field with sufficient accuracy. Besides, many other experimental and theoretical studies have also been performed to study the wave-breakwater interactions, including Brossard et al. [2], Muttray and Oumeraci [3], and Chen et al. [4].

Numerical modeling based on the Navier-Stokes (N-S) equations has the advantage of including the irregular seabed geometries, inhomogeneous porous media, nonlinear waves, and nonlinear friction forces. They are capable of calculating the flows inside the complex geometries to disclose very refined information about the velocity, pressure, turbulence, transport property, and so forth. The numerical models based on the 2D N-S type equations and the Reynolds averaged N-S (RANS) equations are possibly the most common to the study of wave-structure interactions and wave overtopping for engineering purposes, as the computational efforts are reasonably small, and the number of simplifying assumptions is considerably reduced as compared to other existing models. The relevant works include Qiu and Wang [5], Liu et al. [6], Huang and Dong [7], and Garcia et al. [8].

In this paper, we propose an incompressible smoothed particle hydrodynamics (ISPH) model to study the wave interaction and overtopping for different breakwater designs. The SPH method was originally developed for the astrophysics by Monaghan [9] and recently commonly used to the fluid flows [10]. One of the great advantages of the SPH to model free surface flows is that the particles move in a Lagrangian coordinate, and the advection is directly calculated by the particle motion. Thus, free surfaces can be conveniently and accurately tracked by the particles without numerical diffusion, which is usually encountered in the traditional Eulerian approach. In the early simulations of fluid flows by the weakly compressible SPH [10], the incompressibility was realized through an equation of state so that the fluid was assumed to be slightly compressible. In this case, a large sound speed has to be introduced, which could easily cause problems of sound wave reflections at the solid boundaries, and the high sound speed could lead to the crippling CFL time-step constraint. In the ISPH conception [11, 12], the pressure is not a thermodynamic variable obtained from the equation of state, but obtained by way of solving a pressure Poisson equation derived from a semi-implicit algorithm. It has been demonstrated that both the computational efficiency and stability could be improved in the ISPH due to that a relatively larger time step can be used, and the particle fluctuation is reduced [13].

Here we use the ISPH model to study the wave overtopping of a coastal structure with different characteristics: vertical and sloping walls, with and without the protection of porous materials. The flow velocity field, turbulence, and pressure distributions will be compared for the different designs to evaluate their performance. Finally, the flow overtopping discharges will be validated against the available data published in the literature [6]. It is worth to mention that many of the state-of-the-art wave overtopping simulations have been carried out by researchers using either the mesh-based or mesh-free methods, such as in [14–18].

2. Principles of Incompressible SPH Model

2.1. Governing Equations. Following the work of Shao [11], the Lagrangian form of governing equations is used in the ISPH. In an SPH framework, the mass and momentum equations for the flow field are represented as follows:

$$\frac{1}{\rho}\frac{d\rho}{dt} + \nabla \cdot \mathbf{u} = 0, \tag{1}$$

$$\frac{d\mathbf{u}}{dt} = -\frac{1}{\rho}\nabla P + \mathbf{g} + v_0 \nabla^2 \mathbf{u} + \frac{1}{\rho}\nabla \cdot \vec{\tau}, \tag{2}$$

where ρ = density, t = time, $\mathbf{u}$ = velocity, P = pressure, $\mathbf{g}$ = gravitational acceleration, v_0 = laminar viscosity, and $\vec{\tau}$ = turbulence stress. It is noted that both (1) and (2) are written in the form of a full derivative on the left side of equations to enable an SPH formulation.

The turbulence stress $\vec{\tau}$ in (2) needs to be modeled to close the equation. In Liu et al. [6], the effect of turbulence is modeled by an improved $k - \varepsilon$ model. Here a simple and effective eddy-viscosity-based subparticle scale (SPS) turbulence model originally developed by Gotoh et al. [19], which has been widely used in both the coastal and river hydrodynamics, is used to model the turbulence stress as:

$$\frac{\tau_{ij}}{\rho} = 2v_T S_{ij} - \frac{2}{3}k\delta_{ij}, \tag{3}$$

where v_T = turbulence eddy viscosity, S_{ij} = strain rate of the mean flow, k = turbulence kinetic energy, and δ_{ij} =

Kronecker's delta. We use the widely adopted Smagorinsky model [20] to calculate the turbulence eddy viscosity v_T as follows:

$$v_T = (C_s \Delta X)^2 |S|, \tag{4}$$

where C_s = Smagorinsky constant, which is taken as 0.1 in the paper, ΔX = particle spacing, which represents the characteristic length scale of the small eddies, and $|S| = (2S_{ij}S_{ij})^{1/2}$ is the local strain rate.

To apply the above numerical model for the flows inside the porous materials, it is generally not easy to solve the N-S equations directly inside the pores. Thus, by following Gotoh and Sakai [21], the effect of a permeable layer is addressed by taking into account the additional external forces, namely, the drag forces, into the momentum equation (2). The drag forces due to the existence of the permeable layer can be written as follows:

$$\mathbf{F} = -\frac{3C_D}{4\Delta X}|\mathbf{u}|\mathbf{u}, \tag{5}$$

where C_D = drag coefficient due to the existence of porous materials. Shimizu and Tsujimoto [22] estimated the range of values of the drag coefficient to be between 1.0 and 1.5, based on the experiment of flow inside a permeable layer made by the vertical cylinders. In the current paper, a value of 1.25 was taken but we did not test the sensitivity of the value.

Although much more advanced porous flow treatment has been given in Shao [23], it was found that the above simple formulation can well address many kinds of the porous flows with enough accuracy. Besides, this approach was also successfully employed by Gotoh and Sakai [21] to study the plunging wave breaking on a permeable slope using the moving particle semi-implicit (MPS) modeling approach.

2.2. ISPH Solution Procedures. Following Shao [11], the ISPH model employs a two-step prediction/correction solution approach similar to the two-step projection method of Chorin [24] for solving the Navier-Stokes equations.

The prediction step is an explicit integration in time without enforcing the incompressibility. In this step, only the gravitational force, viscous/turbulence, and resistance forces in (2) and (5) are used and an intermediate particle velocity and position are obtained as:

$$\Delta \mathbf{u}_* = \left(\mathbf{g} + v_0 \nabla^2 \mathbf{u} + \frac{1}{\rho}\nabla \cdot \vec{\tau} - \frac{3C_D}{4\Delta X}|\mathbf{u}|\mathbf{u}\right)\Delta t,$$

$$\mathbf{u}_* = \mathbf{u}_t + \Delta \mathbf{u}_*, \tag{6}$$

$$\mathbf{r}_* = \mathbf{r}_t + \mathbf{u}_*\Delta t,$$

where $\Delta \mathbf{u}_*$ = increment of particle velocity during the prediction step, Δt = time increment, $\mathbf{u}_t$ and $\mathbf{r}_t$ = particle velocity and position at time t, and $\mathbf{u}_*$ and $\mathbf{r}_*$ = intermediate particle velocity and position.

In the correction step, the pressure is used to update the particle velocity obtained from the prediction step

$$\Delta \mathbf{u}_{**} = -\frac{1}{\rho_*} \nabla P_{t+1} \Delta t,$$

$$\mathbf{u}_{t+1} = \mathbf{u}_* + \Delta \mathbf{u}_{**}, \tag{7}$$

where $\Delta \mathbf{u}_{**}$ = increment of particle velocity during the correction step, ρ_* = intermediate particle density calculated after the prediction step, and P_{t+1} and $\mathbf{u}_{t+1}$ = particle pressure and velocity at time $t + 1$.

Finally, the positions of particle are centered in time

$$\mathbf{r}_{t+1} = \mathbf{r}_t + \frac{(\mathbf{u}_t + \mathbf{u}_{t+1})}{2} \Delta t, \tag{8}$$

where $\mathbf{r}_t$ and $\mathbf{r}_{t+1}$ = positions of particle at time t and $t + 1$.

The pressure is implicitly calculated from the Poisson equation of pressure as follows:

$$\nabla \cdot \left(\frac{1}{\rho_*} \nabla P_{t+1} \right) = \frac{\rho_0 - \rho_*}{\rho_0 \Delta t^2}, \tag{9}$$

where ρ_0 = initial constant density at each of the particle in the beginning of computation. Equation (9) was derived from the combination of the mass and momentum equations (1) and (2), by enforcing the incompressibility of particle densities. It is analogous to that employed in the moving particle semi-implicit (MPS) method [25] in that the source term of the equation is the variation of particle densities, while it is usually the divergence of intermediate velocity fields in a finite difference method.

2.3. Basic SPH Theories and Formulations. The advantages of the SPH approach arise from its gridless nature. Since there is no mesh distortion, the SPH method can effectively treat the large deformations of free surface and multi-interface in a pure Lagrangian frame. In an SPH framework, the motion of each particle is calculated through the interactions with the neighboring particles using an analytical kernel function. All terms in the governing equations are represented by the particle interaction models, and thus the grid is not needed. For a detailed review of the SPH theories see Monaghan [9]. Among a variety of kernels documented in the literatures the spline-based kernel normalized in 2-D [9] is widely used in the hydrodynamic calculations. We use the following basic formulations for the proposed ISPH model.

The fluid density ρ_a of particle a is calculated by

$$\rho_a = \sum_b m_b W(|\mathbf{r}_a - \mathbf{r}_b|, h), \tag{10}$$

where a and b = reference particle and all of its neighbors; m_b = particle mass, $\mathbf{r}_a$ and $\mathbf{r}_b$ = particle positions, W = interpolation kernel, and h = smoothing distance, which determines the range of particle interactions and is equal to 1.2 times of the initial particle spacing in the paper.

The pressure gradient assumes a symmetric form as:

$$\left(\frac{1}{\rho} \nabla P \right)_a = \sum_b m_b \left(\frac{P_a}{\rho_a^2} + \frac{P_b}{\rho_b^2} \right) \nabla_a W_{ab}, \tag{11}$$

where the summation is over all particles other than particle a and $\nabla_a W_{ab}$ = gradient of the kernel taken with respect to the positions of particle a. In a similar way, the velocity divergence of particle a is formulated by

$$\nabla \cdot \mathbf{u}_a = \rho_a \sum_b m_b \left(\frac{\mathbf{u}_a}{\rho_a^2} + \frac{\mathbf{u}_b}{\rho_b^2} \right) \cdot \nabla_a W_{ab}. \tag{12}$$

The turbulence stress in (2) is formulated by applying the above SPH definition of divergence as

$$\left(\frac{1}{\rho} \nabla \cdot \vec{\vec{\tau}} \right)_a = \sum_b m_b \left(\frac{\vec{\vec{\tau}}_a}{\rho_a^2} + \frac{\vec{\vec{\tau}}_b}{\rho_b^2} \right) \cdot \nabla_a W_{ab}. \tag{13}$$

The Laplacian of pressure and laminar viscosity terms are formulated as a hybrid of a standard SPH first derivative with a finite difference approximation for the first derivative. They are also represented in the symmetrical forms as

$$\nabla \cdot \left(\frac{1}{\rho} \nabla P \right)_a = \sum_b m_b \frac{8}{(\rho_a + \rho_b)^2}$$
$$\times \frac{(P_a - P_b)(\mathbf{r}_a - \mathbf{r}_b) \cdot \nabla_a W_{ab}}{|\mathbf{r}_a - \mathbf{r}_b|^2}, \tag{14}$$

$$\left(\nu_0 \nabla^2 \mathbf{u} \right)_a = \sum_b m_b \frac{2(\nu_a + \nu_b)}{\rho_a + \rho_b}$$
$$\times \frac{(\mathbf{u}_a - \mathbf{u}_b)(\mathbf{r}_a - \mathbf{r}_b) \cdot \nabla_a W_{ab}}{|\mathbf{r}_a - \mathbf{r}_b|^2}. \tag{15}$$

2.4. Treatment of Solid Boundary and Free Surface. In the ISPH computations, the free surface can be easily and accurately tracked by the fluid particles. Since there is no fluid particle existing in the outer region of the free surface, the particle density on the surface should drop significantly. A zero pressure is given to each of the surface particles.

The impermeable solid boundaries such as the horizontal sea bed and sloping sea walls are treated by the fixed wall particles, which balance the pressure of inner fluid particles and prevent them from penetrating the wall. The pressure Poisson equation is solved on these wall particles. The offshore boundary is the incident wave boundary, which is modeled by a numerical wave paddle composed of moving wall particles. In the computations, the frequency and amplitude of the numerical wave paddle are given so as to generate the desired incident waves. Most kinds of the practical waves can be easily generated by the SPH model. For a more detailed description of the boundary treatment in particle models, refer to [11, 21, 25].

3. Wave Overtopping for Different Breakwater Designs

3.1. Model Setup and Numerical Parameters. In this section, we use the developed ISPH model to study a practical engineering problem: the breaking wave overtopping on a caisson breakwater under different conditions, including with and without the protection by a porous armor layer and different slope geometries. We will investigate the overtopping mass rate, pressure, velocity, and turbulence features in front of the breakwater to study the flow characteristics. The computational setting is based on the laboratory experiment of Sakakiyama and Kayama [26] and the numerical computations of Liu et al. [6].

The laboratory experiment used an impermeable caisson breakwater with a dimension of 30 cm × 18 cm and a layer of porous materials in front of the caisson. The effective porosity is 0.5, and the mean diameter of porous materials is 0.05 m. The ISPH model is used to reproduce the experiment of Sakakiyama and Kayama [26] in which the wave period was $T = 1.4$ s, wave height $H = 0.105$ m, and still water depth $d = 0.28$ m. A sketch view of the numerical setup including the caisson breakwater and a layer of porous armor units is shown in Figure 1(a), where the origin of coordinates is chosen at the intersection of the still water level and front wall of the caisson. The free surfaces were measured at several sections for more than 40 seconds in the experiment. The overtopped mass was also weighed to estimate the overtopping rate.

In the ISPH simulations, a smaller computational domain and shorter simulation time are used to reduce the computational effort. The computational domain is 5.3 m long and covers the caisson breakwater and a numerical wave paddle located at the offshore boundary. A uniform particle spacing of $\Delta X = 0.01$ m is used and about 12,000 particles are involved in the computations. The spatial resolution in the ISPH run is similar to that used in the RANS computation of Liu et al. [6], who used a nonuniform grid of $\Delta x_{min} = 0.01$ m and $\Delta y_{min} = 0.007$ m. Because the leading reflected wave from the caisson reached the wave paddle about 7 seconds after the first wave was generated by the paddle, we impose the nonreflecting wave paddle of Hayashi et al. [27] to absorb the reflected waves to ensure that the quasisteady condition can be attained. The total simulation time is $t = 13$ seconds in the incompressible SPH computations.

To further demonstrate the protective role of the porous armor layer, two alternative numerical experiments are also performed to compare the flow velocity, turbulence, and pressure characteristics in front of the caisson breakwater without the protection of the porous layer. The design of the problem follows Liu et al. [6]. In the first case, the porous layer is completely removed so the waves impact directly on the vertical caisson wall. In the second case, the porous layer is replaced by an impermeable material, so the impermeable material and the caisson become a single structure like an impermeable sloping seawall. The numerical settings for the additional two cases are shown in Figures 1(b) and 1(c), respectively.

Here we should emphasize that in the work of Shao [23], the detailed breaking wave running up and overtopping characteristics for the porous case have been discussed and the wave profiles have been validated against the benchmark data. In the current paper, the focus will be the comparisons of flow characteristics for the different design scenarios and especially the flow overtopping rate, which is a key parameter in the practical breakwater design.

3.2. Discussions of Flow Velocity Field for Different Cases. According to the ISPH simulations, the flow velocity fields during the wave interaction and overtopping on the caisson breakwater are shown in Figures 2, 3, and 4, respectively, corresponding to the case of the porous layer, impermeable vertical, and sloping walls. From Figures 2(a) and 2(b), the velocity fields demonstrate the protective role of the porous armor layer in absorbing the wave energy and attenuating the flow impact on the caisson, where it acts as an equivalent buffer layer. From the figure, it is shown that the velocities decrease towards the core of the armor units and become nearly zero near the toe of the caisson. Thus, the scour of the caisson by the continuous wave actions can be greatly reduced due to the use of the armor unit. This phenomenon has also been well described by Liu et al. [6] in their numerical simulations by using an advanced RANS approach [28].

For Figures 3(a) and 3(b) in case of the impermeable vertical wall, due to the absence of the porous armor units, it lacks an efficient mean to absorb the wave energy. Compared with the case by using the porous armor layer in Figure 2, the flow motion in Figure 3 is much stronger. When the wave overtopping starts to occur, both the vertical velocities and vertical accelerations are quite large in front of the caisson. Therefore, the potential scouring at the foot of the caisson becomes severe in this case. In addition, the size of the overtopping jet is also much larger and thus carries a lot of energy, which could threaten the safety of the onshore areas.

For Figures 4(a) and 4(b) in case of the impermeable sloping wall, although this kind of design has a similar physical geometry as that used in Figure 2, the impermeable sloping wall prevents the wave energy dissipation and provides a chance for the wave to run up. Although the size and intensity of the overtopping jet are relatively smaller as compared with the vertical jet generated in Figure 3 in a vertical caisson without any protection, the wave overtopping capacity is actually enhanced in Figure 4, due to that the flow can maintain sufficiently large horizontal momentum for the subsequent overtopping process.

3.3. Discussions of Flow Pressure Field for Different Cases. The impact pressure on the caisson breakwater is another very important topic in the breakwater stability and scouring problems. For studying this, the pressure fields in front of the caisson breakwater are shown in Figures 5, 6, and 7, respectively, corresponding to the three different cases. In the figures, the pressure values have been normalized by ρg to represent the normalized pressure head. For Figure 5, in case of the existence of the porous layer, it is shown that the pressure generally follows a hydrostatic distribution before the wave impacts on the structure as shown in Figure 5(a). However, as the wave front approaches the caisson and overtops on the crest, the fluid particles experience a vertical acceleration that results in a slightly larger pressure than the hydrostatic value. As shown in Figure 5(b), the maximum pressure seems to be resulted from the wave impact because

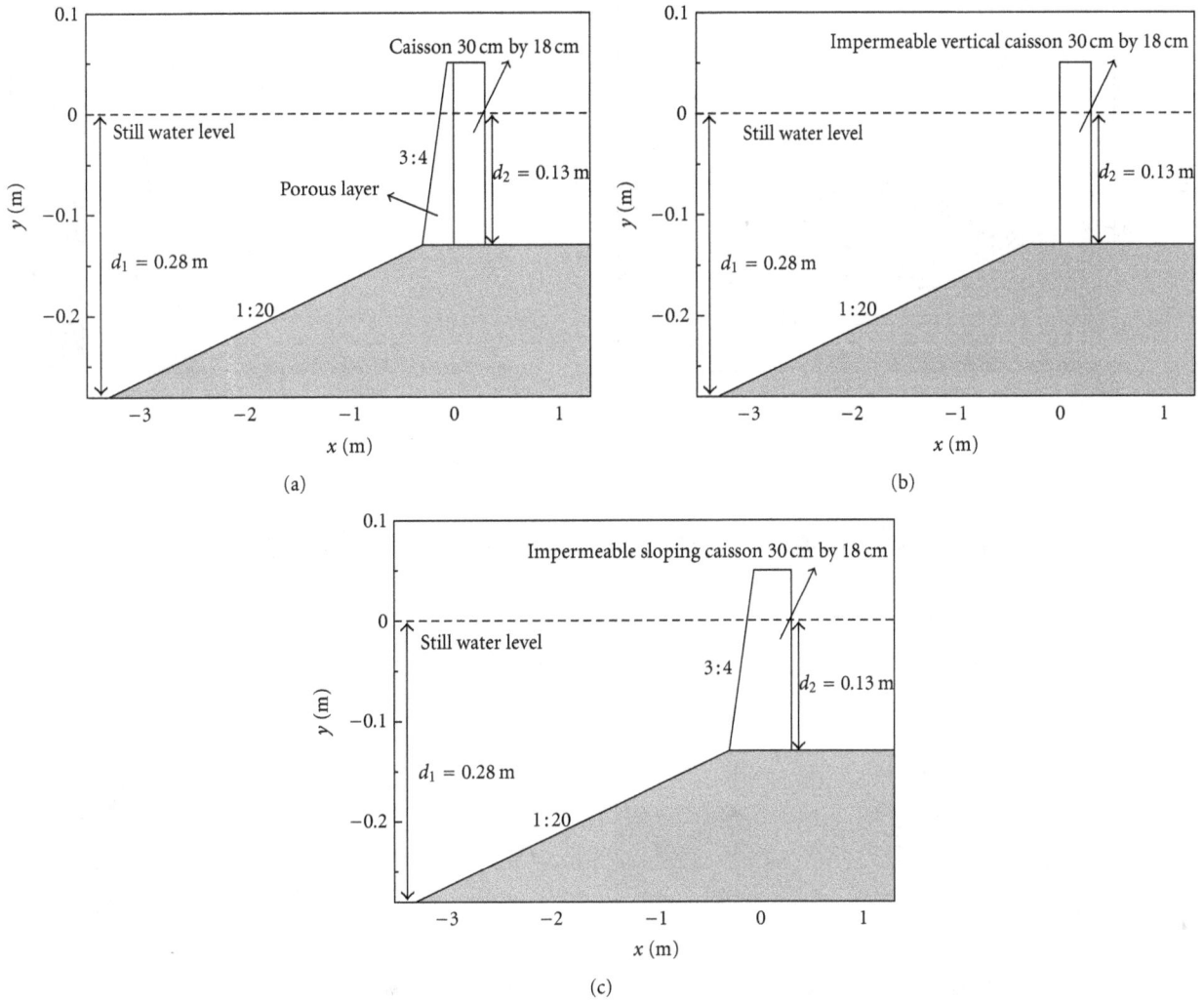

FIGURE 1: Sketch view of numerical setup for wave overtopping of different breakwaters: (a) with a permeable layer, (b) impermeable vertical wall, and (c) impermeable sloping wall.

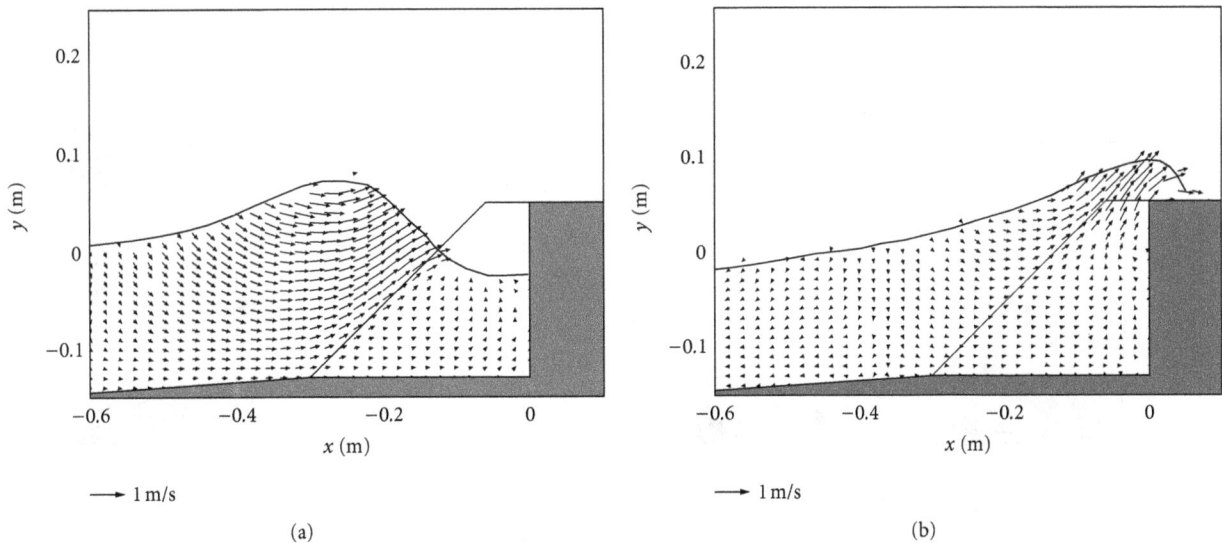

FIGURE 2: Velocity fields during (a) wave interaction and (b) overtopping on caisson breakwater protected by a porous layer.

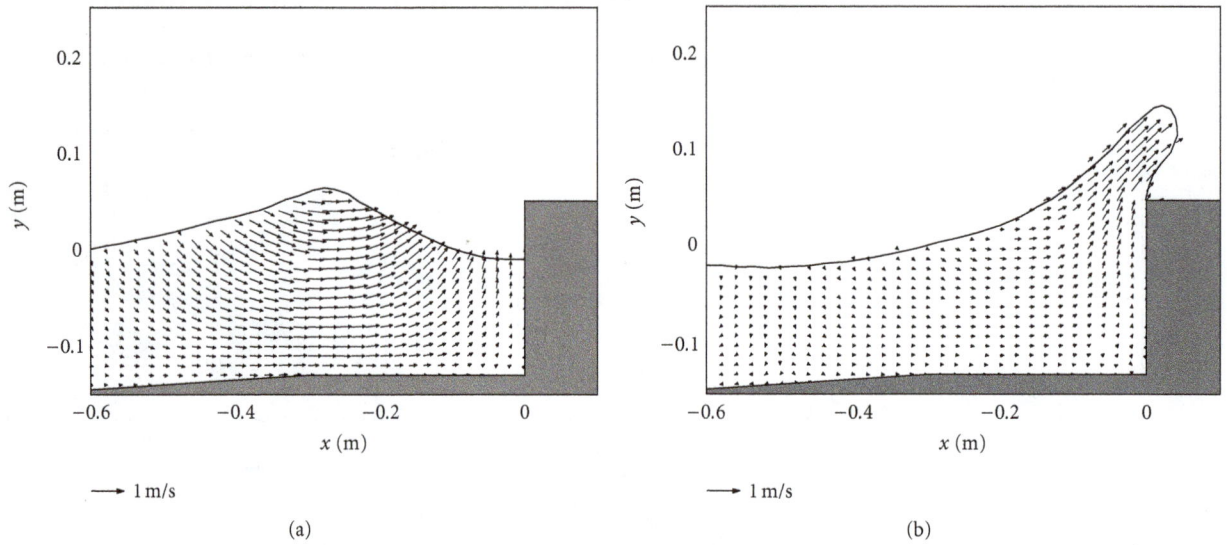

FIGURE 3: Velocity fields during (a) wave interaction and (b) overtopping on impermeable vertical caisson breakwater.

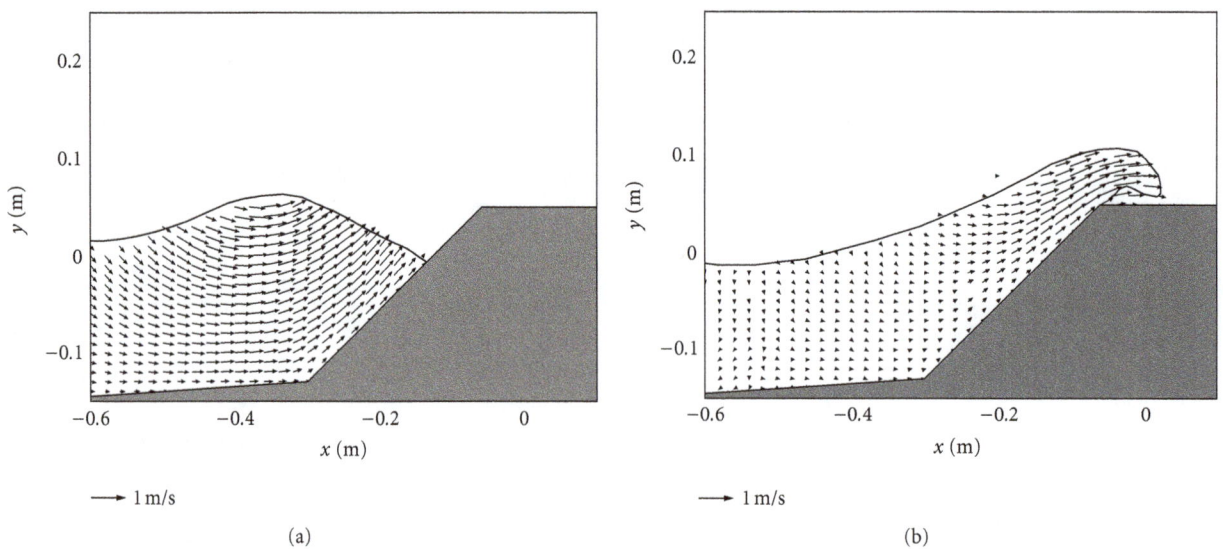

FIGURE 4: Velocity fields during (a) wave interaction and (b) overtopping on impermeable sloping caisson breakwater.

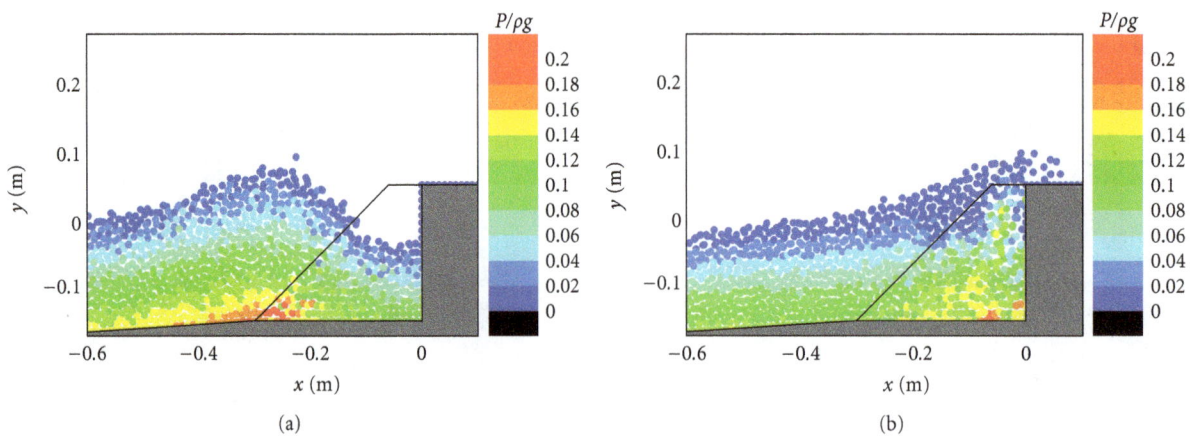

FIGURE 5: Pressure fields during (a) wave interaction and (b) overtopping on caisson breakwater protected by a porous layer.

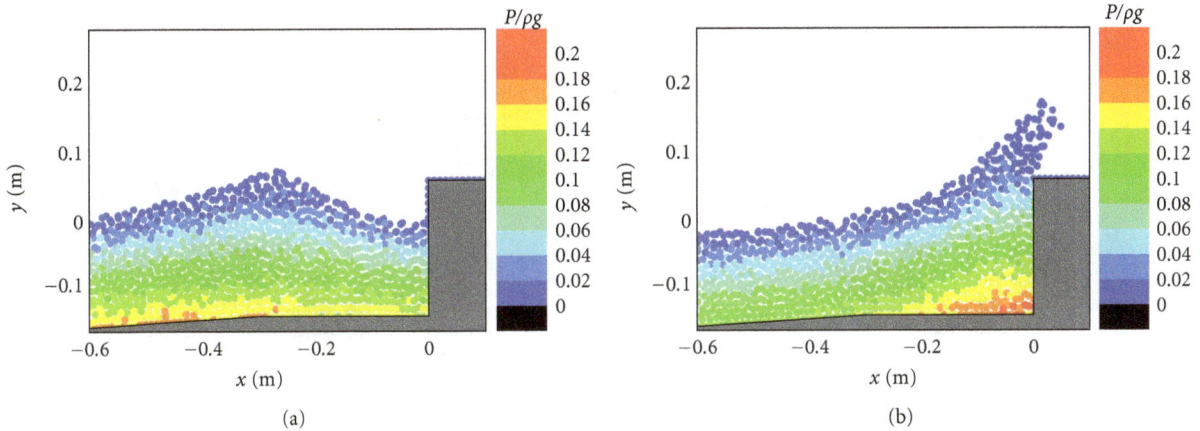

FIGURE 6: Pressure fields during (a) wave interaction and (b) overtopping on impermeable vertical caisson breakwater.

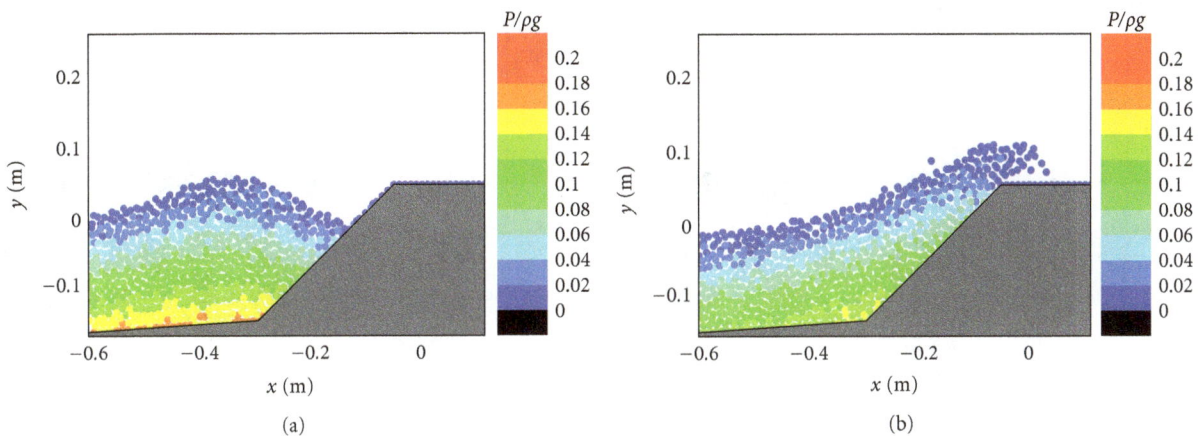

FIGURE 7: Pressure fields during (a) wave interaction and (b) overtopping on impermeable sloping caisson breakwater.

the pressure distributions in front of the caisson do not exactly follow the hydrostatic law. Besides, the range of impact pressures is greatly reduced due to the protection of the porous armor layer. The computed pressure patterns in Figure 5 generally agree with the RANS results of Liu et al. [6].

The pressure fields presented in Figure 6 indicate that stable pressure calculations are also achieved for the simplification of the physical problem without considering the porous flow. Although no much difference in the pressure patterns is observed during the wave approaching the caisson breakwater as shown in Figures 6(a) and 5(a), the pressure patterns seem quite different during the wave overtopping on the caisson as shown in Figures 6(b) and 5(b). Without the protection by the porous materials, the pressures increase more rapidly and the high pressure regions are more widely spread at the foot of the caisson, which can pose a great threat to the structure stability in practice. For Figure 7, in case of the impermeable sloping wall, although the maximum pressure is smaller than that in the porous case in Figure 5, the wave overtopping capacity can significantly increase due to the reasons as mentioned before.

The above stable pressure simulations demonstrated the robustness of the ISPH model presented in the paper. It is well known that the pressure fluctuation is a common problem in most particle modeling approaches, which arises from the particle interactions and inevitably leads to the particle fluctuations. Such a problem has been reported in the widely used weakly compressible SPH approach [9], and additional numerical treatments are needed to address this problem. In the ISPH computation, we can directly obtain very smoothed and reasonable pressure fields without any numerical smoothing. This is due to the fact that, in the ISPH formulation, the pressure is calculated through a strict hydrodynamic formulation. So the incompressible approach could represent a promising particle modeling technique for different hydrodynamic problems.

3.4. Discussions of Flow Turbulence Field for Different Cases. The flow turbulence fields in front of the caisson are shown in Figures 8, 9, and 10, respectively, for the three different design cases. In the figures, the turbulence eddy viscosity values have been normalized by the laminar viscosity to represent the equivalent turbulence intensity. As shown in Figure 8,

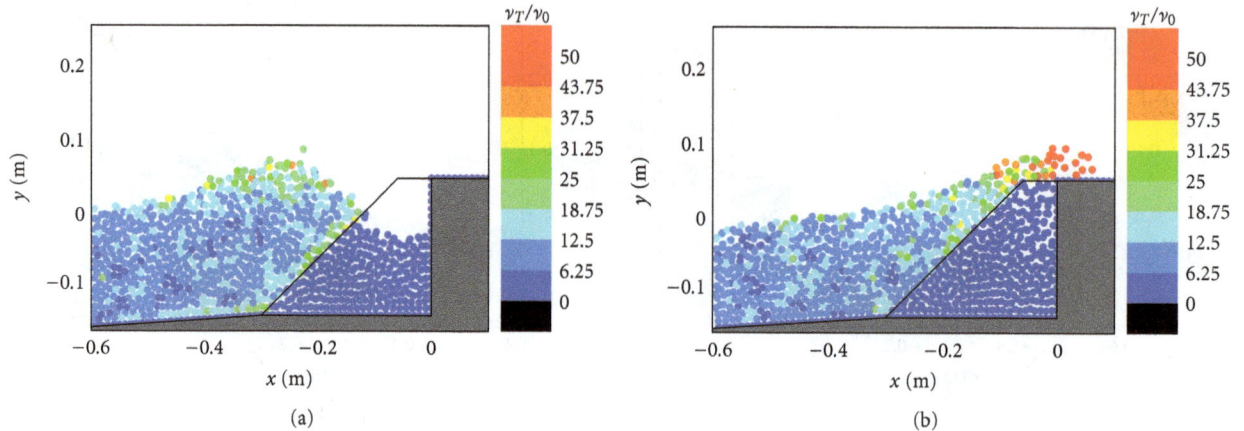

FIGURE 8: Turbulence fields during (a) wave interaction and (b) overtopping on caisson breakwater protected by a porous layer.

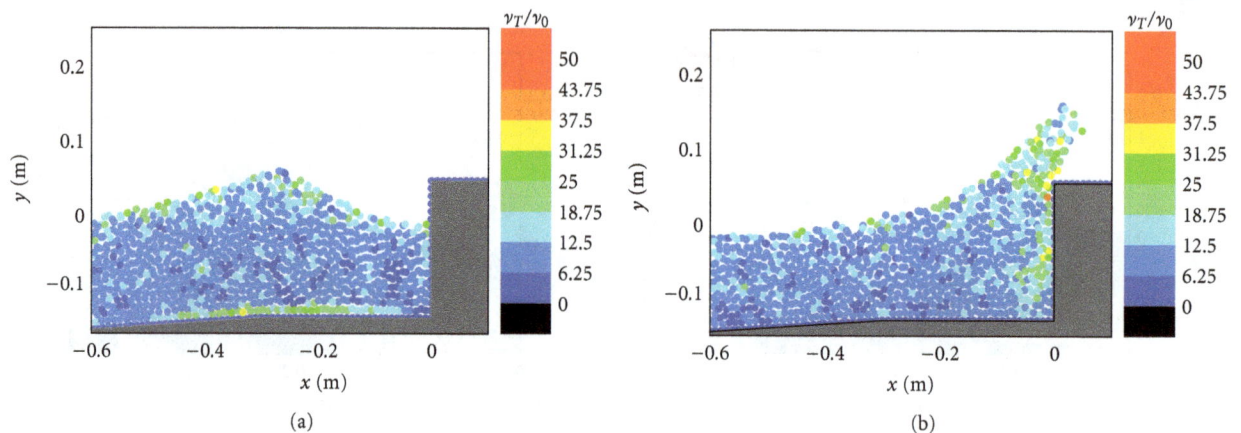

FIGURE 9: Turbulence fields during (a) wave interaction and (b) overtopping on impermeable vertical caisson breakwater.

the turbulence levels inside the porous layer are very low and close to zero, because the porous material dampens most of the flow energy. Figure 8 also indicates that the high turbulence areas almost concentrate on the wave crest and the overtopping wave front, where the flow velocity is also the largest. In Figure 8(b), it is quite obvious that the highest turbulence areas concentrate on the overtopping jet. Since the turbulence is related to the energy dissipation, we can reasonably conclude that most of the kinetic energy of the overtopping wave is dissipated by the turbulence generation, and thus the overtopping intensity can be greatly reduced (We will have the detailed comparison of the overtopping mass rate in the following section to support this).

The computed flow turbulence fields in Figures 9 and 10 for the two impermeable cases have demonstrated a similar evolution pattern as that in the case with the protection of a porous armor layer in Figure 8, that is, the high turbulence regions almost concentrate on the wave crest and overtopping jet, as well as the lower solid boundary. The overall turbulence intensity in the overtopping front for the impermeable sloping wall in Figure 10(b) is higher than that in impermeable vertical wall case in Figure 9(b).

3.5. *Comparisons of Wave-Overtopping Load for Different Cases.* The wave overtopping load is an important parameter to evaluate the performance of the breakwater design in practice. In order to quantitatively analyze the effectiveness of three different designs of the caisson breakwater, the time history of the wave-overtopping load for each design is shown in Figure 11, based on the ISPH computations. As a comparison, the numerical results computed from the RANS model of Liu et al. [6] are also shown. Regardless of some differences in the detailed velocity, turbulence, and pressure fields computed by the two numerical models, the overall agreement in the wave-overtopping mass is quite excellent. The relatively large deviation is found for the computations with the presence of the porous armor layer, which could be attributed to the different treatments of the turbulence boundary conditions and the drag forces by the two models.

Figure 11 indicates that the caisson breakwater protected by the porous armor layer (denoted as "SPH-Porous" in the figure) has the smallest wave overtopping. By using the statistical analysis, it is calculated that the caisson without the armor units (denoted as "SPH-Vertical") increases the wave-overtopping load by about 55% and the use of an impermeable sloping wall (denoted as "SPH-Sloping") increases

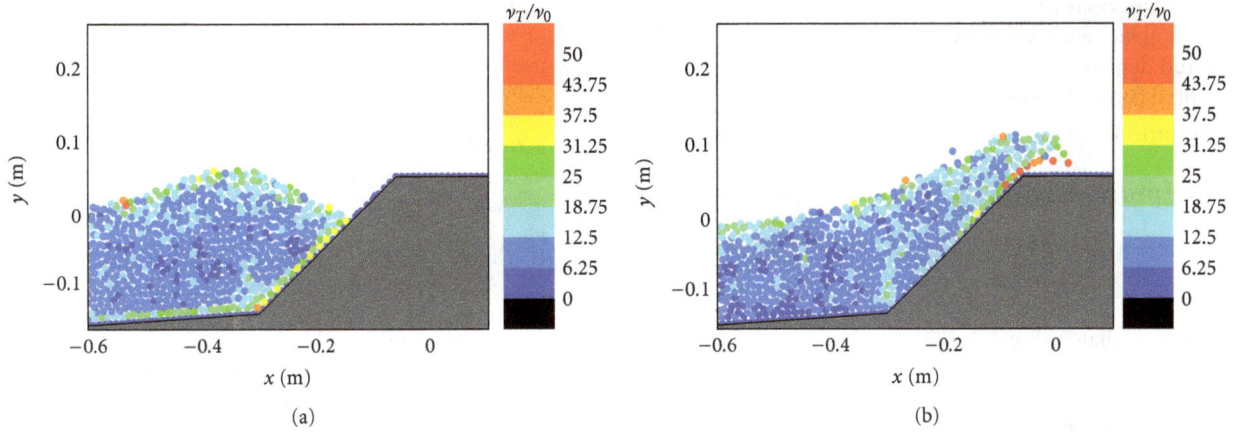

FIGURE 10: Turbulence fields during (a) wave interaction and (b) overtopping on impermeable sloping caisson breakwater.

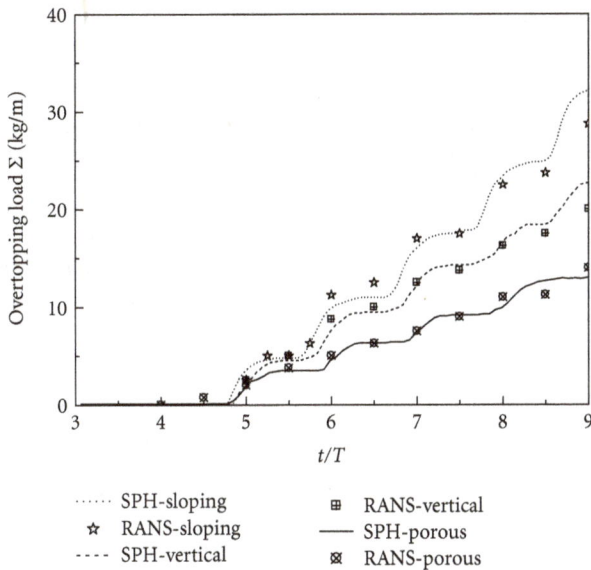

FIGURE 11: Time history of wave overtopping load for different breakwater designs.

the overtopping load by about 105%. This is quite close to the documented values of 45% and 100%, respectively, computed by Liu et al. [6]. The figure has provided a quantitative measurement to show the effectiveness of the porous materials in attenuating the incident wave energy and protecting the coastal structure from severe wave attack.

4. Conclusions

An incompressible smoothed particle hydrodynamics model has been used to evaluate the wave-breakwater interactions and wave overtopping. The SPH numerical scheme has the advantage of treating the free surfaces and complex solid boundaries in an easy and accurate way. All of the computations were made by using a CPU 2.13 G and RAM 1.0 G laptop. A single run was finished within 6–8 hours by employing 12,000 particles for a wave simulation of 13 seconds (with an averaged time step of 0.001 s).

The numerical model was applied to the problem of a breaking wave interacting and overtopping on a caisson breakwater. The computed wave overtopping mass rate is in good agreement with the available numerical results computed from an RANS model. The numerical results of the flow velocities, pressures, and turbulence quantities demonstrated that the armor units play an important role in dissipating the wave energy and stabilizing the caisson breakwater. According to the numerical study of different designs, it is shown that the overtopping mass can be reduced by about 55% and 105%, respectively, as compared with a similar design of the caisson breakwater without any protection, or with an attached impermeable sloping seawall.

Notations

C_D: Drag coefficient
C_s: Smagorinsky constant
d: Still water depth
$\mathbf{F}$:   Drag force caused by porous media
$\mathbf{g}$:   Gravitational acceleration
h: SPH kernel smoothing distance
H: Wave height
k: Turbulence energy
m: Particle mass
P: Pressure
$\mathbf{r}$:   Position vector
$|S|$:   Local strain rate
S_{ij}:   Strain rate of mean flow
T: Wave period
$\mathbf{u}$:   Velocity vector
W: SPH interpolation kernel
δ_{ij}:   Kronecker's delta
Δx_{min}: Horizontal grid spacing in RANS
ΔX: Particle spacing
Δt: Time increment
$\Delta \mathbf{u}$:   Change in velocity
Δy_{min}: Vertical grid spacing in RANS

ν_0: Laminar viscosity
ν_T: Turbulence eddy viscosity
ρ: Fluid density
ρ_0: Initial constant density
$\vec{\tau}$: Turbulence stress.

Subscripts and Symbols

a: Reference particle
b: Neighboring particle
t: Time
$*$: Intermediate value
$**$: Corrected value.

Acknowledgments

This research work is supported by the Nazarbayev University Seed Grant, entitled "Environmental assessment of sediment pollution impact on hydropower plants". S. Shao also acknowledges the Royal Society Research Grant (2008/R2 RG080561). Jaan Hui Pu is currently an assistant professor at Nazarbayev University, Kazakhstan and Songdong Shao is currently a senior lecturer at University of Bradford, UK.

References

[1] European Overtopping Manual_EurOtop, "Wave Overtopping of Sea Defences and Related Structures: Assessment Manual," 2008, http://www.overtopping-manual.com/.

[2] J. Brossard, A. Jarno-Druaux, F. Marin, and E. H. Tabet-Aoul, "Fixed absorbing semi-immersed breakwater," Coastal Engineering, vol. 49, no. 1-2, pp. 25–41, 2003.

[3] M. O. Muttray and H. Oumeraci, "Theoretical and experimental study on wave damping inside a rubble mound breakwater," Coastal Engineering, vol. 52, no. 8, pp. 709–725, 2005.

[4] H. B. Chen, C. P. Tsai, and J. R. Chiu, "Wave reflection from vertical breakwater with porous structure," Ocean Engineering, vol. 33, no. 13, pp. 1705–1717, 2006.

[5] D. Qiu and L. Wang, "Numerical and experimental research for wave damping over a submerged porous breakwater," in Proceedings of the 6th International Offshore and Polar Engineering Conference, pp. 572–576, Los Angeles, Calif, USA, May 1996.

[6] P. L. F. Liu, P. Lin, K. A. Chang, and T. Sakakiyama, "Numerical modeling of wave interaction with porous structures," Journal of Waterway, Port, Coastal and Ocean Engineering, vol. 125, no. 6, pp. 322–330, 1999.

[7] C. J. Huang and C. M. Dong, "On the interaction of a solitary wave and a submerged dike," Coastal Engineering, vol. 43, no. 3-4, pp. 265–286, 2001.

[8] N. Garcia, J. L. Lara, and I. J. Losada, "2-D numerical analysis of near-field flow at low-crested permeable breakwaters," Coastal Engineering, vol. 51, no. 10, pp. 991–1020, 2004.

[9] J. J. Monaghan, "Smoothed particle hydrodynamics," Annual Review of Astronomy and Astrophysics, vol. 30, no. 1, pp. 543–574, 1992.

[10] J. J. Monaghan and A. Kos, "Solitary waves on a cretan beach," Journal of Waterway, Port, Coastal and Ocean Engineering, vol. 125, no. 3, pp. 145–154, 1999.

[11] S. Shao, "SPH simulation of solitary wave interaction with a curtain-type breakwater," Journal of Hydraulic Research, vol. 43, no. 4, pp. 366–375, 2005.

[12] A. Khayyer, H. Gotoh, and S. Shao, "Enhanced predictions of wave impact pressure by improved incompressible SPH methods," Applied Ocean Research, vol. 31, no. 2, pp. 111–131, 2009.

[13] E. S. Lee, C. Moulinec, R. Xu, D. Violeau, D. Laurence, and P. Stansby, "Comparisons of weakly compressible and truly incompressible algorithms for the SPH mesh free particle method," Journal of Computational Physics, vol. 227, no. 18, pp. 8417–8436, 2008.

[14] H. Gotoh, H. Ikari, T. Memita, and T. Sakai, "Lagrangian particle method for simulation of wave overtopping on a vertical seawall," Coastal Engineering Journal, vol. 47, no. 2-3, pp. 157–181, 2005.

[15] S. Shao, "Incompressible SPH simulation of wave breaking and overtopping with turbulence modelling," International Journal for Numerical Methods in Fluids, vol. 50, no. 5, pp. 597–621, 2006.

[16] A. J. C. Crespo, M. Gómez-Gesteira, and R. A. Dalrymple, "3D SPH simulation of large waves mitigation with a dike," Journal of Hydraulic Research, vol. 45, no. 5, pp. 631–642, 2007.

[17] I. J. Losada, J. L. Lara, R. Guanche, and J. M. Gonzalez-Ondina, "Numerical analysis of wave overtopping of rubble mound breakwaters," Coastal Engineering, vol. 55, no. 1, pp. 47–62, 2008.

[18] M. T. Reis, K. Hu, T. S. Hedges, and H. Mase, "A comparison of empirical, semiempirical, and numerical wave overtopping models," Journal of Coastal Research, vol. 24, no. 2, pp. 250–262, 2008.

[19] H. Gotoh, T. Shibahara, and T. Sakai, "Sub-particle-scale turbulence model for the MPS method—lagrangian flow model for hydraulic engineering," Computational Fluid Dynamic Journal, vol. 9, pp. 339–347, 2001.

[20] J. Smagorinsky, "General circulation experiments with the primitive equations, I. The basic experiment," Monthly Weather Review, vol. 91, pp. 99–164, 1963.

[21] H. Gotoh and T. Sakai, "Lagrangian simulation of breaking waves using particle method," Coastal Engineering Journal, vol. 41, no. 3-4, pp. 303–326, 1999.

[22] Y. Shimizu and T. Tsujimoto, "Numerical analysis of turbulent open-channel flow over a vegetation layer using a k-ε turbulence model," Journal of Hydroscience and Hydro Engineering, vol. 11, pp. 57–67, 1994.

[23] S. Shao, "Incompressible SPH flow model for wave interactions with porous media," Coastal Engineering, vol. 57, no. 3, pp. 304–316, 2010.

[24] A. J. Chorin, "Numerical solution of the Navier-Stokes equations," Mathematics of Computation, vol. 22, pp. 745–762, 1968.

[25] S. Koshizuka, A. Nobe, and Y. Oka, "Numerical analysis of breaking waves using the moving particle semi-implicit method," International Journal for Numerical Methods in Fluids, vol. 26, no. 7, pp. 751–769, 1998.

[26] T. Sakakiyama and M. Kayama, "Numerical simulation of wave breaking and overtopping on wave absorbing revetment," in Proceedings of Coastal Engineering, vol. 44, pp. 741–745, Japan Society of Civil Engineers, Tokyo, Japan, 1997.

[27] M. Hayashi, H. Gotoh, T. Memita, and T. Sakai, "Gridless numerical analysis of wave breaking and overtopping at upright seawall," in Proceedings of the 27th International Conference on Coastal Engineering (ICCE '00), pp. 2100–2113, July 2000.

[28] P. Lin and P. L. F. Liu, "A numerical study of breaking waves in the surf zone," Journal of Fluid Mechanics, vol. 359, pp. 239–264, 1998.

2

Simulation of Nonisothermal Consolidation of Saturated Soils Based on a Thermodynamic Model

Zhichao Zhang and Xiaohui Cheng

Department of Civil Engineering, Tsinghua University, Beijing 100084, China

Correspondence should be addressed to Zhichao Zhang; zhangzhichaopt@163.com

Academic Editors: M. Jha and S. Kaewunruen

Based on the nonequilibrium thermodynamics, a thermo-hydro-mechanical coupling model for saturated soils is established, including a constitutive model without such concepts as yield surface and flow rule. An elastic potential energy density function is defined to derive a hyperelastic relation among the effective stress, the elastic strain, and the dry density. The classical linear nonequilibrium thermodynamic theory is employed to quantitatively describe the unrecoverable energy processes like the nonelastic deformation development in materials by the concepts of dissipative force and dissipative flow. In particular the granular fluctuation, which represents the kinetic energy fluctuation and elastic potential energy fluctuation at particulate scale caused by the irregular mutual movement between particles, is introduced in the model and described by the concept of granular entropy. Using this model, the nonisothermal consolidation of saturated clays under cyclic thermal loadings is simulated in this paper to validate the model. The results show that the nonisothermal consolidation is heavily OCR dependent and unrecoverable.

1. Introduction

Since the 80s of last century, the thermo-hydro-mechanical (THM) coupling problem has become an important scientific research focus in many engineering areas such as geothermal resources development, oil exploration, and nuclear waste storage. Researches show that many physical properties such as the permeability, water content, and thermal expansion coefficient of soils are affected by the temperature [1, 2]. Moreover, the water flow in soils, the pore pressure development, and the shear strength are all sensitive to the temperature variation [3–5]. For example, under temperature elevation, significant thermal volumetric deformation (i.e., the nonisothermal consolidation) [5–7] will occur and the shear strength will be changed significantly. Experimental studies show that the nonisothermal consolidation is heavily dependent on the overconsolidation ratio (OCR) [5–7]. During undrained heating process, due to the difference in the thermal expansion coefficients of solid and fluid phases and the nonelastic deformation, the pore pressure of soils will be increased notably so that thermal failure may occur because of the decrease of effective stress [8].

However, there are few models that can represent all of the THM coupling features discussed before for saturated soils. The existing models at present are usually based on the improvement of the classical elasto-plastic model [8–10] or based on empirical formulae of the pore pressure and strain from the fitting of experimental data [11]. These models are phenomenological and cannot represent the real physical mechanisms of the THM coupling processes in soils. On the contrary, models based on the thermodynamics are able to provide an effective and comprehensive approach for complex multifield coupling problems of soils. In this paper, a thermodynamic theoretical framework, which is usually referred to as hydrodynamic method, is adopted to establish a THM coupling model. In this approach, based on the conservation laws and the nonequilibrium thermodynamics, many independent thermodynamic state variables and thermodynamic potential functions (e.g., free energy) are defined and thus many useful thermodynamic identities that provide a plenty of relations between state variables are obtained theoretically. On the other hand, the unrecoverable energy processes in materials are described by the migration coefficient model in linear nonequilibrium thermodynamics.

This approach has been successfully applied in the researches of many different materials such as fluid, crystal, and dry sand [12–16]. Jiang and Liu [14–16] further introduce the concept of granular entropy S_g for granular solid materials to describe the energy process of granular fluctuation, which represents the irregular mutual movement between particles, and derives a constitutive model without the concepts of yield surface, flow rule, and hardening/softening rule.

Based on the thermodynamic theoretical framework discussed before, a THM coupling model is established for the simulation of nonisothermal consolidation of saturated soils. The granular entropy is introduced to describe the granular fluctuation, which includes the kinetic energy fluctuation and the elastic potential energy fluctuation. A modified form of the evolution law of granular fluctuation is given from the perspective of the physical mechanism of nonisothermal processes, in which the conversion between the free pore water and the bound pore water is considered. Moreover, an equivalent nonelastic strain is defined to determine the elastic potential energy fluctuation and better represent the nonisothermal consolidation under repeated thermal loadings. Based on this model, the nonisothermal consolidation tests are simulated to validate the model.

2. Model Formulation

2.1. General Assumptions

(1) Saturated soils can be divided into a solid phase and a liquid phase. The liquid phase is composed of a free water phase and a bound water phase. The bound water can be converted to the free water during temperature elevation, while there is no mass exchange between the solid phase and the liquid phase.

(2) Every phase is continuous in space and all phases have the same temperature.

(3) The liquid phase does not undergo solidification and vaporization in the temperature range considered in this paper, and the soils always maintain saturation.

2.2. Basic Equations.
Denote the porosities of the free water and the bound water as ϕ_{fw} and ϕ_{bw}, respectively. The total porosity is $\phi = \phi_{fw} + \phi_{bw}$. Define the density, the velocity, the momentum density, and the entropy density of each phase as ρ_α, v_i^α, m_i^α, and s_α, respectively, wherein, $\alpha = s$ and f, respectively, represent the solid phase, the free water phase and the bound water phase, the same hereinafter. Assume that the bound water is completely absorbed on the surface of the solid particles and thus its velocity equals the velocity of the solid phase; that is, $v_i^s = v_i^{bw}$.

2.2.1. Mass Conservation Equations.
The mass conservation equations for the solid phase, the free water phase, and the bound water phase are shown in (1a)-(1c), respectively, wherein, d_t is the material derivative using the solid phase as a reference and has a relation with the spatial derivative

∂_t, $d_t = \partial_t + v_k^s \nabla_k$. The material derivative is used in this paper for the simplification of the forms of partial differential equations. $v_i^{fw} - v_i^s$ is the relative velocity of the free water and the solid phase, and thus $\phi_{fw}(v_k^{fw} - v_k^s)$ is just the average water flow velocity. ε_{kk} is the volumetric strain of the solid skeleton and is taken as positive under compression. The assumption of small strain is adopted; that is, the strain rate $d_t\varepsilon_{ij} = -(v_{i,j}^s + v_{j,i}^s)/2$. Q represents the mass converted from bound water to free water per unit time and per unit volume during the elevation of temperature. In this paper, $Q = \rho_{bw}\alpha_{bf}\phi_{bw}d_tT$, wherein, the parameter α_{bf} is the mass of free water converted from per unit mass of bound water per each unit rise in temperature:

$$d_t[\rho_s(1-\phi)] = \rho_s(1-\phi)d_t\varepsilon_{kk}, \tag{1a}$$

$$d_t(\rho_{fw}\phi_{fw}) = -\rho_{fw}\left[\phi_{fw}\left(v_k^{fw} - v_k^s\right)\right]_{,k}$$
$$- \phi_{fw}\left(v_k^{fw} - v_k^s\right)\nabla_k\rho_{fw} \tag{1b}$$
$$+ \rho_{fw}\phi_{fw}d_t\varepsilon_{kk} + Q,$$

$$d_t(\rho_{bw}\phi_{bw}) = \rho_{bw}\phi_{bw}d_t\varepsilon_{kk} - Q. \tag{1c}$$

The average water flow velocity can be determined by a generalized flow formula [10, 17] as shown in (2), where k is the intrinsic permeability of the solid skeleton, μ is the kinematic viscosity of the fluid, g_i is the gravity acceleration vector, and T is the temperature. The term $\theta_{ik}\nabla_kT$ in (2) represents the coupling between the water flow and the thermal conduction, where θ_{ik} is a thermal coupling coefficient [10]:

$$\phi_{fw}\left(v_i^{fw} - v_i^s\right) = -\frac{k}{\mu}\left[\nabla_i p - \rho_{fw}g_i\right] - \theta_{ik}\nabla_kT. \tag{2}$$

In (1a), (1b), (1c), and (2), the density of each phase, the intrinsic permeability of the solid skeleton, and the kinematic viscosity of the fluid are the functions of temperature and stresses, as shown in (3a), (3b), (4a), and (4b). Wherein, p' is the effective mean stress, p is the pore pressure, K_s is the bulk modulus of the solid particle, c_f is the compressibility of the liquid phase, and T_0 is the reference temperature. ρ_{s0} and ρ_{f0} are the densities of solid and liquid phases, respectively, at zero pressure and the temperature $T = T_0$. As shown in (4a), k is the function of the porosity, where k_0 and b_k are material parameters. The kinematic viscosity of the fluid is the function of temperature [18], as in (4b):

$$\rho_s = \rho_{s0}\left[1 - 3\beta_s(T - T_0) + \frac{p'}{(1-\phi)K_s}\right], \tag{3a}$$

$$\rho_{fw} = \rho_{bw} = \rho_{f0}\left[1 + c_f p - \beta_f(T - T_0)\right], \tag{3b}$$

$$k = k_0\exp\left(\frac{b_k}{1-\phi_{fw}}\right), \tag{4a}$$

$$\mu = 1.984 \times 10^{-6}\exp\left(\frac{1825.85}{T}\right)(\text{N} \cdot \text{s/m}^2). \tag{4b}$$

2.2.2. Momentum Conservation Equation.

The momentum conservation for saturated soils is shown in (5). Wherein, σ_{ij}^s, σ_{ij}^{fw}, and σ_{ij}^{bw} are the stresses of the solid phase, the free water phase, and the bound water phase, respectively. All of them are defined as average stresses on the section with a total surface porosity of ϕ and taken as positive under compression. Thus, the pore pressure also contributes to the stress of solid phase. According to the mixture theory [10], stresses of each phase can be expressed as shown in (6a), (6b), and (6c), where σ_{ij}' is the effective stress and K_b is the bulk modulus of the solid skeleton. Thus, the total stress can be written as in (7), where α is the Biot coefficient which takes the compressibility of solid particles into account. Equation (7) is just the generalized effective stress principle in soil mechanics:

$$
\begin{aligned}
&\rho_s \left(1 - \phi\right) d_t v_i^s + \rho_{\text{bw}} \phi_{\text{bw}} d_t v_i^{\text{bw}} \\
&\quad + \rho_{\text{fw}} \phi_{\text{fw}} d_t v_i^{\text{fw}} + \left(v_i^{\text{fw}} - v_i^s\right) \rho_{\text{fw}} \phi_{\text{fw}} \nabla_j v_i^{\text{fw}} \\
&\quad + \nabla_j \left(\sigma_{ij}^s + \sigma_{ij}^{\text{fw}} + \sigma_{ij}^{\text{bw}}\right) \\
&= g_i \left[\rho_s \left(1 - \phi\right) + \rho_{\text{fw}} \phi_{\text{fw}} + \rho_{\text{bw}} \phi_{\text{bw}}\right],
\end{aligned}
\tag{5}
$$

$$
\sigma_{ij}^s = \sigma_{ij}' + \left[1 - \frac{K_b}{K_m \left(1 - \phi\right)}\right] p \left(1 - \phi\right) \delta_{ij}, \tag{6a}
$$

$$
\sigma_{ij}^{\text{fw}} = p \phi_{\text{fw}} \delta_{ij}, \tag{6b}
$$

$$
\sigma_{ij}^{\text{bw}} = p \phi_{\text{bw}} \delta_{ij}, \tag{6c}
$$

$$
\sigma_{ij} = \sigma_{ij}^s + \sigma_{ij}^{\text{fw}} + \sigma_{ij}^{\text{bw}} = \sigma_{ij}' + \alpha p \delta_{ij}, \quad \alpha = 1 - \frac{K_b}{K_m}. \tag{7}
$$

2.2.3. Entropy Increase Equation.

In the thermodynamic framework of the hydrodynamic method [19, 20], the energy conservation laws are fully used in the derivation of the constitutive relations. Therefore, the entropy increase equation is adopted as the governing equation of the thermal field. According to the thermodynamics, the change of entropy should consist of the entropy production and the entropy flow induced by water flow, thermal conduction, and thermal convection. Define the specific entropy of the solid phase, the free water phase, and the bound water phase as v_s, v_{fw}, and v_{bw}, respectively. The entropy increase equation of saturated soils is

$$
\begin{aligned}
&\rho_s \left(1 - \phi\right) d_t v_s + \rho_{\text{fw}} \phi_{\text{fw}} d_t v_{\text{fw}} \\
&\quad + \rho_{\text{bw}} \phi_{\text{bw}} d_t v_{\text{bw}} = \frac{R}{T} + \nabla_k \frac{\kappa \nabla_k T}{T} \\
&\quad - \left(v_k^{\text{fw}} - v_k^s\right) \rho_{\text{fw}} \phi_{\text{fw}} \nabla_k v_{\text{fw}},
\end{aligned}
\tag{8a}
$$

$$
v_s = C_s \ln\left(\frac{T}{T_0}\right) + v_{s0}, \tag{8b}
$$

$$
v_{\text{fw}} = v_{\text{bw}} = C_f \ln\left(\frac{T}{T_0}\right) + v_{f0}, \tag{8c}
$$

where C_s and C_f are the specific heat capacity of solid and fluid phases, respectively; v_{s0} and v_{f0} are the specific entropy of solid and fluid phases, respectively, at $T = T_0$; κ is the effective thermal conductivity of saturated soils; R is the energy dissipation rate induced by all unrecoverable process, including the viscosity of each phase, the thermal conduction, the pore water flow, the transient elasticity, and the granular fluctuation [14–16, 21]. R/T is called entropy production rate. In this paper, only the dissipations corresponding to the transient elasticity and the granular fluctuation are considered for the purpose of simplification, as they are the most important dissipation mechanisms for saturated soils.

The transient elasticity, which corresponds to the relaxation of elastic potential energy and means that the materials are no longer elastic as long as external loadings begin, is an important mechanism of nonelastic deformation of solid materials [14–16]. For granular solid materials, under external loadings, there are random fluctuation movements at the granular scale around the macroaverage movements, such as the slide, roll, and collision between particles, corresponding to the kinetic energy fluctuation. Along with the kinetic energy fluctuation, there is also an elastic potential energy fluctuation at granular scale since that the elastic potential energy is stored through the interactions between particles. The granular fluctuation is always accompanied by the energy dissipation and is the important source of nonelastic deformation for granular solid materials. If there are no continuous incentives, the granular fluctuation will be attenuated by the form of macroenergy dissipation (i.e., heat generation) until the fluctuation disappears. Unlike the discrete approach in granular mechanics, this paper adopts a continuum method called "double entropy" theory proposed by Jiang and Liu [14–16] who introduce the concepts of granular entropy density s_g and the granular entropy temperature T_g to give a unified consideration of the granular fluctuation.

According to the nonequilibrium thermodynamics [19, 20], energy dissipation rate R can be expressed as the sum of multiplying all dissipative forces (generalized force that represents the driving actions making a system deviate from the thermodynamic equilibrium state) and corresponding dissipative flows. The dissipative forces of transient elasticity and granular fluctuation are the thermodynamic conjugate state variable of the elastic strain and the granular entropy temperature [16, 21], denoted as π_{ij} and T_g, respectively. T_g is the conjugate variable of the granular entropy density s_g and presents the severe degree of the granular fluctuation. It can be proved that π_{ij} approximatively equals the effective stress of saturated soils; that is, $\pi_{ij} \approx \sigma_{ij}'$ [21]. Define the dissipative flows corresponding to π_{ij} and T_g as Y_{ij} and I_g, respectively. Thus,

$$
R = Y_{ij} \pi_{ij} + I_g T_g. \tag{9}
$$

According to the linear nonequilibrium thermodynamics, the dissipative flows can be expressed as the functions of dissipative forces, as in (10a) and (10b). Wherein, λ_{ijkl} and γ are called migration coefficients:

$$Y_{ij} = \lambda_{ijkl}\pi_{kl}, \tag{10a}$$

$$I_g = \gamma T_g. \tag{10b}$$

2.3. Constitutive Relations of Mechanical Field.
As long as evolution laws of the strain of solid skeleton and the effective stress are determined, a completed THM coupling model for saturated soils can be established, using (1a)–(1c)–(10a) and (10b). It is worthwhile mentioning that in the framework of hydrodynamic method [12–16], all of these evolution laws can be derived theoretically. The derivation process of these laws can be found in [16], and the results are given directly in this paper.

2.3.1. Hyperelastic Relation.
In this paper, the effective stress is defined as the function of elastic strain ε_{ij}^e of the solid skeleton through the elastic potential energy density function ω_e. This hyperelastic relation can be expressed as

$$\sigma_{ij}' \approx \pi_{ij} = \frac{\partial \omega_e}{\partial \varepsilon_{ij}^e}. \tag{11}$$

Introducing a nonlinear term to the linear model of ω_e, Jiang and Liu [14–16] proposed a model for dry sands that can represent the effect of effective mean stress on the elastic modulus and provide a maximum stress ratio in effective stress space. Based on the Jiang and Liu model, this paper proposes a new model of ω_e as shown in (12a) and (12b), considering the cohesion of soils and the thermoelastic coupling:

$$\omega_e = \frac{2}{5}B\left(\varepsilon_v^e + c\right)^{1.5}\left(\varepsilon_v^e\right)^2 + B\xi\left(\varepsilon_v^e + c'\right)^{1.5}\left(\varepsilon_s^e\right)^2$$
$$+ \int 3K_e\beta_T\left(T - T_0\right)d\varepsilon_v^e, \tag{12a}$$

$$B = B_0 \exp\left(B_1\rho_d\right), \quad \varepsilon_v^e = \varepsilon_{kk}^e, \quad \varepsilon_s^e = \sqrt{e_{ij}^e e_{ij}^e}. \tag{12b}$$

In (12a) and (12b), ε_v^e and ε_s^e are the elastic volumetric strain and the second invariant of the elastic strain; $e_{ij} = \varepsilon_{ij} - \varepsilon_{kk}\delta_{ij}/3$ is the deviatoric strain tensor; B_0 is a parameter with the same dimension as stresses; ρ_d is the dry density of the saturated soils, that is, $\rho_d = \rho_s(1 - \phi)$; B_1 is a parameter that represents the effect of the dry density on the elastic modulus and shear strength of soils; c is a parameter related to the cohesion and should be taken as zero for cohesionless materials like sands; ξ is a parameter related to the shear behavior of soils; c' is a parameter relevant to the critical shear strength. The last term in (12a) and (12b) is the thermoelastic coupling term, where β_T is the elastic expansion coefficient of the solid skeleton and K_e is the secant elastic bulk modulus of

the solid skeleton. K_e is defined by $\pi_{kk} = 3K_e[\varepsilon_v^e + \beta_T(T - T_0)]$. Thus, from (11), (12a), and (12b),

$$K_e = 0.6B\left(\varepsilon_v^e + c\right)^{0.5}\varepsilon_v^e$$
$$+ \frac{1.5B\xi\left(\varepsilon_v^e + c'\right)^{0.5}\left(\varepsilon_s^e\right)^2}{\varepsilon_v^e} + 0.8B\left(\varepsilon_v^e + c\right)^{1.5}, \tag{13}$$

provided that β_T is only dependent on the thermal expansion of the solid phase and the bound water phase, as the thermal expansion of the free water does not contribute to the elastic deformation of the solid skeleton. Thus,

$$\beta_T = \beta_s\left(1 - \phi\right) + \beta_f\phi_{bw}. \tag{14}$$

2.3.2. "Granular Entropy Increase" Equation.
Define the specific granular entropy as $v_g = s_g/\rho_s$. Similar to the entropy increase equation, the "granular entropy increase" equation is

$$\rho_s\left(1 - \phi\right)d_t v_g = \frac{R_g}{T_g} - I, \tag{15a}$$

$$R_g = \sigma_{ij}^g d_t\varepsilon_{ij} + M d_t T, \tag{15b}$$

$$I = \frac{I_g T_g + Y_{ij}^g\pi_{ij}}{T_g}, \tag{15c}$$

$$\sigma_{ij}^g = \eta_g T_g d_t e_{ij} + \zeta_g T_g d_t\varepsilon_{kk}\delta_{ij}, \tag{15d}$$

$$M = \frac{\psi_g T_g\pi_{kk}\alpha_{bf}\phi_{bw}}{3\left(1 - \phi\right)}, \quad \left(\psi_g > 0 \ (d_t T > 0)\right);$$
$$\psi_g = 0 \ (d_t T \leq 0)\right), \tag{15e}$$

where R_g is the granular fluctuation energy production rate induced by external stimulations and can be determined by (15b) using a similar method described in (9) ($R_g \geq 0$). Note that both mechanical loadings and thermal loading can stimulate the reorganized movement of the particles, such as the mechanical consolidation and the nonisothermal consolidation [5–7]. In this paper, the "dissipative forces" for granular fluctuation are the strain rate and the temperature rate, as in (15b). The "dissipative flows" corresponding to $d_t\varepsilon_{ij}$ and $d_t T$ are denoted as σ_{ij}^g and M, respectively. σ_{ij}^g is expressed as a linear function of the strain rate, as in (15d), where η_g and ζ_g are material parameters also called migration coefficients. M can be theoretically determined by the conversion process from bound water to free water during temperature elevation [21], as in (15e), where ψ_g is a material parameter.

Different from the entropy, the granular entropy can be converted to the "real" entropy. In (15a)–(15e), I is the conversion rate of the granular entropy to the real entropy and should satisfy $I \geq 0$. In (15c), $I_g T_g$ and $Y_{ij}^g\pi_{ij}$ are the dissipation rates of the kinetic energy fluctuation and the elastic potential energy fluctuation into real entropy, respectively. Y_{ij}^g is the elastic potential energy fluctuation "dissipative flow."

The energy dissipation corresponding to the elastic potential energy fluctuation described by $Y_{ij}^g \pi_{ij}$ will be transformed to be a part of the transient elasticity dissipation described by $Y_{ij}\pi_{ij}$ in (9). The energy dissipation rate $I_g T_g$ described in (9) represents only the energy dissipation of granular kinetic energy fluctuation.

By defining a granular fluctuation energy density function, the relation between the specific granular entropy v_g and the granular entropy temperature T_g can be obtained [16, 21], as shown as follows (16):

$$v_g = bT_g. \tag{16}$$

2.3.3. Evolution Laws of Elastic and Nonelastic Strains. Divide the strain rate into the elastic strain rate $d_t\varepsilon_{ij}^e$ and the nonelastic strain $d_t\varepsilon_{ij}^D$; that is,

$$d_t\varepsilon_{ij}^e = d_t\varepsilon_{ij} - d_t\varepsilon_{ij}^D. \tag{17}$$

In this paper, the nonelastic strain evolution is quantitatively determined by the dissipations induced by the transient elasticity and the granular fluctuation described by (9), (10a), (10b), (15a)–(15e), and (16). Using the thermodynamic principles proposed by Jiang and Liu [16], it can be proved that the nonelastic strain rate has a relation with the dissipative flow Y_{ij} and Y_{ij}^g, as shown in (18a). Here, a simplified model for the dissipative flow Y_{ij} described in (10a) has been proposed as the so-called relaxation time model [16, 21], as in (18b). λ_v, λ_s, and a are material parameters. The transient elasticity must be stimulated by the granular fluctuation. Therefore, in (18b), $Y_{ij} = 0$ when $T_g = 0$:

$$d_t\varepsilon_{ij}^D = Y_{ij} - Y_{ij}^g, \tag{18a}$$

$$Y_{ij} = \lambda_s(T_g)^a e_{ij}^e + \lambda_v(T_g)^a \varepsilon_{kk}^e \delta_{ij}. \tag{18b}$$

Following with (18b), an equivalent nonelastic strain ε_{ij}^h is introduced to determine the value of Y_{ij}^g, as in (19a). The evolution of ε_{ij}^h is defined by the nonelastic strain rate, as shown in (19b). w and h are material parameters. The second term on the right side of (19b) restricts the value of $\sqrt{\varepsilon_{ij}^h \varepsilon_{ij}^h}$ within a maximum value h:

$$Y_{ij}^g = \lambda_s(T_g)^a e_{ij}^h + \lambda_v(T_g)^a \varepsilon_{kk}^h \delta_{ij}, \tag{19a}$$

$$d_t\varepsilon_{ij}^h = d_t\varepsilon_{ij}^D - w\frac{d_t\varepsilon_{kl}^D \cdot \varepsilon_{kl}^h}{h\sqrt{\varepsilon_{mn}^h \varepsilon_{mn}^h}}$$

$$\times \varepsilon_{ij}^h \quad \begin{cases} w = 1 & d_t\varepsilon_{kl}^D \cdot \varepsilon_{kl}^h > 0 \\ 0 < w < 1 & d_t\varepsilon_{kl}^D \cdot \varepsilon_{kl}^h < 0. \end{cases} \tag{19b}$$

Setting $\tilde{T}_g = \lambda_s^{1/a} T_g$, $m_1 = \lambda_v/\lambda_s$, $m_2 = (\lambda_s)^{1/a}\eta_g/\gamma$, $m_3 = \zeta_g/\eta_g$, $m_4 = \gamma/b$ and $m_5 = (\lambda_s)^{1/a}\psi_g/b$ and combining with (10b), (15a)–(15e), and (16), the "granular entropy increase" equation expressed with T_g can be written in (20). The term

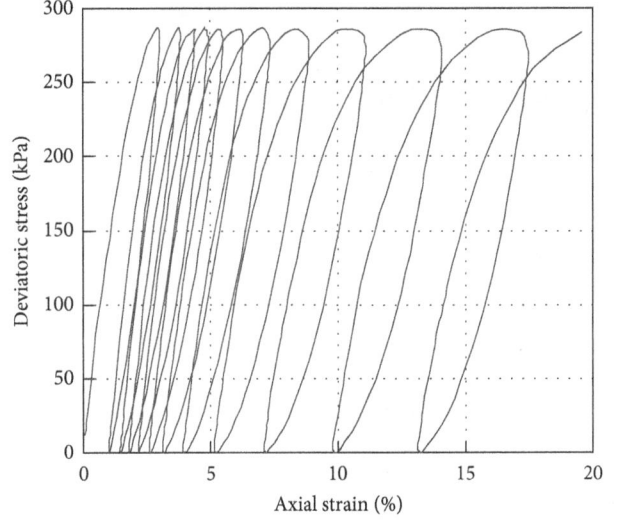

FIGURE 1: Simulation result of one-way cyclic undrained triaxial shear test using the constitutive model presented in this paper.

$Y_{ij}^g \pi_{ij}$ is ignored in (20) for the simplification of the model calculation and parameter calibration:

$$d_t\tilde{T}_g = \frac{m_2 m_4 d_t e_{ij} d_t e_{ij} + m_2 m_3 m_4 (d_t \varepsilon_{kk})^2}{\rho_d}$$

$$+ \frac{m_5 \pi_{kk} \alpha_{bf} \phi_{fw}}{3\rho_d(1-\phi)} d_t T - \frac{m_4}{\rho_d}\tilde{T}_g. \tag{20}$$

2.4. Model Summary. The mass conservation equations (1a)–(1c), the generalized water flow formula (2), the momentum conservation equation (5), the effective stress principle (7), the entropy increase equation (8a)–(8c)–(10a), and (10b), and the constitutive model of the mechanical field form a THM coupling model for saturated soils. The constitutive model in this paper does not need the concepts such as yield surface and flow rule. It contains the following three parts: (1) hyperelastic relation, (11), (12a), and (12b); (2) evolution laws of elastic and nonelastic strains, (17)–(19a), and (19b); (3) "granular entropy increase" equation, (20). If the strain rate $d_t\varepsilon_{ij}$ is given, the granular entropy temperature T_g can be determined using (20) and the elastic strain value can be obtained from (17)–(19a) and (19b). Thus, the effective stress can be calculated by (11), (12a), and (12b). Using this constitutive model, many important mechanical features of soils can be simulated, such as the hysteresis behavior of soils under cyclic shear loadings, as shown in Figure 1. In the following, this model will be applied to the analysis of nonisothermal consolidation, which is a kind of THM coupling behavior.

3. Simulation and Discussion

Based on the model presented before, in this section, the nonisothermal consolidation for silty clay under cyclic thermal loadings will be simulated in order to validate the model.

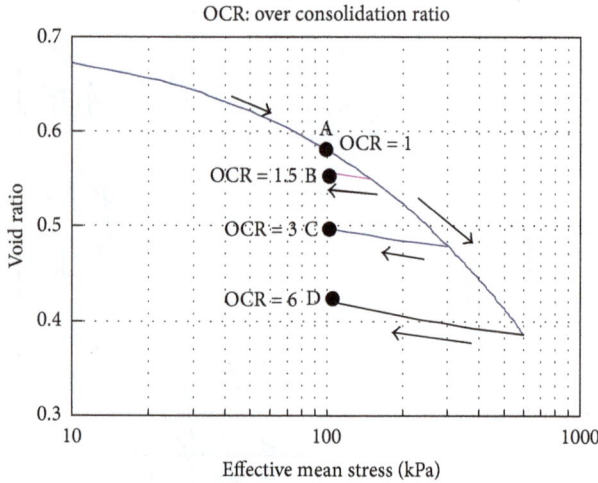

FIGURE 2: Mechanical consolidation simulation results before non-isothermal consolidation.

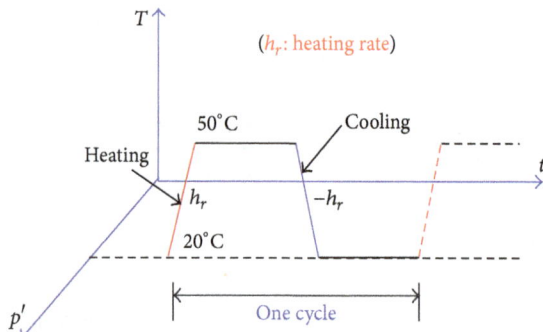

FIGURE 3: Repeated nonisothermal consolidation paths.

The elemental measured results of nonisothermal consolidation for silty clay provided by Bai and Su [11] will be used in this paper.

3.1. Simulation Paths. As shown in Figure 2, before the non-isothermal consolidation, the clay samples are isotropically consolidated to different initial states: (1) normally consolidated state (point A in Figure 2) and (2) overconsolidated state (points B, C, and D in Figure 2). When the mechanical consolidation is finished, keeping the confining pressure constant, repeated thermal loading is applied to the clay samples, as shown in Figure 3. In each thermal loading cycle, under undrained condition, the samples are heated at a constant heating rate h_r to a maximum temperature and cooled at a cooling rate $-h_r$ to the initial temperature. After the heating and cooling processes are finished, under drained condition, the temperature is kept constant for a period of time so that the pore pressure induced by the thermal loadings can fully dissipate.

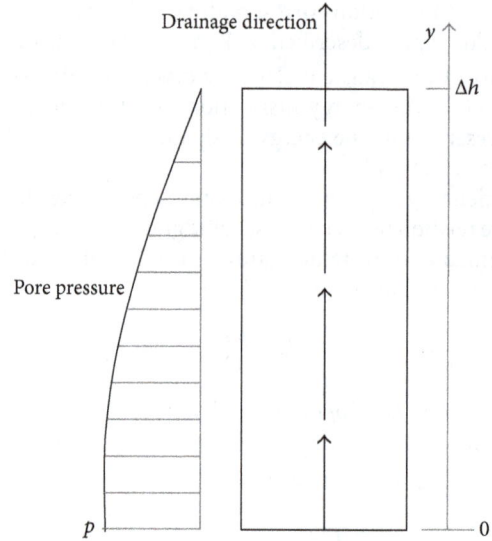

FIGURE 4: Schematic diagram of drainage in a sample.

For isotropic consolidated state, $\varepsilon_s^e = 0$. Thus, from (11), (12a), and (12b), the effective mean stress p' and the confining pressure $\overline{\sigma} = \sigma_{kk}/3$ are, respectively,

$$p' = 0.6B(\varepsilon_v^e + c)^{0.5}(\varepsilon_v^e)^2 + 0.8B(\varepsilon_v^e + c)^{1.5}\varepsilon_v^e$$
$$+ 3K_b\beta_T(T - T_0),$$
(21a)

$$\overline{\sigma} = p' + p.$$
(21b)

In (21b), the Biot coefficient α is set at 1. The clay samples are assumed to be with uniform deformation and temperature in order to perform calculations at the elemental scale. Thus, the gradient of fluid density in (1b) and the gradient of temperature in (2) are ignored. An example of uniform upward drainage is shown in Figure 4, with a unit height of Δh, a pore water pressure of zero at the top, and a pore water pressure of $p(y = 0)$ at the bottom, provided that the pore pressure in the samples is uniform along the horizontal direction and the change rate of pore pressure along the height of the samples is zero at the bottom. Thus, the pore pressure along the height of the samples and the term $[\phi_{\text{fw}}(v_k^{\text{fw}} - v_k^s)]_{,k}$ in (1b) can be simplified as

$$p(y) = -\frac{p(y=0)}{\Delta h^2}y^2 + p(y=0),$$
(22a)

$$\left[\phi_{\text{fw}}\left(v_k^{\text{fw}} - v_k^s\right)\right]_{,k} = -\frac{k}{\mu}\nabla_k(\nabla_k p) = 2\frac{k}{\mu\Delta h^2}p(y=0).$$
(22b)

Due to uniform deformation being assumed within the samples, the deformation state at the bottom of the samples is used to represent the overall deformation state. In the following, the pore pressure p represents the pore pressure value at the bottom.

(a)

(b)

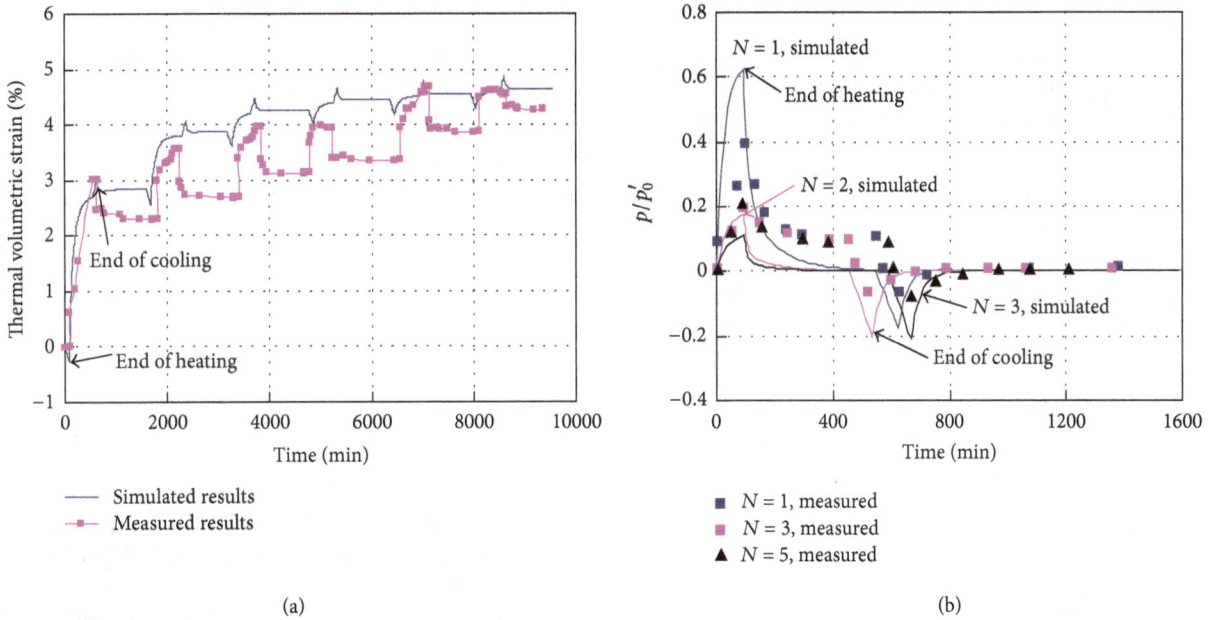

FIGURE 5: Responses of thermal volumetric strain (a) and pore pressure (b) during the repeated nonisothermal consolidation (OCR = 1; $\bar{\sigma} = 100\,\text{kPa}$; N represents the cycle number; measured results are from [11]).

In the simulations, the change rates of the temperature and the confining pressure are controlled by tests; that is, $d_t(p' + p) = f_1(t)$ and $d_t T = f_2(t)$ are known conditions. In the mechanical consolidation, $f_2(t) = 0$; in the nonisothermal consolidation, $f_1(t) = 0$. Using these known conditions, the mass conservation equations (1a)–(1c), (3a), (3b), (4a), (4b), (22a), (22b), (17)–(21a), and (21b), the pore pressure development, and the strain evolution during the mechanical consolidation and nonisothermal consolidation can be calculated. Because both the total stress state and the temperature are known, the momentum conservation equation and the entropy increase equation are not used in the simulations. The main model parameters used in this paper are listed in Table 1.

FIGURE 6: Simulation results of the granular entropy temperature evolution under repeated thermal loadings.

3.2. Simulation Results. The mechanical consolidation simulation results have been shown in Figure 2 and the simulation results of nonisothermal consolidation will be discussed in this section. Figure 5 shows the responses of thermal volumetric strain and pore pressure during the repeated nonisothermal consolidation for normally consolidated silty clay. Figure 6 shows the evolution of granular entropy temperature during the repeated nonisothermal consolidation. In the simulations, the initial temperature is 20°C, the maximum temperature is 50°C, and the heating rate $h_r = 0.27$°C/min. From Figures 5 and 6, in the undrained heating processes, the pore pressure and the granular entropy temperature increase significantly. After the heating process is finished, the pore pressure begins to dissipate and thermal volumetric deformation develops significantly until the pore pressure and the granular entropy temperature decrease to zero. The negative pore pressure is generated in the undrained cooling process,

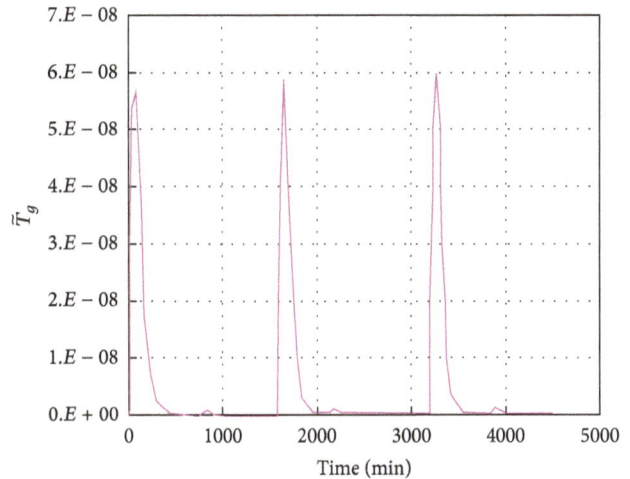

followed by a water-absorbing process and a volume expansion after the cooling process is finished. Figure 5 shows that the nonisothermal consolidation is an unrecoverable process. With the increase of cycle number, the thermal volumetric strain is accumulated gradually. After four cycles, the volume shrinkage induced by the heating process basically equals the volume expansion induced by the cooling process. Thus, the accumulation of thermal volumetric strain disappears after certain cycles of thermal loading. Correspondingly, the maximum pore pressure in each cycle decreases gradually, while no significant change in the minimum pore pressure is observed in all cycles. This feature is attributed to the

TABLE 1: List of the main parameters used in this paper.

Parameter type	Elastic potential energy density function parameters			Migration coefficient		Hysteric parameters	Non-isothermal consolidation	Thermal expansion coefficient ($^\circ$C^{-1})	
	B_0/Pa	B_1/(m^3/kg)	c/[—]	m_1/[—]	m_2/min$^{-(1-2a)/a}$	h/[—]	α_{bf}/$^\circ$C^{-1}	β_s	β_f
	c'/[—]	ξ/[—]		m_3/[—]	m_4/(kg/m^3/min)	w/[—]	m_5/(kg·min/m^4)		
Silty clay	510	0.0116	0.02	0.558	1.79×10^3	0.022	0.02	1×10^{-6}	3×10^{-4}
	0.12	0.276		0.447	48.4	0.98	3.2×10^{-6}		

Note: the two upper and lower values correspond to the upper and lower parameters for the corresponding clay. The parameter a in (18b) is taken as 0.455; the initial value of porosity of bound water is 0.1. Parameters B_0, B_1, c, c', and ξ can be seen in (12a) and (12b); parameters m_1, m_2, m_3, m_4, and m_5 can be seen in (20); parameters h and w can be seen in (19b); parameter α_{bf} can be seen in Section 2.2.1 and (15e). In (3) and (4), $K_s = 10^{12}$ Pa, $c_f = 10^{-10}$ Pa^{-1}, $k_0 = 6.4 \times 10^{-18}$ m^2, and $b_k = 3.25$.

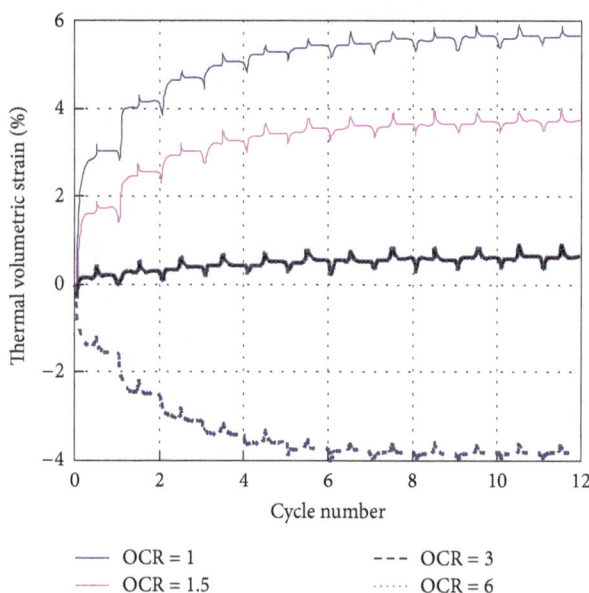

FIGURE 7: Thermal volumetric strain responses of silty clays with different OCR values under repeated thermal loadings.

of nonisothermal consolidation has been proved in many experimental studies [5–7, 11].

In summary, the THM coupling model presented in this paper is effective for the simulations of nonisothermal consolidation, which shows obvious OCR dependency and unrecoverability. In the model, the nonisothermal consolidation is described by the granular fluctuation stimulated by the conversion process between the bound water and free water phases during the thermal loadings; see (15b) and (15e). As long as the granular fluctuation begins, the nonelastic deformation will develop due to the transient elasticity of soils. That is the physical mechanism of the nonisothermal consolidation. The ability of the model to represent the features of nonisothermal behavior of saturated soils discussed before is very important and useful for engineering areas like the shallow geothermal engineering, in which repeated thermal loadings are applied perennially, possibly resulting in the development of excess pore pressure and the significant unrecoverable deformation of the ground.

4. Conclusions

A THM coupling model based on nonequilibrium thermodynamics has been established in this paper, including a constitutive model of the mechanical field without such concepts as yield surface and flow rule. The dependency of the permeability and the density of each phase on the deformation and temperature are taken into account. The entropy increase equation, in which the energy dissipation is described by the concepts of dissipative force and dissipative flow in nonequilibrium thermodynamics, is introduced as a basic governing equation of the model. In the model, the effective stress is defined as the function of the elastic strain and the dry density through the elastic potential energy density function, which considers the cohesion and thermoelastic coupling effect of soils. On the other hand, the nonelastic deformation evolution is determined by two important dissipation mechanisms called transient elasticity and granular fluctuation, which is described by the concept of granular entropy.

In the model, the granular fluctuation is linked with the conversion process between the bound water and free water phases under thermal loadings. Therefore, this model is able to represent the nonisothermal consolidation of

unrecoverable energy processes stimulated by the thermal loading, that is, the granular fluctuation and the triggered transient elasticity dissipation described by (18a), (18b), (19a), (19b), and (20). The simulation results shown in Figure 5 are basically consistent with the measured results provided by Bai and Su [11]. However, residual pore pressure in the nonisothermal consolidation is observed in the measured results. This may be due to the failure in considering the effect of temperature on the readings of pore pressure sensors in the experiments.

Figure 7 shows the simulation results of nonisothermal consolidation under repeated thermal loadings for silty clays with different OCR values. Obviously, the nonisothermal consolidation is heavily OCR dependent. For normally consolidated or slightly overconsolidated clay, the accumulation of volume shrinkage is observed under repeated thermal loadings. On the contrary, for heavily overconsolidated clays, the accumulation of volume expansion will be generated under repeated thermal loadings. This OCR dependency saturated soils under repeated thermal loadings. Simulation results show that the nonisothermal consolidation is heavily OCR dependent and unrecoverable. Under repeated thermal loadings, volume shrinkage will be generated for normally consolidated or slightly overconsolidated clay, while volume expansion will develop for heavily overconsolidated clay. The residual thermal volumetric strain is gradually accumulated until a maximum value is reached after several cycles of thermal loading. These simulation results are consistent with the experimental results of the nonisothermal consolidation.

Acknowledgment

This study was supported by the Tsinghua-Cambridge-MIT Low Carbon Energy University Alliance (TCM-LCEUA) seed funding project, to which the authors hereby express their sincere gratitude.

References

[1] Y. J. Cui, Y. F. Lu, P. Delage, and M. Riffard, "Field simulation of in situ water content and temperature changes due to ground-atmospheric interactions," *Geotechnique*, vol. 55, no. 7, pp. 557–567, 2005.

[2] P. Delage, N. Sultan, and Y. J. Cui, "On the thermal consolidation of Boom clay," *Canadian Geotechnical Journal*, vol. 37, no. 2, pp. 343–354, 2000.

[3] P. Kuntiwattanakul, I. Towhata, K. Ohishi, and I. Seko, "Temperature effects on undrained shear characteristics of clay," *Soils and Foundations*, vol. 35, no. 1, pp. 147–162, 1995.

[4] L. Moritz, "Geotechnical properties of clay at elevated temperature," Report 47, Swedish Geotechnical Institute, Linkioping, Sweden, 1995.

[5] C. Cekerevac and L. Laloui, "Experimental study of thermal effects on the mechanical behaviour of a clay," *International Journal for Numerical and Analytical Methods in Geomechanics*, vol. 28, no. 3, pp. 209–228, 2004.

[6] R. E. Passwell, "Temperature effects on clay soil consolidation," *Soil Mechanic and Foundations Division, ASCE*, vol. 93, pp. 9–22, 1967.

[7] M. Tidfors and G. Sallfors, "Temperature effect on preconsolidation pressure," *Geotechnical Testing Journal*, vol. 12, no. 1, pp. 93–97, 1989.

[8] T. Hueckel, A. Peano, and R. Pellegrini, "A constitutive law for thermo-plastic behaviour of rocks: an analogy with clays," *Surveys in Geophysics*, vol. 15, no. 5, pp. 643–671, 1994.

[9] L. Laloui and C. Cekerevac, "Thermo-plasticity of clays: an isotropic yield mechanism," *Computers and Geotechnics*, vol. 30, no. 8, pp. 649–660, 2003.

[10] F. G. Tong, L. R. Jing, and R. W. Zimmerman, "A fully coupled thermo-hydro-mechanical model for simulating multiphase flow, deformation and heat transfer in buffer material and rock masses," *International Journal of Rock Mechanics and Mining Sciences*, vol. 47, no. 2, pp. 205–217, 2010.

[11] B. Bai and Z. Q. Su, "Thermal Responses of Saturated Silty Clay During Repeated Heating-Cooling Processes," *Transport in Porous Media*, vol. 93, no. 1, pp. 1–11, 2012.

[12] L. D. Landau and E. M. Lifshitz, *Theory of Elasticity*, Pergamon Press, Oxford, UK, 3rd edition, 1986.

[13] P. C. Martin, O. Parodi, and P. S. Pershan, "Unified hydrodynamic theory for crystals, liquid crystals, and normal fluids," *Physical Review A*, vol. 6, no. 6, pp. 2401–2420, 1972.

[14] Y. M. Jiang and M. Liu, "Energetic instability unjams sand and suspension," *Physical Review Letters*, vol. 93, no. 14, pp. 1–148001, 2004.

[15] Y. Jiang and M. Liu, "From elasticity to hypoplasticity: dynamics of granular solids," *Physical Review Letters*, vol. 99, no. 10, Article ID 105501, 2007.

[16] Y. Jiang and M. Liu, "Granular solid hydrodynamics," *Granular Matter*, vol. 11, no. 3, pp. 139–156, 2009.

[17] S. M. Hassanizadeh, "Derivation of basic equations of mass transport in porous media, Part 2. Generalized Darcy's and Fick's laws," *Advances in Water Resources*, vol. 9, no. 4, pp. 207–222, 1986.

[18] V. Guvanase and T. Chan, "A three-dimensional numerical model for thermohydromechanical deformation with hysteresis in a fractured rock mass," *International Journal of Rock Mechanics and Mining Sciences*, vol. 37, no. 1-2, pp. 89–106, 2000.

[19] S. R. de Groot and P. Mazur, *Non-Equilibrium Thermodynamics*, North-Holland publishing company, Amsterdam, The Netherlands, 1962.

[20] L. Onsager, "Reciprocal relations in irreversible processes. I," *Physical Review*, vol. 37, no. 4, pp. 405–426, 1931.

[21] Z. C. Zhang and X. H. Cheng, "Formulation of Tsinghua-Thermosoil model: a fully coupled THM model based on non-equilibrium thermodynamic approach," in *Multiphysical Testing of Soils and Shales, SSGG*, L. Laloui and A. Ferrari, Eds., pp. 155–161, Springer, Berlin, Germany, 2013.

3

A Case Study on Stratified Settlement and Rebound Characteristics due to Dewatering in Shanghai Subway Station

Jianxiu Wang,[1,2] Tianrong Huang,[1,2] and Dongchang Sui[1,2]

[1] *Key Laboratory of Geotechnical and Underground Engineering of Ministry of Education, Tongji University, Shanghai 200092, China*
[2] *College of Civil Engineering, Tongji University, Shanghai 200092, China*

Correspondence should be addressed to Jianxiu Wang; jianxiu_wang@yeah.net

Academic Editors: S. Kaewunruen and X. Yan

Based on the Yishan Metro Station Project of Shanghai Metro Line number 9, a centrifugal model test was conducted to investigate the behavior of stratified settlement and rebound (SSR) of Shanghai soft clay caused by dewatering in deep subway station pit. The soil model was composed of three layers, and the dewatering process was simulated by self-invention of decompressing devise. The results indicate that SSR occurs when the decompression was carried out, and only negative rebound was found in sandy clay, but both positive and negative rebound occurred in the silty clay, and the absolute value of rebound in sandy clay was larger than in silty clay, and the mechanism of SSR was discussed with mechanical sandwich model, and it was found that the load and cohesive force of different soils was the main source of different responses when decompressed.

1. Introduction

Located in the Yangtze Delta, Shanghai is a typical area of soft soil distribution. As one of the most dynamic economic center in the world, metro railway construction is now being developed on a large scale, and the location of metro stations are often in densely populated districts. The dewatering measures were often adopted in metro station constructions, which often cause soil layer compression, land subsidence, and foundation's deformation. Therefore, it is of vital importance to conduct research on it and find out the corresponding measures to ensure the safety of construction and protection of the environment. For this reason, various researches have been conducted to investigate the relationship between land subsidence and water withdrawal by dewatering [1–11]. These studies discussed the effects of soil consolidation and land subsidence induced by dewatering, mostly in a perspective of mechanics as groundwater pumping leading to a decline in the groundwater level.

In the fields related to deformation of clay, Yuan et al. established a modified Cam-clay model with nonassociated flow rule considering S-D effect on deviatory plane and cohesion on meridian plane [12]. Yu et al. proposed a constitutive model to fully describe the mechanical behavior of the boom clay by combining the transversely isotropic elastic model and the modified Mohr-Coulomb criterion [13]. Taiebat et al. extended the SANICLAY model to include restructuration by conducting two distinct types of restructuration (isotropic and frictional) [14]. Liu and Carter described that the behavior of natural clay is proposed in a new four-dimensional space, consisting of the current stress state, stress history, the current voids ratio, and a measure of the current soil structure [15]. Cotecchia and Chandler suggested that the structure of clays may be distinguished simply as either sedimentation or postsedimentation, depending on whether gross yield in one-dimensional compression in void ratio-vertical effective stress plane occurs on the original sedimentation compression curve or to the right of this curve [16]. Mutman investigated the properties of bentonite clay stabilized by the burned olive waste and proposed a solution to the problem of the olive oil waste which are incriminated for a high quantity of pollution [17]. These works have made people gain more understanding about the characteristic of clay, but few were concerned about the special phenomenon (stratified settlement and rebound) in soft clay.

The stratified settlement means different settlement with different location in one layer, and the rebound includes positive and negative rebound, which was expansion and compression in the soil, respectively. The SSR was a brand new deformation behavior of soil; it was first discovered by Shanghai Tunnel Construction Corporation and reported by Wang et al. [18]; but this behavior had not been paid enough attention in engineering application. As the accelerating of the city's development, metro stations are now being conducted in more dense districts with considerable surrounding tall buildings, and the control of land subsidence become mores strictly. For example, the land subsidence needs to be controlled within 7 millimeters in Yishan metro station of number 3 Ming Zhu subway in Shanghai, PR, China [19]. Therefore, the research of stratified settlement and rebound is of great significance, for it affects the accurate calculation and control of land subsidence, which has also great potential in engineering application. In previous study [18], the SSR was mainly supported and explained by the field monitoring data, and no model tests have been conducted to do the relevant research, which prohibits the profound recognition of this phenomenon and potential application, so it is very necessary to conduct the centrifuge model test to research the mechanism of SSR.

This paper is to analyze the mechanism of SSR of Shanghai clay often found in the metro station construction by the centrifugal model test. For this purpose, the soil samples were obtained from the Yishan Metro Station Project of Shanghai Metro Line 9. Then, the model test was conducted by using the centrifuge of Tongji University and self-invention of decompressing devise, and the results are explicitly discussed. The relation between the distance from the center of pit and land subsidence is also discussed. Finally, the mechanism of SSR of Shanghai clay is analyzed based on a mechanical sandwich model. It brings useful reference to designers and is helpful to analyze and control ground subsidence in this situation.

2. Centrifugal Model Test

2.1. Engineering Background. Yishan Road station was located at Yishan Road with four storied island platform structure and the terminal station of Shanghai subway number 9. It extended from the west (West Zhongshan Road) to the east (Kaixuan Road) with a total distance of 297.40 m. The width of its standard part was 21.2 m, and the environmental conditions around Yishan Road station were very complex, and it required high level of environmental protection. The middle part of the south side of Yishan station was Shanghaiqijian decoration building with 17 storied concrete frame structure, which was 14 m away from the boundary of the enclosure protection of the pit; in the north of the station are Jianshijia building and Jinyindao material hall, which were 13 m away from the pit. In its east, there were Mingzhu viaduct and the Yishan Road station of Mingzhu subway. The minimum distance was 7 m between the pile cap of Mingzhu viaduct and the pit while that between the Yishan Road station of Mingzhu subway and the pit was 23 m; the west of the station was Zhong Shan viaduct with a minimum distance of 25 m between its center and the pit. All the buildings and roads had a strict requirement for settlement control which was mainly caused by dewatering of the confined aquifer. The

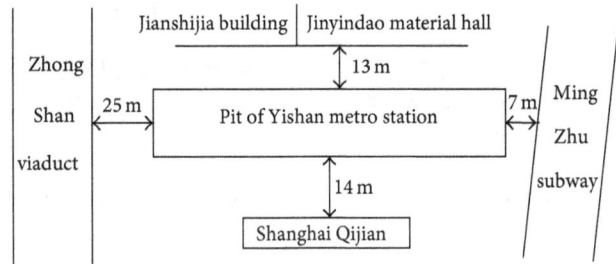

FIGURE 1: Layout of Yishan metro station.

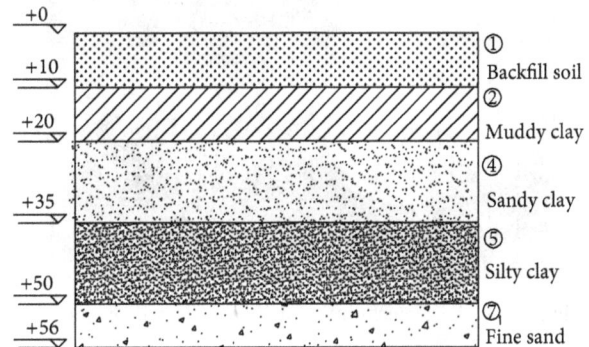

FIGURE 2: Geological section of Yishan metro station.

location and neighboring building of Yishan Road station are presented in Figure 1.

According to the geotechnical investigation report, the simplified geological section of Yishan Road station was shown in Figure 2.

2.2. The Geotechnical Centrifuge. This geotechnical centrifuge of Tongji University was shown in Figure 3, whose capacity is $150 \, g \cdot t$ and maximum acceleration is $200 \, g$, and it had been successfully applied to related research by a group of researchers in Tongji University [20–22].

2.3. Soil Properties and Preparation. Since the bearing stratum of metro stations is often located at the layer number 4 of sandy clay and layer number 5 of silty clay in Shanghai [23], this paper focus on this two kinds of soft clay in Shanghai. In this paper, in order to simulate the SSR, self-invention of decompressing devise was placed at the bottom of layer number 5 to simulate the decompression induced by dewatering in deep pit, and a topsoil layer of silty clay was placed on top of the two layers to stand for equal gravity. The properties of test soil are listed in Table 1.

The soil samples were demonstrated in Figure 4, and they were silty clay and sandy clay, respectively.

To minimize side friction, the wall of model was covered with a thin layer of smooth plastic membrane. The procedures of preparing soil layer were as follows:

(1) the silt clay obtained from the site was used to construct the bottom silt clay (layer number 5), 150 mm in thickness;

TABLE 1: The properties of the soil samples.

Layer number	Name	Density (g/cm^3)	Water content (%)	Porosity ratio	Cohesive force (kPa)	Internal friction angle (°)
Topsoil*	Silty clay	18.4	39.8	1.13	19	27.5
Layer number 4	Sandy clay	18.1	36.9	1.1	7	31
Layer number 5	Silty clay	18.4	39.8	1.13	20	27

*Means that layer topsoil is the layer with same gravity of all layers above layer number 4 and number 5.

FIGURE 3: The geotechnical centrifuge of Tongji University.

(2) layer No. 4 (150 mm in thickness) was constructed by the sandy clay obtained from the site;

(3) last, 30-mm-thick topsoil layer was constructed on top of layer No. 4 and No. 5.

The model was consolidated at 150 g for about 3 h after soil layers were constructed.

At the bottom of layer number 5, a self-invention of decompressing devise was adopted to simulate the decompression induced by dewatering in confined aquifer. It was composed by a plate of low-density polyethylene with 70 mm in thickness, 660 mm in length, and 660 mm in width, and a taper was on top of this plate with a 3 mm-thick silicone rubber membrane. The diameter of this taper was 600 mm, and the bottom of the taper was connected to the pressure control system by a gas transmission line. The decompression was simulated and controlled by the pressure control system. The soil layers and self-invention of decompressing devise were illustrated in Figure 5.

2.4. Instrument and Test Procedures. Confined to the size of test box, the model scale was taken as n = 20. The model box was $700 \times 900 \times 700$ mm in width, length, and height, respectively. To measure the stratified settlement and rebound, nine displacement meters were placed as Figure 6 indicates.

In Figure 6, WY1 were set in the centre of the model and reached the top of the decompressing device, and the displacement meters WY8, WY9, and WY10 were placed on top of the topsoil, which was to measure the displacement on topsoil, and their distance from the centre of the model was 90 mm, 180 mm, and 270 mm, respectively. In the same way, WY5, WY6, and WY7 were located on top of layer number 4, and WY2, WY3, and WY4 were located on top

of layer number 5. Therefore, WY8-WY5, which means the subtraction value of displacement WY8 and WY5, represents the settlement of layer number 4 with a distance of 90 mm from the model center, and WY9-WY6 and WY10-WY7 represented the settlement of layer number 4 with a distance of 180 mm and 270 mm from the model center, respectively. In the same case, WY5-WY2, WY6-WY3, and WY7-WY4 represented the stratified settlement of layer number 5 with a distance of 90 mm, 180 mm and 270 mm from the model center, respectively.

The centrifuge model test was conducted to study SSR of Shanghai clay. It has five accelerations, including 10 g, 20 g, 30 g, 40 g and 50 g. During the test, it needed 30 minutes to accelerate the centrifuge before it reached the stable stage. And the whole running time of the centrifuge was 2 h.

3. Test Results and Analysis

To express concisely, the displacement on different layer surface in 30 g were first chosen to be discussed, then the SSR in layer No. 4 and No. 5 was investigated respectively, and then the SSR in different accelerations was also discussed. Finally the mechanism of SSR was investigated with a mechanical sandwich model.

3.1. Displacement on Different Layer Surface in 30 g. The displacement on each layer's surface was drawn in Figures 7, 8, and 9, respectively.

As indicated, there existed no displacements in the soil at the time of 400 seconds, during which the WY1 displacement meter remained zero and mean no decompression. However, the displacement of all layers came to a sudden change when displacement of WY1 increased to 28 mm, but the displacement of WY1 remained constant from 400 to 1000 seconds, which also mean the constant decompression at the bottom of layer No. 5, and then the displacement began to decrease with the unloading of centrifuge when the time exceeded 1000 seconds. This demonstrated that the decompression had a strong impact on soil deformation and surface movement. Comparing the three figures, it can also be found that the displacement on top of topsoil was the smallest among the three layers, which showed that the decompression exerted the greatest influence on the nearby layer number 5 but the least impact on the surface of topsoil.

3.2. Stratified Settlement and Rebound

3.2.1. Stratified Settlement and Rebound in 30 g. Based on the data in Figures 7 and 8, the settlement of layer number 4 was drawn in Figure 10.

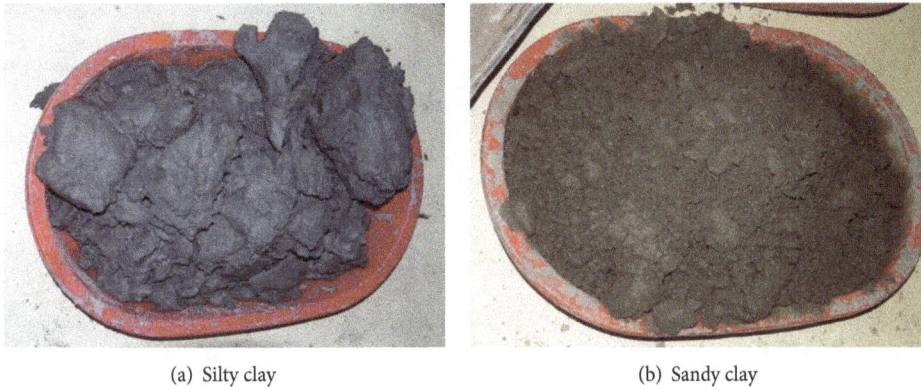

(a) Silty clay

(b) Sandy clay

FIGURE 4: Pictures of soil samples.

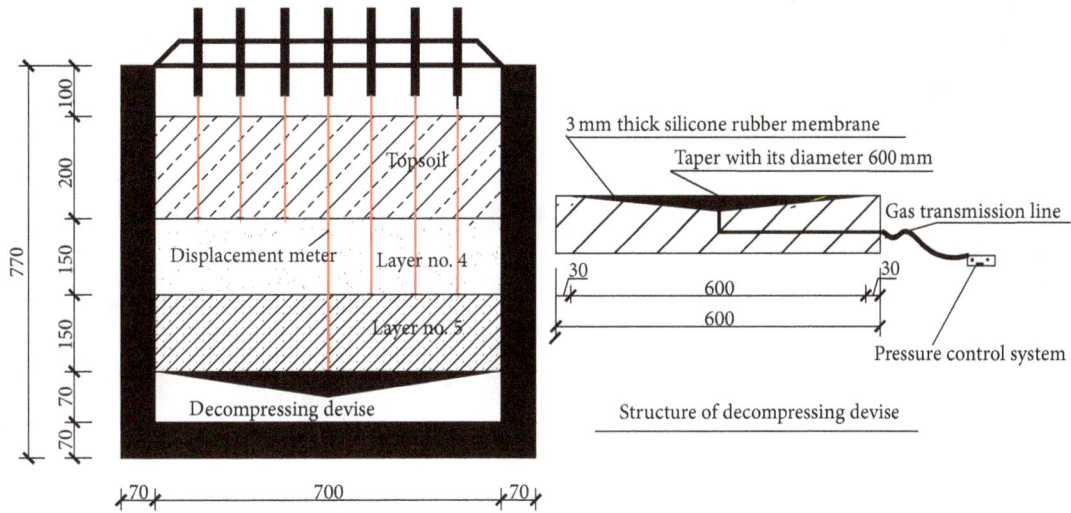

FIGURE 5: The distribution of stratum and decompressing devise in test model (unit: mm).

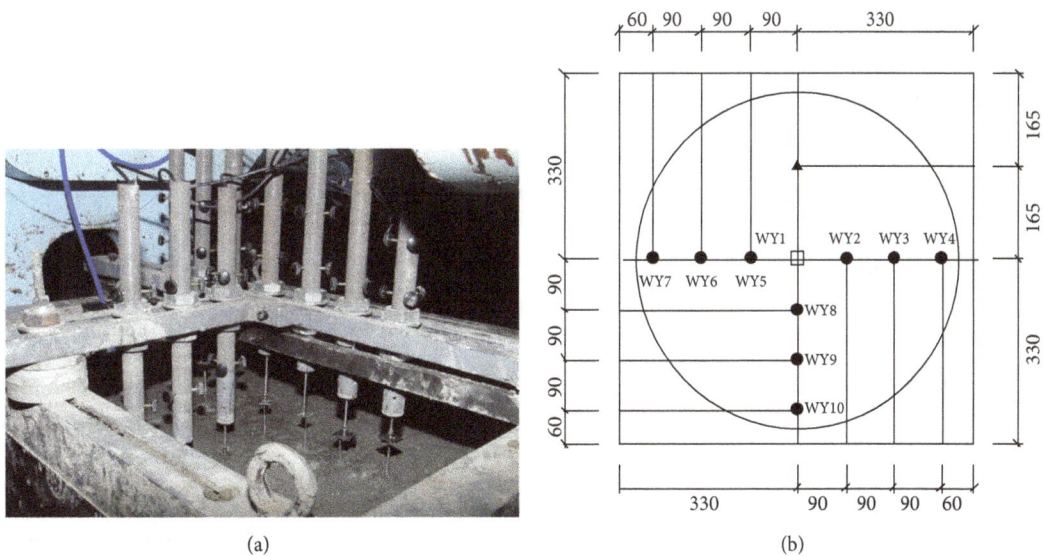

(a)

(b)

FIGURE 6: The distribution of displacement meters (unit: mm).

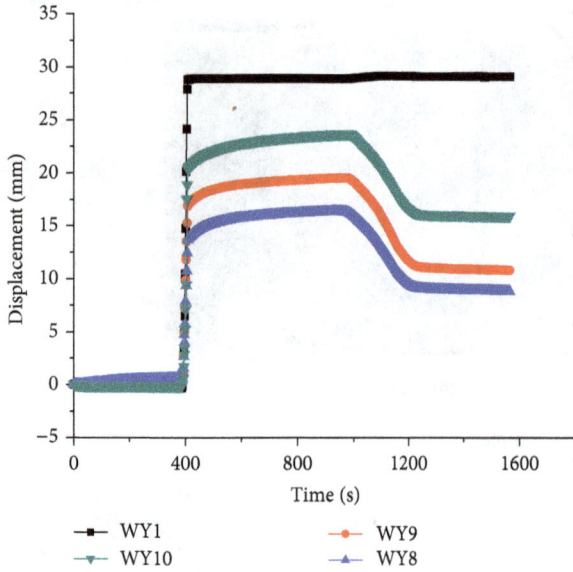

FIGURE 7: The displacement on top of topsoil.

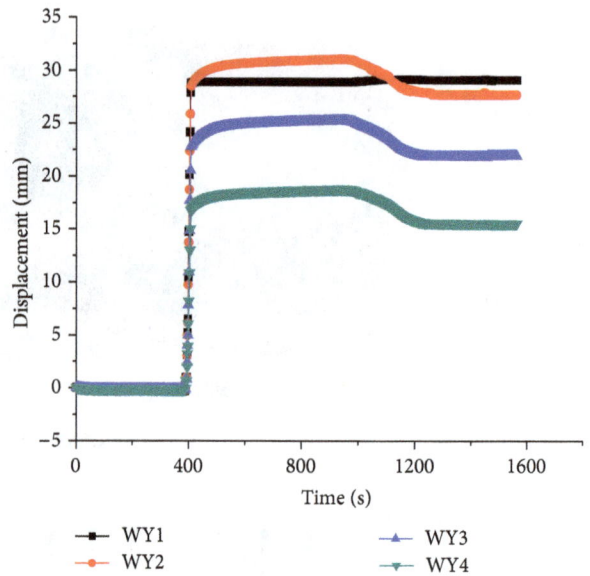

FIGURE 9: The displacement on top of layer number 5.

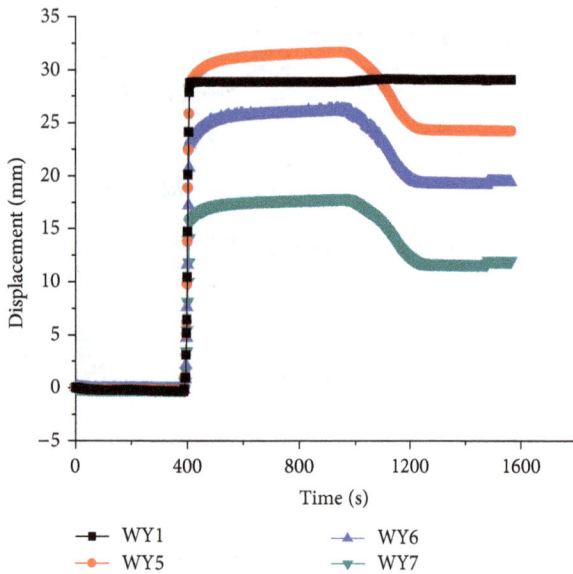

FIGURE 8: The displacement on top of layer number 4.

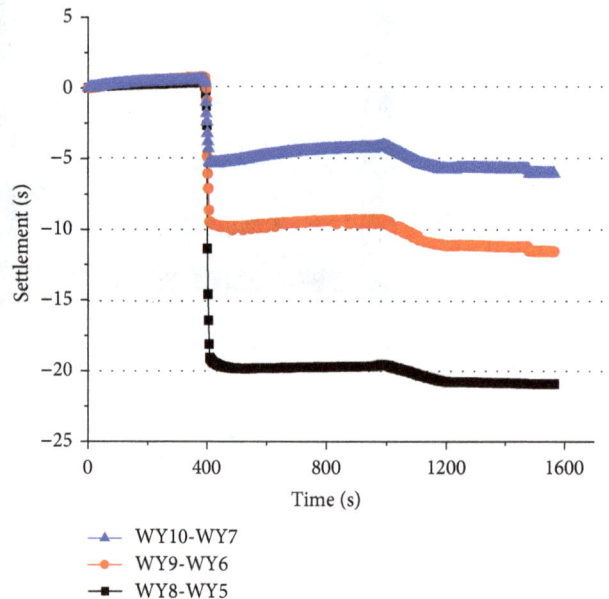

FIGURE 10: The settlement of layer number 4.

As mentioned in Section 2.3, WY8-WY5, WY9-WY6, and WY10-WY7 represented the settlement of layer No. 4 with a distance of 90 mm, 180 mm, and 270 mm from model center, respectively. From Figure 8, it was found that the settlement of WY8-WY5, WY9-WY6, and WY10-WY7 remained zero before the time of 400 seconds. After the decompression began at 400 seconds, the settlement come to a sudden decrease but was of different value, respectively. The settlement of WY10-WY7 dropped from 0 mm to −5 mm, and the settlement of WY8-WY5 and WY9-WY6 decreased from 0 mm to −10.5 mm and −20.0 mm, respectively, so the settlement of layer number 4 was not unanimous with different location, which was called stratified settlement, and

the absolute value of stratified settlement decreased with increasing distance from the model center.

On the other hand, the settlement after decompression also stands for the rebound occurred in the layer number 4. This rebound was negative and often called compression. And it was found that the rebound decreased with the distance from the model center too, and the rebound only occurred after decompression began in layer number 5, which showed that decompression was the main source of rebound.

In the same way, according to the data of Figures 8 and 9, the settlement of layer number 5 was shown in Figure 11.

In Figure 11, WY5-WY2, WY6-WY3, and WY7-WY4 represent the settlement with a distance of 90 mm, 180 mm,

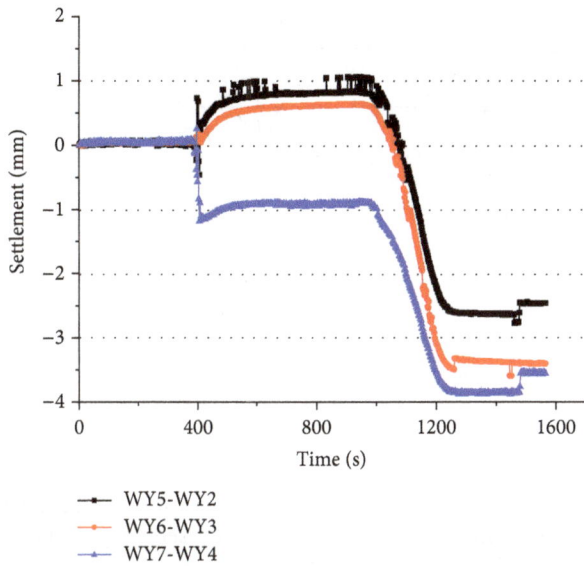

FIGURE 11: The settlement of layer number 5.

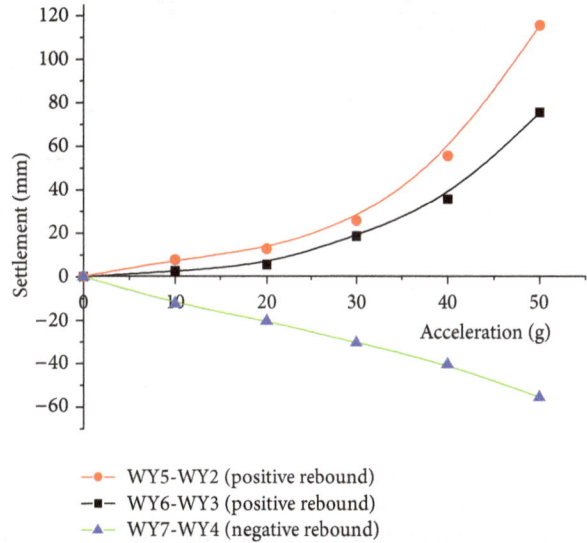

FIGURE 13: The settlement of layer number 5 in different accelerations.

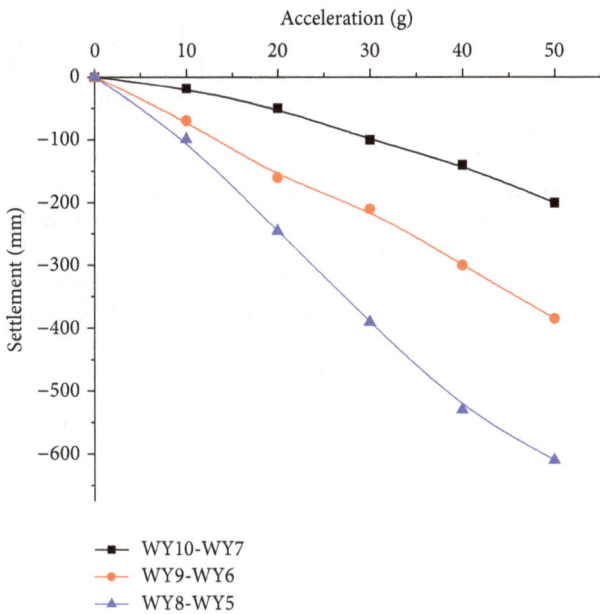

FIGURE 12: The settlement of layer number 4 in different accelerations.

and 270 mm in layer number 5 respectively. Similarly, the SSR was also found in layer number 5 with the decompression. However, there existed two kinds of rebound in layer number 5, positive rebound and negative rebound, which are also called expansion and compression, respectively; for the WY5-WY2 and WY6-WY3, the rebound was positive, so the soil was expanded when the decompression began, and it was also found that the value of expansion decreased with the increasing distance from model center, and for WY7-WY4, the rebound was negative which shows that the soil was compressed. In addition, comparing with Figure 7, we find

that the absolute value of positive and negative rebound in layer number 5 was smaller than those in layer number 4.

In previous study [18], it was reported that the stratified settlement and rebound existed and decreased with the distance from the center of the pit when the dewatering measure was taken in the confined water. Therefore the centrifuge model test is feasible and keeping well confirmation with the field monitoring information.

3.2.2. Stratified Settlement and Rebound in Different Acceleration. According to the test scheme, the settlement of layer number 4 and number 5 in different acceleration was drawn in Figures 12 and 13 respectively (The value of SSR has multiplied with the scale 20 to make it more clearly).

As it was found in Figure 12 that, the settlement of layer number 4 increased with the acceleration; the bigger acceleration mean larger t settlement in the sandy clay, which may be interpreted that larger acceleration makes the soil more compact just like the soil's consolidation, and it was also found that the absolute value of settlement of layer number 4 also decreased with the increasing distance from model center for different series, which also means that the decompression exerted its impact on nearby zone in a more profound way. These results were similar to the field monitoring information, which demonstrates the validity of the SSR found in actually engineering.

From Figure 13, the stratified settlement and positive or negative rebound were found in different accelerations for the silty clay of layer number 5. Even in different accelerations, the soil was expanded in the location with a distance of 90 mm; 180 mm from the model center; while it was compressed in the location with a distance of 270 mm. The deformation behavior of soil was varying with the distance from the model center, which was called the stratified settlement in Section 3.1. Moreover, the value of positive and negative rebound increased with the acceleration, and the bigger

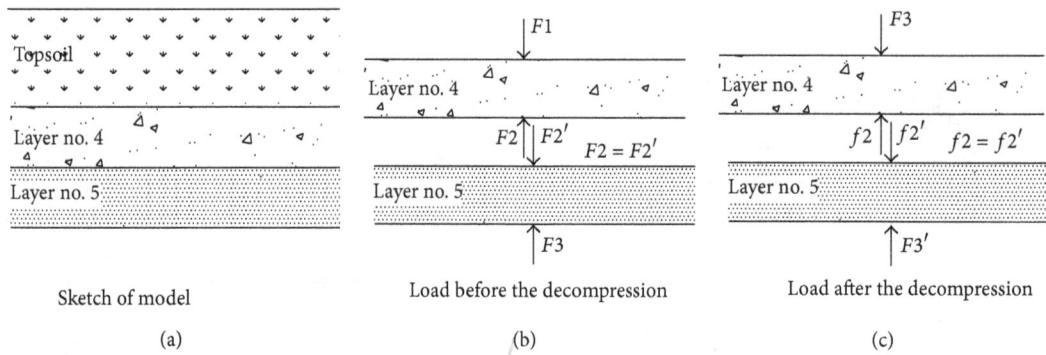

FIGURE 14: The sandwich model of soil.

acceleration meant the larger positive rebound and smaller negative rebound. This may be explained that the higher acceleration means longer time of consolidation for the centrifuge which has a good amplification effect [20–22]. Therefore, the existence of the SSR was not temporarily, and it was the fresh new deformation behavior of silty clay. To make the land subsidence calculation more precisely, further investigation needs to be conducted in a quantitative way.

3.3. The Mechanism of Stratified Settlement and Rebound.
Generally speaking, the soil layers were located in a field of self-weight stress in normal condition. For the soil layers being of certain consolidation and stress, rebound would occur in the soil no matter the unloading was induced in the top or bottom of the layer, which was also proved by the rebound often observed at the bottom of the pit during deep pit excavation [24]. However, the soil was made by nature and always heterogeneous aggregate; its properties may differ with location and result in their different response of same loading. So the settlement varied with different distance from the model center in a specific layer. In addition, comparing with the properties index of layer number 4 and layer number 5, it is found that the cohesive force of layer number 4 is 14 kPa bigger than layer number 5. Because the cohesive force means the stronger internal bond and connection, the response of the two kinds of soil layers showed a different trend. In a whole, the property of the soil including the type and constitution of soil was the major effect leading to the stratified settlement.

In order to reveal this mechanism, the mechanical sandwich model was drawn in Figure 14.

As Figure 14 indicates, the load of each layer was different both before and after decompression. Layer number 4 and number 5 remained in a balance under the gravity of topsoil before the decompression. When the decompression began, the supporting force was decreasing at the bottom of layer number 5; it decreased from F3 to F3' which lead to deformation occurred in layer number 5 before a new balance was attained. The bottom of layer number 5 began to move downward when decompression began. Meanwhile the top of layer number 5 also moved downward, but it moved at a smaller speed. And it was because the layer number 5 was silty clay with high cohesive force of 20 kPa. In the model center,

the effect of the decompression was the biggest. All these made the settlement in layer number 5 become different with location, which was so called stratified settlement, and for WY5-WY2 and WY6-WY3, the displacement at the bottom was big enough to exceed on the top, so their rebound is positive which means expansion in the center nearby, but their value was small; it was 0.5 mm or 0.75 mm, respectively, and for the WY7-WY4, it was located at a distance of 270 mm from the test model. The effect of decompression was relatively smaller so the settlement of the top exceeded the bottom, so WY7-WY4 were negative rebounds which means compression at the zone far from the center.

For layer number 4, the gravity of topsoil remained the same on the top. As the decompression began at the bottom of layer number 5, layer number 4 also was influenced by the decreasing support, which led to an increasing load on top of layer number 4 correspondingly; its value was the subtraction of F1 to F2, which was smaller than the subtraction of F1 to F2. Moreover, the layer number 5 had higher cohesive force and some positive rebound the bottom of layer number 4 would also receive some of the expansion force. All these resulted in the negative rebound in layer number 4, which led to the compression in layer number 4. Similar to layer number 5, the rebound also decreased with the distance from the model center for the effect of decompression reduced gradually with the distance.

4. Conclusions

Through the centrifuge model test composed of three layers of soil and self-invention of decompressing devise, the decompression in pit dewatering was simulated successfully. The conclusion may be summarized as the following.

(1) The centrifuge model test is feasible to simulate the SSR induced by decompression induced by dewatering in confined water during metro station construction.

(2) The stratified settlement induced by decompression was found both in sandy clay and silty clay, and the absolute value of SSR decreased with increasing distance from the model center.

(3) When decompression was conducted, only negative rebound was found in the sandy clay, but both positive

and negative rebound were found in silty clay, and the absolute value of SSR in the sandy clay was bigger than the silty clay.

(4) The stratified settlement and rebound in other acceleration was similar to the acceleration of 30 g; the absolute value of SSR in bigger acceleration was larger than the smaller acceleration, which showed larger acceleration made the soil more compact.

(5) The mechanism of stratified settlement and rebound lies in the load in the soil layer and type of soil. The decompression exerted more profound impact on the bottom of layer number 5, and the different properties of sandy clay and silty clay also contributed to the different response of same decompression.

(6) Owing to the model scale and the complexity of soil, current work is mainly to simulate the stratified settlement and rebound qualitatively; quantitative researches are needed to be conducted to make more profound recognition in future.

Conflict of Interests

The authors of the paper declare that they have no direct financial relation with the commercial identities mentioned in the paper that might lead to a conflict of interests.

Acknowledgments

This work was supported by the National Natural Science Funding of China (no. 41072205) and Natural Science Funding of Shanghai Municipality (no. 10ZR1431500), which are gratefully acknowledged. The authors would like to thank Dongchang Sui for his contribution to the realization of the experimental works, and comments from the reviewers are greatly appreciated.

References

[1] Y. Q. Xue, Y. Zhang, S. J. Ye, J. C. Wu, and Q. F. Li, "Land subsidence in China," *Environmental Geology*, vol. 48, no. 6, pp. 713–720, 2005.

[2] Y. Q. Tang, Z. D. Cui, J. X. Wang, C. Lu, and X. X. Yan, "Model test study of land subsidence caused by high-rise building group in Shanghai," *Bulletin of Engineering Geology and the Environment*, vol. 67, no. 2, pp. 173–179, 2008.

[3] Y. Q. Tang, Z. D. Cui, J. X. Wang, L. P. Yan, and X. X. Yan, "Application of grey theory-based model to prediction of land subsidence due to engineering environment in Shanghai," *Environmental Geology*, vol. 55, no. 3, pp. 583–593, 2008.

[4] X. Shi, J. Wu, S. Ye et al., "Regional land subsidence simulation in Su-Xi-Chang area and Shanghai City, China," *Engineering Geology*, vol. 100, no. 1-2, pp. 27–42, 2008.

[5] J. Wang, L. Hu, L. Wu, Y. Tang, Y. Zhu, and P. Yang, "Hydraulic barrier function of the underground continuous concrete wall in the pit of subway station and its optimization," *Environmental Geology*, vol. 57, no. 2, pp. 447–453, 2009.

[6] S. L. Shen and Y. S. Xu, "Numerical evaluation on land subsidence induced by groundwater pumping in Shanghai," *Canadian Geotechnical Journal*, vol. 48, no. 9, pp. 1378–1392, 2011.

[7] Y. S. Xu, L. Ma, Y.-J. Du, and S.-L. Shen, "Analysis of urbanisation-induced land subsidence in Shanghai," *Natural Hazards*, vol. 63, no. 2, pp. 1255–1267, 2012.

[8] L. Y. Ding, X. G. Wu etc, H. Li, H. B. Luo, and Y. Zhou, "Study on safety control for Wuhan metro construction in complex environments," *International Journal of Project Management*, vol. 29, no. 7, pp. 797–807, 2011.

[9] S. Jiang, X. Kong, H. Ye, and N. Zhou, "Groundwater dewatering optimization in the Shengli no. 1 open-pit coalmine, Inner Mongolia, China," *Environmental Earth Science*, vol. 69, no. 1, pp. 187–196, 2013.

[10] J. Wang, B. Fen, Y. Liu et al., "Controlling subsidence caused by de-watering in a deep foundation pit," *Bulletin of Engineering Geology and Environment*, vol. 71, no. 3, pp. 545–555, 2012.

[11] J. Wang, B. Feng, H. Yu, T. Guo, G. Yang, and J. Tang, "Numerical study of dewatering in a large deep foundation pit," *Environmental Earth Science*, vol. 69, no. 3, pp. 863–872, 2013.

[12] K. Yuan, W. Chen, H. Yu et al., "Modified Cam-clay model considering cohesion and S-D effect and its numerical implementation," *Chinese Journal of Rock Mechanics and Engineering*, vol. 31, no. 8, pp. 1574–1580, 2012 (Chinese).

[13] H. Yu, W. Chen, X. Li, and X. Sillen, "A transversely isotropic damage model for boom clay," *Rock Mechanics and Rock Engineering*, 2013.

[14] M. Taiebat, Y. F. Dafalias, and R. Peek, "A destructuration theory and its application to SANICLAY model," *International Journal for Numerical and Analytical Methods in Geomechanics*, vol. 34, no. 10, pp. 1009–1040, 2010.

[15] M. D. Liu and J. P. Carter, "Volumetric deformation of natural clays," *International Journal of Geomechanics ASCE*, vol. 3, no. 2, pp. 236–252, 2003.

[16] F. Cotecchia and R. J. Chandler, "A general framework for the mechanical behaviour of clays," *Geotechnique*, vol. 50, no. 4, pp. 431–447, 2000.

[17] U. Mutman, "Clay improvement with burned olive waste ash," *The Scientific World Journal*, vol. 2013, Article ID 127031, 4 pages, 2013.

[18] J. Wang, L. Wu, Y. Zhu, Y. Tang, P. Yang, and R. Lou, "Mechanism of dewatering-induced ground subsidence in deep subway station pit and calculation method," *Chinese Journal of Rock Mechanics and Engineering*, vol. 28, no. 5, pp. 1010–1019, 2009 (Chinese).

[19] Y. Q. Tang, C. Q. Luan, J. X. Wang, Y. F. Zhu, and W. Q. Pan, "Analysis of the effects of environments for dewatering in a metro station in shanghai," *Journal of Wuhan University of Technology*, vol. 30, no. 8, pp. 147–151, 2008 (Chinese).

[20] Z. D. Cui, Y. Q. Tang, and X. X. Yan, "Centrifuge modeling of land subsidence caused by the high-rise building group in the soft soil area," *Environmental Earth Sciences*, vol. 59, no. 8, pp. 1819–1826, 2009.

[21] Z. D. Cui and Y. Q. Tang, "Land subsidence and pore structure of soils caused by the high-rise building group through centrifuge model test," *Engineering Geology*, vol. 113, no. 1–4, pp. 44–52, 2010.

[22] Y. Q. Tang, X. Ren, B. Chen, S. Song, J. X. Wang, and P. Yang, "Study on land subsidence under different plot ratios through centrifuge model test in soft-soil territory," *Environmental Earth Sciences*, vol. 66, no. 7, pp. 1809–1816, 2012.

[23] S. L. Gong, "The influence of urban construction in Shanghai to the land subsidence," *Chinese Journal of Geological Hazard Control*, vol. 9, no. 2, pp. 108–111, 1998 (Chinese).

[24] Y. Pan and Z. Wu, "Experimental study on the resilience of pit unloading," *Chinese Journal of Geotechnical Engineering*, vol. 24, no. 1, pp. 101–106, 2002 (Chinese).

4

3D-Web-GIS RFID Location Sensing System for Construction Objects

Chien-Ho Ko

Department of Civil Engineering, National Pingtung University of Science and Technology, 1 Shuefu Road, Neipu, Pingtung 912, Taiwan

Correspondence should be addressed to Chien-Ho Ko; fpecount@yahoo.com.tw

Academic Editors: J. Mander and B. Uy

Construction site managers could benefit from being able to visualize on-site construction objects. Radio frequency identification (RFID) technology has been shown to improve the efficiency of construction object management. The objective of this study is to develop a 3D-Web-GIS RFID location sensing system for construction objects. An RFID 3D location sensing algorithm combining Simulated Annealing (SA) and a gradient descent method is proposed to determine target object location. In the algorithm, SA is used to stabilize the search process and the gradient descent method is used to reduce errors. The locations of the analyzed objects are visualized using the 3D-Web-GIS system. A real construction site is used to validate the applicability of the proposed method, with results indicating that the proposed approach can provide faster, more accurate, and more stable 3D positioning results than other location sensing algorithms. The proposed system allows construction managers to better understand worksite status, thus enhancing managerial efficiency.

1. Introduction

The construction industry is characterized by intensive manual labor and is prone to errors [1], creating significant challenges for providing a clear understanding of construction site activity [2]. A perennial issue facing construction site managers is object positioning including assets, personnel, material, and equipment [3–6]. Several attempts have been made to facilitate the location of objects on construction sites. Meade and Chignell [7] used ground penetrating radar to locate buried piping without excavation. Grau [8] presented an epistemic model based on belief functions to monitor the positions of mobile sensing nodes. His research demonstrates that the epistemic functions can correctly filter location uncertainties and effectively monitor the movements of mobile sensing nodes. Song et al. [9] tracked construction material to improve project performance and reduce the effort needed to derive project performance indicators. In their approach, materials are fitted with radio frequency identification (RFID) tags to allow for automatic identification and tracking on construction sites. Razavi and Moselhi [10] developed a construction equipment and supply location system using passive RFID tags. Skibniewski and Jang [11]

introduced an architecture for construction asset tracking using wireless sensor modules to track objects via the time-of-flight method. Shahi et al. [12] presented an Ultra Wide Band positioning system as a material and activity tracking tool for indoor construction projects. Global Positioning System (GPS) is frequently used for tracking objects outdoors. Pradhananga and Teizer [13] used GPS devices to automate the assessment of construction site equipment operations by continuously logging time-stamped equipment locations for analysis.

Given the alternatives available for object positioning in construction sites, Jiang et al. [14] and Nasir et al. [15] suggested methodologies for selecting appropriate technologies for various types of projects and objects and suggested that RFID is an appropriate solution for object positioning in indoor construction sites. Razavi and Moselhi [10] also demonstrated the potential for RFID as a method for object tracking in indoor construction sites.

Accurate object positioning offers the possibility of improved object visibility. However, previous studies have either treated positioning algorithms as separate from display systems or have tended to use one- or two-dimensional maps to indicate object locations, thus obscuring object visibility. Displaying objects in three-dimensional (3D) space requires a corresponding location sensing algorithm. Ko [16] proposed an RFID 3D location sensing algorithm, but the algorithm had trouble deriving smooth convergences while searching for the target objects.

The present study develops a 3D-Web-GIS RFID location sensing system to locate objects in indoor construction sites. An improved RFID 3D location sensing algorithm is established, combining Simulated Annealing (SA) and the gradient descent method to overcome the convergence problem while locating objects. The 3D-Web-GIS RFID location sensing system manipulates the location sensing algorithm to better relate the display of construction objects to the real world and help managers better understand construction site activity. This study begins by introducing indoor sensing networks and then explains the evolutionary process of the proposed RFID 3D location sensing algorithm. Section four describes the development of the 3D-Web-GIS RFID location sensing system. Section five describes a demonstration of the system on a construction site. Finally, the paper concludes with suggestions for future research directions.

2. Indoor Sensing Networks

Construction sites require both indoor and outdoor sensing [17, 18], but these sensing contexts require different types of networks and technologies [19]. GPS, a relatively mature location sensing technique, is frequently used in outdoor construction sites [20], but indoor location remains a challenge. This study thus focuses on improving the accuracy and efficiency of the location of indoor construction objects including material, equipment, personnel, and machinery. Construction site activity can only be understood through the simultaneous tracking of multiple objects. In a passive location system, RFID antennas are distributed at reference coordinates within the location space [21], and the target objects are fitted with passive RFID tags [22]. This provides a relatively low-cost solution as compared to active RFID systems in which the target objects have to be equipped with antennas. A passive location mode that attaches an active RFID tag on the target object [23] is therefore selected.

The indoor sensing networks are constructed using RFID readers and tags, as shown in Figure 1. An active RFID system was used to expand the sensing space. Four RFID antennas were set at four corners of the hexahedron space. Nine reference tags were uniformly distributed in the space to build a location sensing network. Through the network topology, signal strengths of the RFID tags from different directions with diverse distances can be collected. The collected signal strength was then analyzed using a 3D location sensing algorithm to calculate target object locations.

3. Location Sensing Algorithms

3.1. Location Concept. This research develops an RFID 3D location sensing algorithm using a trilateration method [24–26] that calculates the target object location using distances

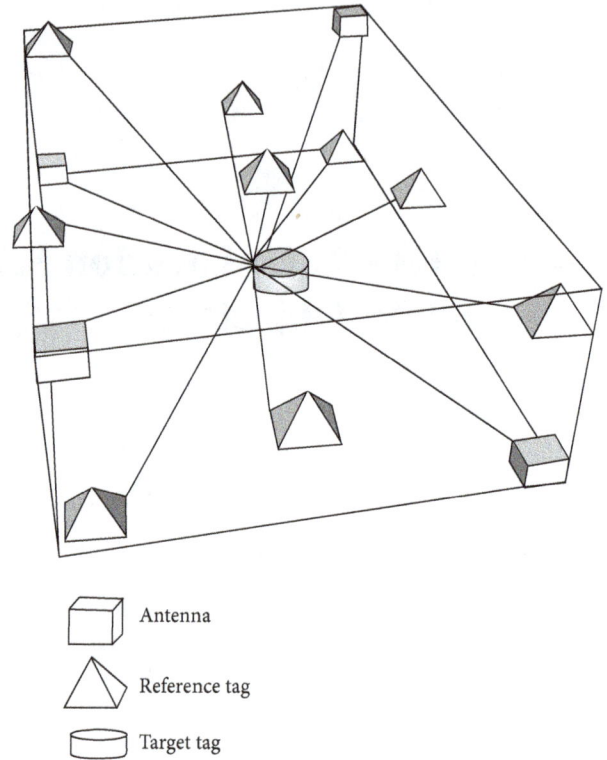

	Antenna
	Reference tag
	Target tag

FIGURE 1: Indoor sensing networks.

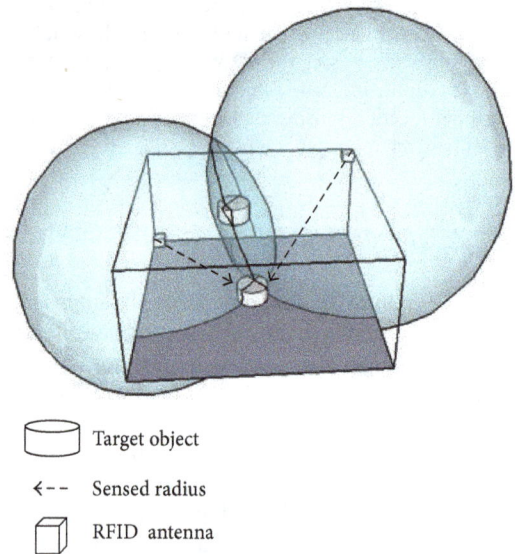

	Target object
	Sensed radius
	RFID antenna

FIGURE 2: 3D trilateration location concept.

from the RFID antennas to the target object. In a 3D space, a single RFID antenna can sense its distance to an RFID tag using Received Signal Strength Indication (RSSI). The possible location of the target object could be expressed as a sphere with a radius of the sensed distance. Adding a second antenna, the solution space is an intersection of the two spheres with the two sensed radii, as shown in Figure 2. Using the same concept, the target object location could be further narrowed to few points using three antennas. Four antennas can be expected to produce a highly specific location.

3.2. Location Algorithm.

Ko [16] developed an RFID 3D location sensing algorithm using the gradient descent method. However, that method used fixed adjustment coefficients to search for the target object location. In a large space, the adjustment coefficient has to be increased to reduce the amount of computational time required, but this may make convergence difficult. A small space, on the other hand, needs smaller adjustment coefficients to converge. Selection of the adjustment coefficients appropriate for the dimensions of the given search space is achieved by trial and error. Simulated Annealing (SA), a technique analogizing the annealing of metals for stable global search [27], could potentially solve this problem. This study thus hybridizes SA and gradient descent methods to locate construction-related objects in a 3D space. The improved RFID 3D location sensing algorithm is shown and explained in Figure 3.

3.2.1. Initializing Location.

The first step of the algorithm is to initiate a location search for the target object. The location of target object i in 3D space is noted as (x_i, y_i, z_i).

3.2.2. Sensing Distances.

The trilateration location method requires the distances from each antenna to the target object. The proposed method senses the distance using RSSI, with an example shown in Figure 4. An antenna receives a signal with a given strength level from the active RFID tag attached to the target object. Through the RSSI curve, the received signal strength can be converted to a distance.

3.2.3. Calculating Error.

This step calculates a positioning error between the initial location and the sensed location, which will be used to adjust the target object location in the next step. The positioning error of target object i and antenna k (e_{ik}) is calculated using the following equation:

$$e_{ik} = \left(S_{ik} - \bar{S}_{ik} \right), \qquad (1)$$

where S_{ik} is a sensed distance between target object i and antenna k converted using RSSI; $\bar{S}_{ik}$ is the distance in 3D space between target object i at (x_i, y_i, z_i) and antenna k at (x_k, y_k, z_k) calculated using the following equation:

$$\bar{S}_{ik} = \sqrt{(x_i - x_k)^2 + (y_i - y_k)^2 + (z_i - z_k)^2}. \qquad (2)$$

3.2.4. Refining Coordinates.

SA is combined with the gradient descent method to narrow the potential location of the target object. SA is used to gradually decrease the adjustment (i.e., cooling down) to help the algorithm converge, while the gradient descent method is used to reduce location error.

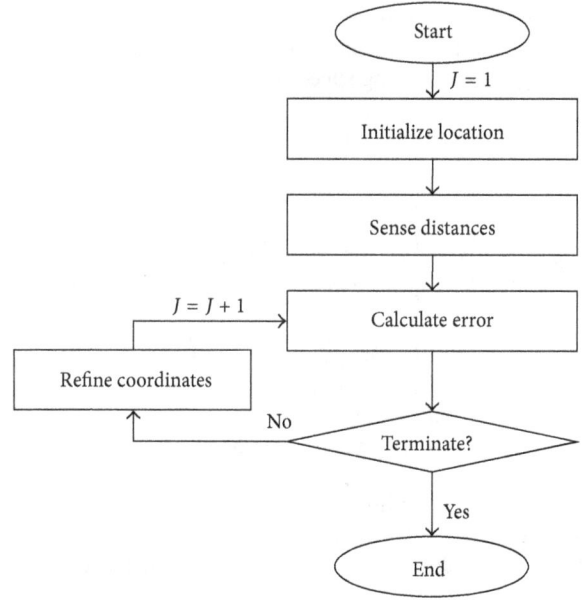

FIGURE 3: RFID 3D location sensing algorithm.

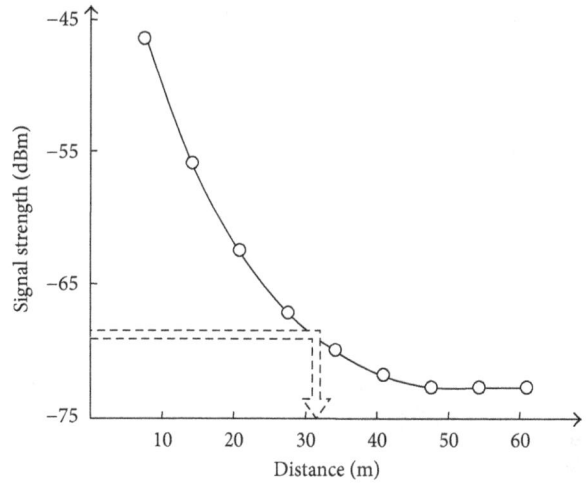

FIGURE 4: Received signal strength indication.

The target object's location at the epoch (j) is adjusted using the following equation:

$$(x_i(j+1), y_i(j+1), z_i(j+1))$$
$$= \begin{cases} x_i(j+1) = x_i(j) + \Delta x_i(j) \\ y_i(j+1) = y_i(j) + \Delta y_i(j) \\ z_i(j+1) = z_i(j) + \Delta z_i(j), \end{cases} \qquad (3)$$

where $(\Delta x_i(j), \Delta y_i(j), \Delta z_i(j))$ is the amount of adjustment for the x-, y-, and z-axes. SA is applied to cool down the adjustment, as shown in the following equation:

$$(\Delta x_i(j), \Delta y_i(j), \Delta z_i(j)) = \begin{cases} \Delta x_i(j) = \alpha_x \beta_x \gamma_x \\ \Delta y_i(j) = \alpha_y \beta_y \gamma_y \\ \Delta z_i(j) = \alpha_z \beta_z \gamma_z, \end{cases} \qquad (4)$$

where $(\alpha_x, \alpha_y, \alpha_z)$ is adjustment rate for the x-, y-, and z-axes; $(\beta_x, \beta_y, \beta_z)$ represents the temperature formulated in (5); $(\gamma_x, \gamma_y, \gamma_z)$ is the cooling speed represented in (6). Consider

$$
(\beta_x, \beta_y, \beta_z) = \begin{cases} \beta_x = \dfrac{(S_{ik} - \bar{S}_{ik})}{|S_{ik} - \bar{S}_{ik}|} \exp\left(\dfrac{-1}{\alpha_x |x_i \delta_{ik}|}\right), \\[2mm] \beta_y = \dfrac{(S_{ik} - \bar{S}_{ik})}{|S_{ik} - \bar{S}_{ik}|} \exp\left(\dfrac{-1}{\alpha_y |y_i \delta_{ik}|}\right), \\[2mm] \beta_z = \dfrac{(S_{ik} - \bar{S}_{ik})}{|S_{ik} - \bar{S}_{ik}|} \exp\left(\dfrac{-1}{\alpha_z |z_i \delta_{ik}|}\right), \end{cases} \quad (5)
$$

$$
(\gamma_x, \gamma_y, \gamma_z) = \exp(\mu). \quad (6)
$$

In (5), the k RFID antenna gradient for target i (δ_{ik}) can be calculated using (7). The μ shown in (6) is a parameter simulating the cooling process, which is formulated using (8). Consider

$$
\delta_{ik} = \bar{s}_{ik} \times e_{ik}, \quad (7)
$$

$$
\mu = -0.2 \times (j - 1). \quad (8)
$$

The k RFID antenna adjusts the target object's location using (3) to (8). The error of the target object i for m antennas (ε_i) is calculated using the Root Mean Square Error (RMSE), as shown in the following equation:

$$
\varepsilon_i = \sqrt{\dfrac{\sum_{k=1}^{m}\left((S_{ik} - \bar{S}_{ik})/S_{ik}\right)^2}{m}}. \quad (9)
$$

3.2.5. Terminating Conditions. The algorithm locates the target object's location using an iterative adjustment process. The termination conditions can be met if epoch number (j) reaches a predetermined criterion and/or the RMSE (ε_i) is smaller than an assigned number. The predetermined epoch number can ensure that the location algorithm is completed within a specified duration, while the preassigned RMSE ensures the location's accuracy.

4. System Development

4.1. Use Case. The Rational Unified Process (RUP) [28] and Unified Modeling Language (UML) [29] were used to develop the 3D-Web-GIS RFID location sensing system. To identify system requirements, Use Case, which is regarded as a high-level system descriptor, is used for system analysis and design. Figure 5 shows the Use Case diagram of the 3D-Web-GIS RFID location sensing system. The diagram shows how the system can be used by construction managers to locate objects in construction sites and by project stakeholders to easily understand the system.

The Use Case used in this study is explained as follows:

(i) Use Case: 3D positioning,

(ii) actor: construction managers,

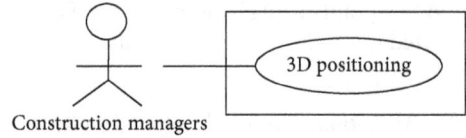

FIGURE 5: Use Case diagram.

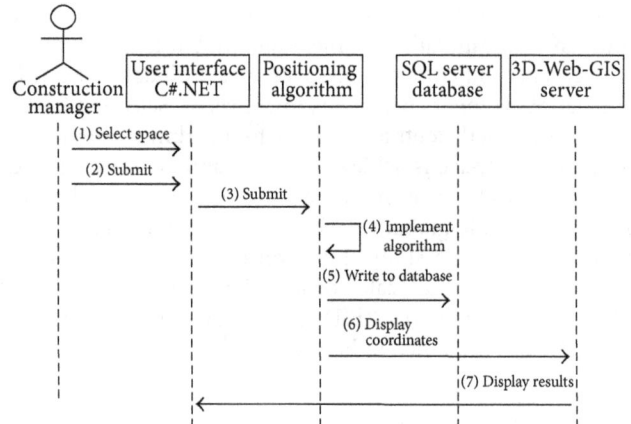

FIGURE 6: Positioning sequence diagram.

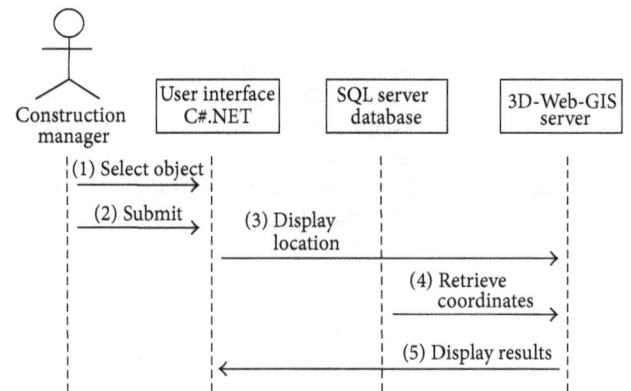

FIGURE 7: Browsing sequence diagram.

(iii) type: primary,

(iv) Descriptions:

(1) users select an object, and the system then displays the object's location in 3D space,

(2) users click the displayed object to retrieve information about the object.

The Use Case identifies two system functions: position and browse. These functions are explained in Table 1, while their sequence diagrams are shown in Figures 6 and 7.

4.2. System Architecture. Figure 8 displays the architecture of the 3D-Web-GIS RFID location sensing system in three tiers. Location sensing algorithm parameters and object coordinates are stored in the storage layer. The 3D positioning algorithm is in the application logic layer that implements system

TABLE 1: Location sensing system functions.

Function	Explanation	Category
Position	Users select a construction site and space. The system calculates and displays object locations using the 3D-Web-GIS system.	Evident
Browse	Users select a construction-related object. The system displays information about the selected object using the 3D-Web-GIS system.	Evident

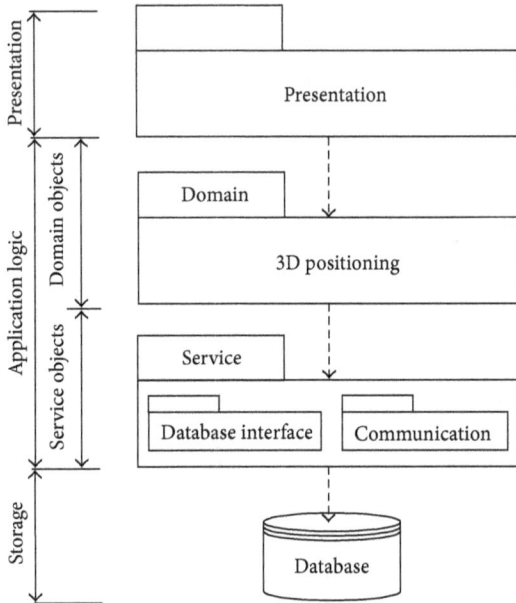

FIGURE 8: System architecture.

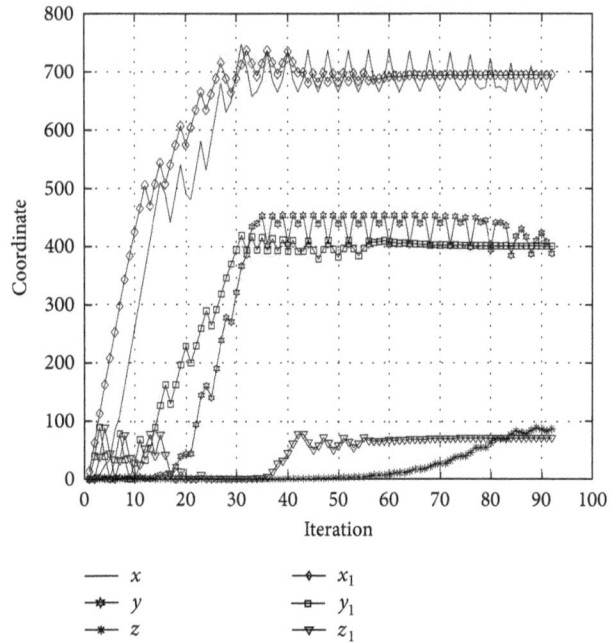

FIGURE 9: Location convergence comparison.

functions. The presentation layer provides user interfaces allowing users to interact with the application logic layer. The system was developed using Microsoft Visual Studio. Net (C#.NET) with SQL server database. Construction object geographic information is displayed by integrating an ESRI ArcGIS Server with the 3D extension module via C#.NET.

5. Verification

To validate its feasibility, the proposed algorithm was applied to a real construction site: a 926 cm × 535 cm × 211 cm indoor space on the third floor of a construction site. The target object is located at (694, 400, 75). An adjustment rate of (0.5) is used for $(\alpha_x, \alpha_y, \alpha_z)$.

Setting the initial location of the target object at (1, 1, 1), Figure 9 compares the convergence trend of the proposed method with that developed by Ko [16]. In the figure, x, y, and z are convergence trends of the previous method, while x_1, y_1, and z_1 are those of the proposed method. The proposed method locates the target object at iteration 60, as opposed to iteration 90 for the previous method. The previous method adjusts the target object's location using gradient decent method. Thus, although adjustments move in the right direction, the convergence becomes spiky in later stages. By contrast, the proposed method combines SA and the gradient decent method to adjust the target object's location.

In the early stages of positioning, both methods display the same conspicuous adjustments. In the cooling down stage, smoother adjustments are applied to locate the target object. Figure 10 compares the error convergence between the two methods. Figure 11 shows the locus of the two methods in 3D space while positioning the target object. As discussed, the proposed method locates the target object faster, more accurately, and more stably in 3D space. Finally, the location of the target object is displayed using the developed 3D-Web-GIS system, as shown in Figure 12, and the location of the construction objects can be visualized. By clicking the located objects, object information is retrieved from the database, as shown in Figure 13.

6. Conclusions

This study hybridizes SA and the gradient descent method to develop a 3D RFID location sensing algorithm. A 3D-Web-GIS system is developed to run the algorithm and display the location of target objects. The proposed method is validated by application to a real construction site, with performance comparisons to the previous best method.

In the proposed algorithm, SA is used to stabilize the search process while the gradient descent method is used to

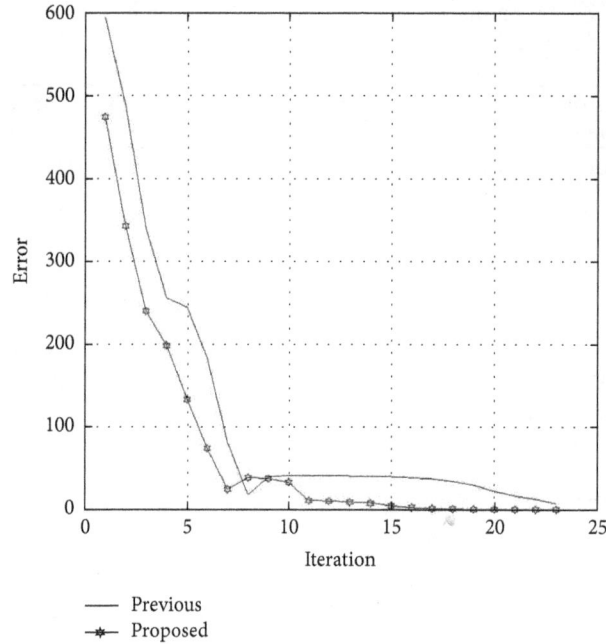

FIGURE 10: Error convergence comparison.

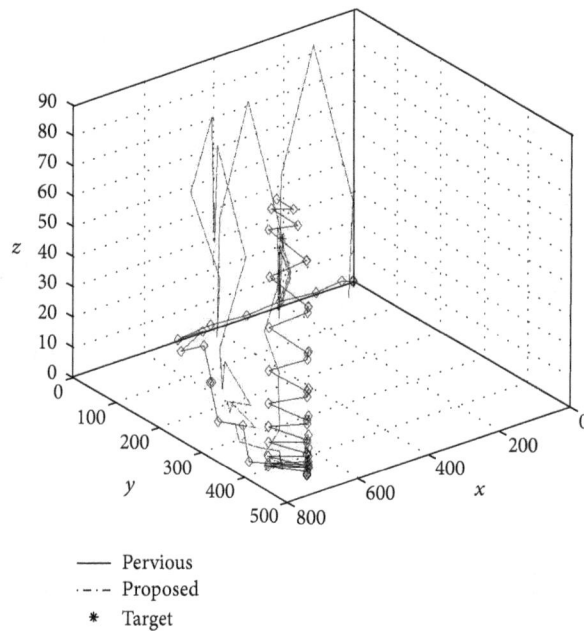

FIGURE 11: 3D locus comparison.

increase location accuracy. At the beginning of the search, while the temperature is high, large location adjustments are made, thus saving time for positioning. The temperature gradually cools as the positioning process enhances the convergence. Application to a real construction site validates that combining SA with the gradient descent method improves the speed, accuracy, and stability of results over those obtained using the previous location sensing algorithm in 3D positioning. Furthermore, the previous 3D positioning algorithm needs to determine algorithm parameters according to location space dimension. The proposed positioning algorithm, however, frontloads target object searching to the beginning of the search process and gradually cools down over time and thus may not need to predetermine algorithm parameters due to the size of the spatial dimensions.

The developed 3D-Web-GIS RFID location sensing system can be accessed through the Internet. Positioning results are visualized in 3D environment, and users can browse information related to the located construction objects. The system allows construction managers to locate construction objects at any time, from anywhere using any operating systems, thus enhancing managerial efficiency.

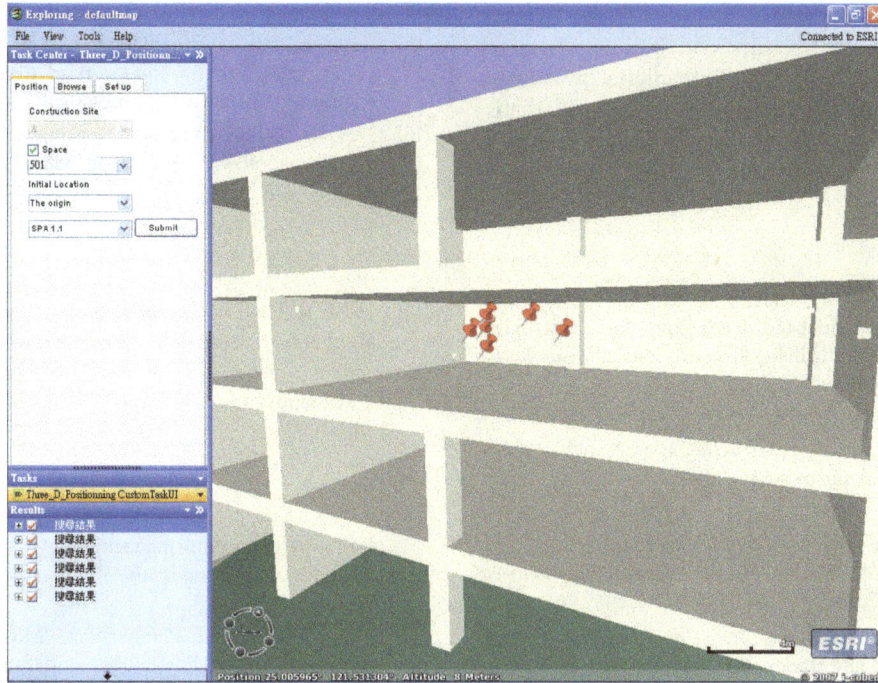

FIGURE 12: Target object location display.

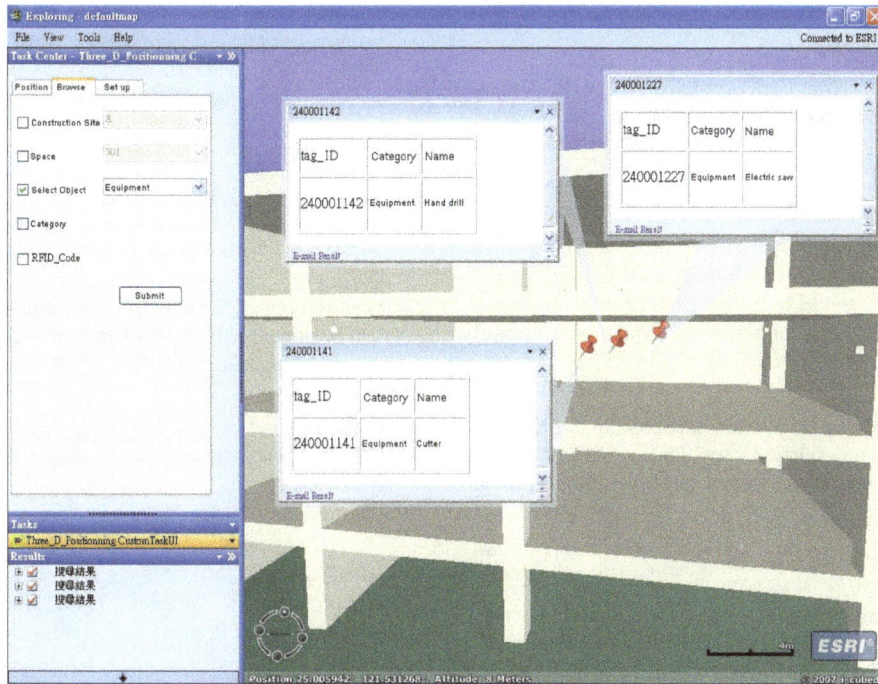

FIGURE 13: Browsing construction object information.

This study did not take into account the RFID signal attenuation effect caused by environmental factors, and future studies could modify the proposed algorithm to consider signal attenuation. Future work could also use this 3D location sensing technology to develop mechanisms for tracking the movement of construction objects within construction sites.

Acknowledgments

This research was funded by the National Science Council (Grant no. NSC 99–2622-E-020-006-CC3) and Architecture and Building Research Institute (Taiwan), whose supports are gratefully acknowledged. Any opinions, findings, conclusions, or recommendations expressed in this paper are those of the author and do not reflect the views of the National Science Council and Architecture and Building Research Institute. The writer would also like to thank the investigated construction site for supporting this study. Special thanks are due to research assistant Tai-Ru Hong who helped in programming.

References

[1] T. Cheng, M. Venugopal, J. Teizer, and P. A. Vela, "Performance evaluation of ultra wideband technology for construction resource location tracking in harsh environments," *Automation in Construction*, vol. 20, no. 8, pp. 1173–1184, 2011.

[2] R. Navon and O. Berkovich, "An automated model for materials management and control," *Construction Management and Economics*, vol. 24, no. 6, pp. 635–646, 2006.

[3] K. Morioka, S. Kovacs, J.-H. Lee, and P. Korondi, "A cooperative object tracking system with fuzzy-based adaptive camera selection," *International Journal on Smart Sensing and Intelligent Systems*, vol. 3, no. 3, pp. 338–358, 2010.

[4] S. Cook, J. Bittner, and T. Adams, "Developing an asset management tool to collect and track commitments on environmental mitigation features," *Transportation Research Record*, no. 2160, pp. 21–28, 2010.

[5] J. Teizer and P. A. Vela, "Personnel tracking on construction sites using video cameras," *Advanced Engineering Informatics*, vol. 23, no. 4, pp. 452–462, 2009.

[6] A. H. Behzadan, Z. Aziz, C. J. Anumba, and V. R. Kamat, "Ubiquitous location tracking for context-specific information delivery on construction sites," *Automation in Construction*, vol. 17, no. 6, pp. 737–748, 2008.

[7] R. B. Meade and R. J. Chignell, "Tool advances pipe location and construction planning," *Pipeline and Gas Journal*, vol. 224, no. 4, p. 3, 1997.

[8] D. Grau, "Epistemic model to monitor the position of mobile sensing nodes on construction sites with rough location data," *Journal of Computing in Civil Engineering*, vol. 26, no. 1, pp. 141–150, 2012.

[9] J. Song, C. T. Haas, and C. H. Caldas, "Tracking the location of materials on construction job sites," *Journal of Construction Engineering and Management*, vol. 132, no. 9, pp. 911–918, 2006.

[10] S. N. Razavi and O. Moselhi, "GPS-less indoor construction location sensing," *Automation in Construction*, vol. 28, pp. 128–136, 2012.

[11] M. J. Skibniewski and W.-S. Jang, "Simulation of accuracy performance for wireless sensor-based construction asset tracking," *Computer-Aided Civil and Infrastructure Engineering*, vol. 24, no. 5, pp. 335–345, 2009.

[12] A. Shahi, A. Aryan, J. S. West, C. T. Haas, and R. C. G. Haas, "Deterioration of UWB positioning during construction," *Automation in Construction*, vol. 24, pp. 72–80, 2012.

[13] N. Pradhananga and J. Teizer, "Automatic spatio-temporal analysis of construction site equipment operations using GPS data," *Automation in Construction*, vol. 29, pp. 107–122, 2013.

[14] S. Jiang, W. S. Jang, and M. J. Skibniewski, "Selection of wireless technology for tracking construction materials using a fuzzy decision model," *Journal of Civil Engineering and Management*, vol. 18, no. 1, pp. 43–59, 2012.

[15] H. Nasir, C. T. Haas, D. A. Young, S. N. Razavi, C. Caldas, and P. Goodrum, "An implementation model for automated construction materials tracking and locating," *Canadian Journal of Civil Engineering*, vol. 37, no. 4, pp. 588–599, 2010.

[16] C.-H. Ko, "RFID 3D location sensing algorithms," *Automation in Construction*, vol. 19, no. 5, pp. 588–595, 2010.

[17] X. Shen and M. Lu, "A framework for indoor construction resources tracking by applying wireless sensor networks," *Canadian Journal of Civil Engineering*, vol. 39, no. 9, pp. 1083–1088, 2012.

[18] Y. Zhu, X. Ding, Z. Li, and S. Zhou, "Discussion on the application of GPS using in Marine construction survey," *Journal of Computers*, vol. 7, no. 7, pp. 1663–1670, 2012.

[19] H. M. Khoury and V. R. Kamat, "Evaluation of position tracking technologies for user localization in indoor construction environments," *Automation in Construction*, vol. 18, no. 4, pp. 444–457, 2009.

[20] T.-H. Yi, H.-N. Li, and M. Gu, "Recent research and applications of GPS-based monitoring technology for high-rise structures," *Structural Control and Health Monitoring*, vol. 20, no. 5, pp. 649–670, 2013.

[21] F. Gustafsson and F. Gunnarsson, "Mobile positioning using wireless networks: Possibilities and fundamental limitations based on available wireless network measurements," *IEEE Signal Processing Magazine*, vol. 22, no. 4, pp. 41–53, 2005.

[22] W. Gueaieb and S. Miah, "An intelligent mobile robot navigation technique using RFID technology," *IEEE Transactions on Instrumentation and Measurement*, vol. 57, no. 9, pp. 1908–1917, 2008.

[23] R. Want, A. Hopper, V. Falcao, and J. Gibbons, "Active badge location system," *ACM Transactions on Information Systems*, vol. 10, no. 1, pp. 91–102, 1992.

[24] T. F. Bechteler and H. Yenigün, "2-D localization and identification based on SAW ID-tags at 2.5 GHz," *IEEE Transactions on Microwave Theory and Techniques*, vol. 51, no. 5, pp. 1584–1590, 2003.

[25] D. E. Manolakis, "Efficient solution and performance analysis of 3-D position estimation by trilateration," *IEEE Transactions on Aerospace and Electronic Systems*, vol. 32, no. 4, pp. 1239–1248, 1996.

[26] F. Thomas and L. Ros, "Revisiting trilateration for robot localization," *IEEE Transactions on Robotics*, vol. 21, no. 1, pp. 93–101, 2005.

[27] W. H. Press, S. A. Teukolsky, W. T. Vetterling, and B. P. Flannery, *The Art of Scientific Computing*, Cambridge University Press, New York, NY, USA, 2007.

[28] P. Kruchten, *The Rational Unified Process: An Introduction*, Addison-Wesley Professional, Reading, Mass, USA, 2003.

[29] G. Booch, J. Rumbaugh, and I. Jacobson, *The Unified Modeling Language User Guide*, Addison-Wesley Professional, Reading, Mass, USA, 2005.

5

Study on the Application of the Kent Index Method on the Risk Assessment of Disastrous Accidents in Subway Engineering

Hao Lu,[1] Mingyang Wang,[1] Baohuai Yang,[2] and Xiaoli Rong[1]

[1] *College of Defense Engineering, PLA University of Science and Technology, Nanjing 210007, China*
[2] *Nanjing KunTuo Civil Engineering Technology Co., Ltd., Nanjing 210007, China*

Correspondence should be addressed to Hao Lu; haohaoluweifeng@gmail.com

Academic Editors: R. Degenhardt and X. F. XU

With the development of subway engineering, according to uncertain factors and serious accidents involved in the construction of subways, implementing risk assessment is necessary and may bring a number of benefits for construction safety. The Kent index method extensively used in pipeline construction is improved to make risk assessment much more practical for the risk assessment of disastrous accidents in subway engineering. In the improved method, the indexes are divided into four categories, namely, basic, design, construction, and consequence indexes. In this study, a risk assessment model containing four kinds of indexes is provided. Three kinds of risk occurrence modes are listed. The probability index model which considers the relativity of the indexes is established according to the risk occurrence modes. The model provides the risk assessment process through the fault tree method and has been applied in the risk assessment of Nanjing subway's river-crossing tunnel construction. Based on the assessment results, the builders were informed of what risks should be noticed and what they should do to avoid the risks. The need for further research is discussed. Overall, this method may provide a tool for the builders, and improve the safety of the construction.

1. Introduction

Creating a perfect design in subway engineering is difficult because of complex geological environment and difficulties in completely obtaining basic information. The influence of current large-scale subway construction, the limited construction period, and poor management caused by the lack of skilled personnel contribute to the increase in the occurrence of accidents in subway construction. Thus, the issue of safety has become very serious [1, 2]. Accidents indicate that subway construction affects the ambient environment (ground buildings, transportation, underground structures, underground pipes, etc.), endangers people's lives, compromises property security, and causes serious economic losses [2]. Several typical subway construction accidents are shown in Table 1.

Plenty of new urgent tasks are being proposed because of the serious safety issue in subway engineering. One of these tasks is to study the safety risk management method. In recent years, the utilization of risk assessment in subway

engineering has significantly increased and has provided particular economic benefits and research results [3–5].

The book "Code for Risk Management of Underground Works in Urban Rail Transit" [6] published in 2011 provides a reference for the application of risk management in subway engineering and considers the classification standard of probability and consequence. However, in the application process, the risk factors that influence scope, occurrence mechanism, and potential damage mechanism in subway construction are very complex. Risk management involves many disciplines such as natural science, social science, engineering technology, system science, and management science. Thus, determining if a probability distribution hypothesis is appropriate becomes difficult when tunnel and underground engineering risks are studied with the probability method [7, 8]. Thus, obtaining the "real" probability value of an accident is difficult [9].

Kent used the index method to study pipeline accidents. He believes that pipeline accidents cannot be accurately

TABLE 1: Several typical subway construction accidents in China.

Time	City	Loss
2004-3-17	Guangzhou	1 death [10]
2007-3-28	Peking	6 deaths [11]
2007-2-5	Nanjing	The ground collapsed, and many residents were affected [12]
2008-11-15	Hangzhou	21 deaths [13]

predicted, and risk assessment does not provide an accurate calculation based on the probability theory. Insufficient sample size or calculation quantity is usually regarded as the reason for the inaccurate calculation, but in truth, the main reason is that too many assumptions are made in the computation or collection of samples, which leads to the inaccuracy of the assessment result. Kent's method does not consider the "real" probability; the indexes in Kent's method contain the probability and are not tied to the "real" probability, which is very persuasive [9].

By adopting advanced techniques from the Kent index method and considering the limitation of the application of Kent's method in subway engineering, a model that can be applied to risk assessment of disastrous accidents in subway engineering is developed in this paper. This paper also provides a reference for quantitative evaluation of disastrous risks involved in subway engineering and other similar fields.

2. Improvement of the Kent Index Method

Kent's method does not intentionally evade the subjective factors in risk assessment. In fact, his method adopted several feasible measures to reduce the negative influence of these factors, thereby providing a good reference for risk assessment in subway engineering. Several researchers have questioned the expert scoring method because of its alleged subjectivity. In truth, opinion, experience, intuition, and other unquantifiable resources are used if knowledge on the matter is limited. Thus, risk assessment becomes at least partially subjective [9]. Moreover, subjectivity is found in any and all risk assessment methodologies. However, experts also have limitations. The assessment results obtained through the expert scoring method could be inconsistent because of the discrepancy between individual and diffused thought. In the same way that experts need the guidance of a risk manager to normalize their thoughts, several research methods (including theoretical research, value simulated, test demonstration, etc.) should be used to minimize subjective influences.

Kent's method has many advantages. Thus, it is extensively applied in pipeline risk assessment. Unfortunately, its disadvantages restrict its application in other fields.

One of the disadvantages of Kent's method is that no specific method or train of thought is used to determine weight, which is very important in risk assessment. Weight is mainly obtained through the experience of an expert, and no perfect solution exists to solve this problem. In the opinion of the author, determining the weight value is an iterative process whose result can be perfected through repeated application.

Another disadvantage is the assumption regarding the independence of the indexes. In risk analysis, the indexes are assumed to be independent. This assumption could have different effects on the analysis of different objects. Thus, the method is not suitable for all cases.

Pipeline risk is a relatively simple problem, only one accident, the bursting of the pipeline. In underground engineering, for example, the common risk factors include collapse, water gushing, surface subsidence, and so on, and a significant correlation exists among these factors. Thus, ensuring construction safety by merely applying Kent's method is difficult. The method must be modified before being applied in underground engineering safety risk assessment.

Several suggestions have been proposed for the application of Kent's method in underground engineering risk assessment.

(a) Practical engineering should be the basis of research; the basic concept of risk must be utilized to make risk assessment more feasible.

(b) In determining the weight value and other data, existing research results should be considered together with an expert scoring method.

(c) An effective assessment cannot be achieved by mere calculation of the summing indexes because of the complex relationship among strata condition, design, and construction (they affect one another). Thus, the model needs to be improved first.

3. Improved Index Method Model

3.1. Index Categories. Based on the characteristics of subway construction, indexes are divided into four categories, namely, basic, design, construction, and consequence indexes. The diversity of subway construction methods causes the indexes to be different in the different construction methods. In this paper, the shield method of construction is utilized as an example.

The basic indexes mainly refer to the attributes of engineering that cannot be routinely changed and are beyond the control of the operators. These attributes are determined immediately after line selection, which considers existing hydrological conditions [14, 15], geological conditions [16], surrounding environment conditions, and tunnel parameters including the size of the tunnel, slope, turning radius, and so on [17].

"Design" in this paper refers to the idea that can prevent the risks caused by the basic indexes and provide convenience and guidance in the construction process. It is the precondition of construction. In view of the characteristics of shield tunneling, the main design indexes include the following: reinforced design, environmental protection design, precipitation design, construction method design, shield machine design, and segment design [18].

The construction indexes mostly focus on management and operation during the period of construction. They mainly include the selection of construction methods and level of construction technology, analysis of the influence of the construction period, specification of the external construction environment, and control measures.

The consequence indexes include five aspects, namely, environmental loss, economic loss, social loss, casualties, and construction time loss.

The definition and relationship of these four kinds of indexes are shown in Figure 1.

3.2. Risk Assessment Model.
The most commonly accepted definition of risk is often expressed as a mathematical equation:

$$R = f(P, c), \qquad (1)$$

where R: risk; P: event probability; c: event consequence.

Considering the basic model, the index method model can be expressed as

$$R = P \times c,$$
$$P = f_1(B, D, C), \qquad (2)$$
$$c = f_2(\text{consequences index}),$$

where B: basic index; D: design index; C: construction index.

In the above expression, P does not represent the exact probability, only the probability index. The scope of P is not from 0 to 1; however, its value contains the meaning of probability and has a positive correlation with the real probability (the higher the value, the greater the probability of risk).

The probability index is composed of the basic, design, and construction indexes. The consequence index represents losses caused by an accident or accidents, as shown in Figure 2.

3.3. Probability Index Model.
The probability index is an important part of risk. The combined form of the probability index is obtained in this study. Different accidents have different risk occurrence modes because accidents occur with different mechanisms. The indexes that can cause accidents are related and cannot be studied independently.

Considering the characteristics of subway construction, the occurrence modes of disastrous accidents can be concluded as follows.

3.3.1. Design Indexes-Insensitive Mode.
This mode considers the basic index as a risk source, together with improper construction management and operation, which leads to the occurrence of the risk. The design index does not often work in accidents. For example, the factors of shield axis risk control such as uniformity of formation, tunnel slope, and turning radius are mainly basic indexes; the construction indexes including the construction level, experience, and design indexes have a very insignificant relationship with the risk.

3.3.2. Basic Index-Insensitive Mode.
The risk of the occurrence of accidents in this mode has no relationship with the basic indexes; the occurrence is mainly caused by improper design and construction. For example, the deformation or damage to the base in the shield moving-out construction is mainly caused by the poor design safety coefficient and improper operation.

3.3.3. Comprehensive Mode.
The above two models involve only two types of indexes; however, most accidents occur because of the combined action of the attributes of the tunnel (strata, environment, tunnel diameter, buried depth, and others), design factors, and construction factors. These three kinds of indexes compose a relationship chain. The probability value changes when one of the indexes changes, such as the risk of collapse, during shield moving-out construction.

The above three occurrence modes are the bases of risk assessment for an accident. An accident may involve one or more modes. The fault tree analysis method can be used to determine the accident occurrence mode.

The probability calculation model is obtained based on the above three modes.

(a) *Design Indexes-Insensitive Mode.* The probability index is affected by the basic and construction indexes. In view of the influence of the construction indexes on probability, the construction coefficient C_k is introduced:

$$\text{construction coefficient } C_k$$
$$= \frac{\text{construction index C}}{\text{construction standard value Cs}}, \qquad (3)$$

where the construction standard value is a constant set in advance. It represents the construction level in general and does not reduce nor increase the disastrous accident probability.

The risk probability is given by

$$P = B \times C_K = B \times \left(\frac{C}{C_S}\right). \qquad (4)$$

The value of the construction coefficient is about 1. If the value is greater than 1, the construction increases the risk of accidents and vice versa.

(b) *Basic Index-Insensitive Mode.* Similar to the design index-insensitive mode, the construction coefficient C_k and the construction standard value C_s are introduced. The calculation model is defined as

$$P = D \times C_K = D \times \left(\frac{C}{C_S}\right). \qquad (5)$$

(c) *Comprehensive Mode.* The design coefficient D_K and construction coefficient S_k are introduced:

$$\text{design coefficient } D_K = \frac{\text{design index } D}{\text{design standard value } D_s}. \qquad (6)$$

The calculation model of this mode is given by

$$P = B \times D_k \times C_k. \qquad (7)$$

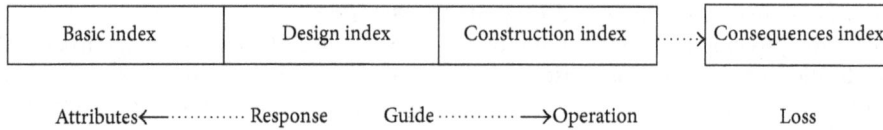

FIGURE 1: Relationship between indexes.

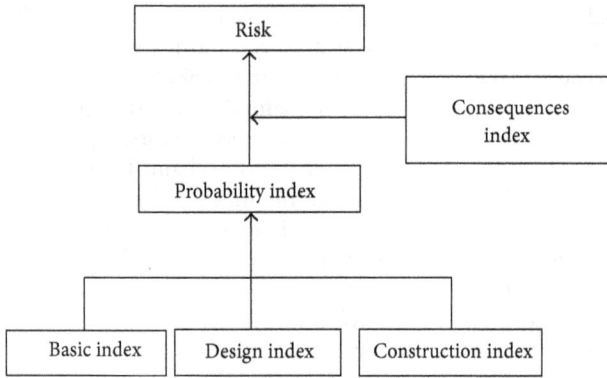

FIGURE 2: Improved index method model.

3.4. *Combining the Basic Indexes and Design Indexes.* The simplest method to combine the basic and design indexes is to consider them separately and then establish the basic index vector. One has

$$B = [B_1, B_2, B_3, B_4 \cdots B_m], \tag{8}$$

where B_i represents the ith basic index.

To establish the design index vector, the following equation is utilized:

$$D = [D_1, D_2, D_3, D_4 \cdots D_n], \tag{9}$$

where D_j represents the jth design index.

The results of the calculation can be expressed as

$$\frac{\text{sum}(B) \times \text{sum}(D)}{(n \times D_s)}. \tag{10}$$

In cases where the basic and design indexes are interrelated, this simple calculation method is feasible. However, in practical engineering, several design indexes may not affect all basic indexes. Considering the foundation consolidation risk in the shield moving-out construction as an example [18], different reinforcement designs have different effects. Several of these designs do not reduce the risk caused by several basic indexes, as shown in Figure 3. When the above calculation is used, the risk may appear to be lower or higher. Thus, the matrix $K_{m,n}$ is introduced to solve this problem:

$$K(i, j) = \begin{cases} 0, & \text{the } i\text{th basic index does not} \\ & \text{affect the } j\text{th design index,} \\ 1, & \text{the } i\text{th basic index affects} \\ & \text{the } j\text{th design index,} \end{cases} \tag{11}$$

where $i = 1, 2, 3, \ldots, m$, $j = 1, 2, 3, \ldots, n$.

TABLE 2: Indicators of economic losses.

Disaster grade	Disastrous	Very serious	Serious	Moderate	Slight
Economic loss ($¥ \times$ million)	>10	3–10	1–3	0.3–1	<0.3
Score	20	16	12	8	4

The results of the calculation can be expressed as

$$B_m \times K_{m,n} \times \left(\frac{D}{D_s}\right). \tag{12}$$

3.5. *Consequence Indexes.* The indicators of the five kinds of losses can be obtained according to the literature [6]. The economic losses are utilized as an example, as shown in Table 2.

4. Risk Assessment Process

Disastrous accidents are risk assessment subjects. The initial steps are to identify the disastrous accidents that happen in subway engineering, to understand the risk occurrence mechanism, and to conduct risk assessment. The fault tree analysis method is a good way to recognize the accident occurrence mechanism. It can be used with other risk assessment methods. The basic events in an accident can be obtained by utilizing the fault tree method. Through the study of these basic events, the relationship between the accident and the indexes could be determined, the risk occurrence mode could be obtained, and the calculation model would be provided. The risk assessment process is shown in Figure 4.

5. Risk Acceptance Criteria

In accordance with the scoring rules and characteristics of the index method, the risk acceptance criteria are determined. The risk level is divided into four grades, namely, unacceptable, unwilling to accept, acceptable, and negligible, as shown in Table 3.

6. Case Study

The risk assessment conducted for the river-crossing tunnel construction of Nanjing subway line 10 was utilized as an example. The basic introduction of the tunnel is shown in Table 4.

6.1. *Risk Assessment.* The first step in risk assessment is risk identification, which aims to predict potential accidents and

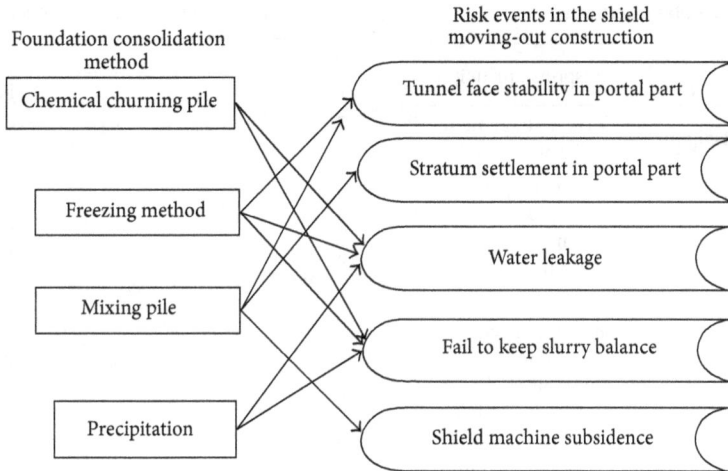

FIGURE 3: Relationship between formation reinforcement design and risk events.

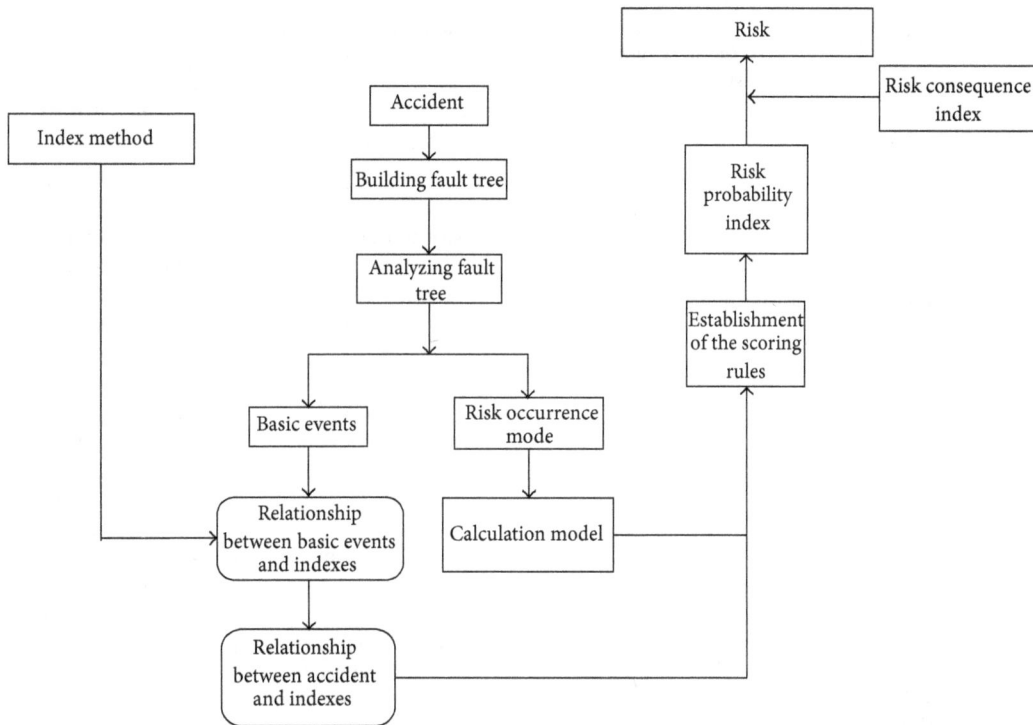

FIGURE 4: Risk assessment process diagram.

to determine the factors of these accidents. A longer tunnel may be divided into several sections according to the strata conditions before risk identification. The conditions of the tunnel's environment are affected by shield tunneling.

As the conditions along the subway line's route change, so does risk. Risk is not constant. Therefore, examining a long tunnel in shorter sections is more efficient. The risk evaluator must decide on a strategy to create sections to obtain an accurate risk value. Each section has its own risk assessment results. Breaking the subway line into many short sections increases the accuracy of the assessment for each section, but may result in higher costs of data collection, handling, and maintenance. Longer sections (fewer in number), on

the other hand, may reduce data costs, but may also reduce accuracy because the average or worst case characteristics dominate if the conditions change within the section [9].

The considered attributes in sectioning or segmenting the tunnel include

(i) strata characteristics,

(ii) environment conditions,

(iii) buried depth.

The next step is to identify the risk and risk factors. The risk identification process is shown in Figure 5. The main steps are collecting data, making the questionnaire,

TABLE 3: Risk acceptance criteria.

Risk value	Risk grade	Risk acceptance level	Response principle
0–1600	IV	Negligible	Risk management can be implemented.
1600–3600	III	Acceptable	Risk management can be implemented, and risk management measures can be taken.
3600–4800	II	Unwilling to accept	Risk management should be implemented to reduce the risk, but the cost of reducing the risk must not be higher than the risk loss.
>4800	I	Unacceptable	Risk management measures must be taken to reduce the risk to grade II.

TABLE 4: Basic introduction of the tunnel.

Tunnel mileage	K11 + 251–K14 + 857
Tunnel design	Single tunnel, double track
Tunnel diameter (outer/inner)	11.2/10.2
Tunnel buried depth	5 m to 35 m
Shield type	Slurry shield
Stratum	silt layer, fine sand; 4-4e1 round gravel layer
Depth of river	0 m to 25.5 m
Permeability coefficient	5.43×10^{-7} to 8.87×10^{-4} cm·s

TABLE 5: Risks or accidents identified in Section 3.

No.	Risk/accident
1	Face instability
2	Cutter head and tool wear
3	Large size bearing breaking
4	Failure of bearing seal
5	Failure of hoisting jack
6	Mud cake
7	Clogging at the exit of slurry
8	Failure of tail skin brush
9	Failure of pushing axis control
10	leakage water at the segment
11	Failure of segment erection
12	Segment uplift
13	Jammed grouting pipe
14	Bad grouting effect

TABLE 6: Risk occurrence mode of face instability and the related indexes.

Occurrence mode	Comprehensive mode
Related indexes	
Basic indexes (no.)	Affects the limit support stress ratio (B1); Change of strata (B2); Overbreak of the affected strata (B3)
Design indexes (no.)	Shield-excavating equipment design (D1); Shield-pushing equipment design (D2); Strata adaptability of the shield (D3)
Construction indexes (no.)	Level of the construction technology (C1); Construction period (C2); Construction environment (C3); Control measures (C4)

The limit support stress ratio is the ratio of the value of limit support stress to original lateral geostress, where the limit support stress refers to the minimum stress that can support face stability.

identifying the risk, and giving opinions and suggestions [19]. Three persons are involved in the risk identification process, namely, the risk evaluator, the expert, and the technology person or operators. The risk evaluators, who function as guides, collect data, make the questionnaires, and send the questionnaires to the experts for risk identification. The experts, as important participants, utilize their experience and finish the questionnaires. The technology person or operators, as the personnel in charge of practical engineering, give their opinions and suggestions on the identified risks, which are obtained by synthesizing the opinions of the experts. Based on the investigation, environment survey, shield selection, and preliminary design data, the risks in sections are identified and are shown in Table 5.

The fault tree and index methods are utilized to analyze and assess the accidents. Face instability, which is the most notable accident, is used as an example.

Based on the fault tree, the risk occurrence mode of face instability and the related basic, design, and construction indexes can be obtained. The results are shown in Table 6.

6.2. Key Parameters in the Probability Index

(a) Standard value. The score scope, design standard value, and construction standard value are determined and shown in Table 7 in accordance with the risk calculation model.

(b) $K_{m,n}$. Based on the construction experiences and the comprehensive scoring of the expert, the weights of each basic and design index and the relationship between the basic and design indexes are determined and shown in Figure 6.

In Figure 6, the meaning of B_1, B_2, B_3, D_1, D_2, D_3, D_4, C_1, C_2, C_3, and C_4 can be got from Table 6, and the length of the column represents the weights. The

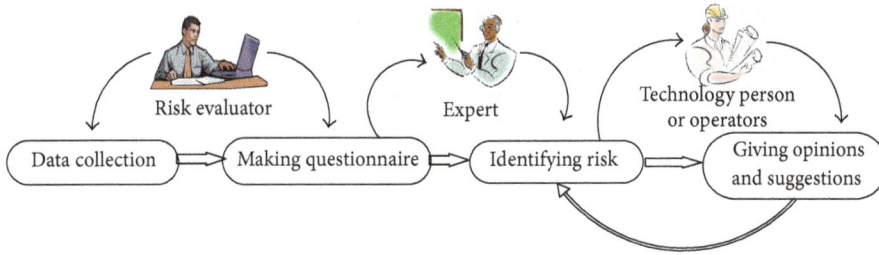

FIGURE 5: Risk identification process.

TABLE 7: Several key parameters.

Indexes	Score scope	Notes
Basic index	0–100	If the score is lower, the probability index is also lower. The construction index is unlimited, which shows that even though the basic index is not large, the possibility of an accident happening would still be high in poor construction conditions.
Design index	0–100	
Construction index	>0	
D_k	50	
S_k	50	

connecting line indicates the correlation between the design and basic indexes.

Thus, we obtain

$$K_{m,n} \text{ is } \begin{bmatrix} 1 & 0 & 1 \\ 0 & 1 & 1 \\ 1 & 0 & 0 \end{bmatrix}. \tag{13}$$

(c) Weights. Obtaining the weights of the indexes is a very important task in risk assessment. The results of existing research and advanced study methods are fully utilized to obtain the weights. The factors that affect the limit support stress ratio, according to the results of existing research, mainly include the buried depth of the tunnel, friction angle and cohesion of the soil, and depth of the groundwater and river. However, no data provide the weights of these factors.

The numerical simulation method is used to obtain the weights in this study. The simulation model is shown in Figure 7, which also contains the effects of the abovementioned factors on the limit support stress ratio.

From the numerical simulation results, increasing the buried depth, friction angle, and cohesion reduces the limit support stress ratio. However, increasing the depth of the underground water and the river increases the limit support stress ratio. Groundwater depth has the biggest impact on limit support stress ratio, followed by river depth, buried depth, cohesion, and friction angle. The weights of these factors are shown in Table 8.

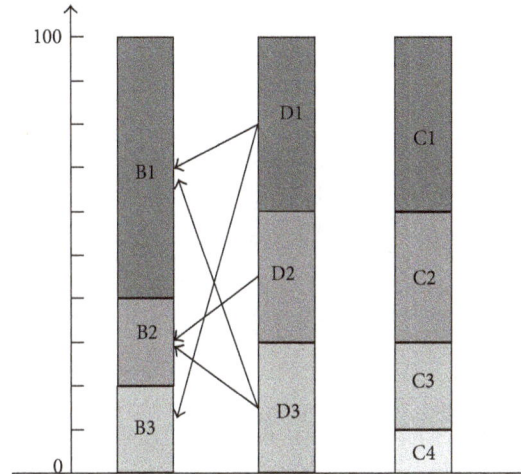

FIGURE 6: Relationship between the basic and design indexes.

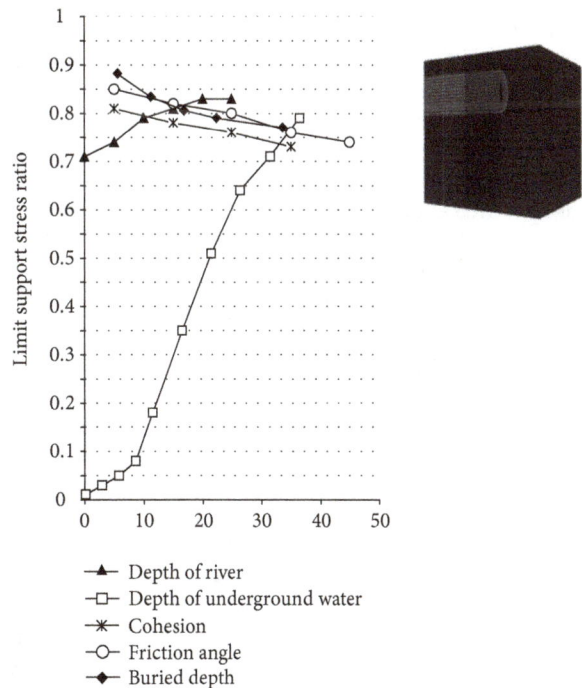

- Depth of river
- Depth of underground water
- Cohesion
- Friction angle
- Buried depth

FIGURE 7: Simulation model and results.

TABLE 8: Weights of the factors that affect the limit support stress ratio.

Factor	Weight
River depth	0.13
Buried depth	0.11
Friction angle	0.05
Cohesion	0.07
Depth of underground water	0.64

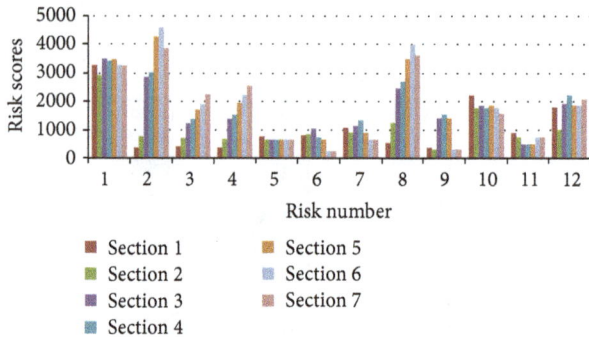

FIGURE 8: Accidents and risk scores in the different sections.

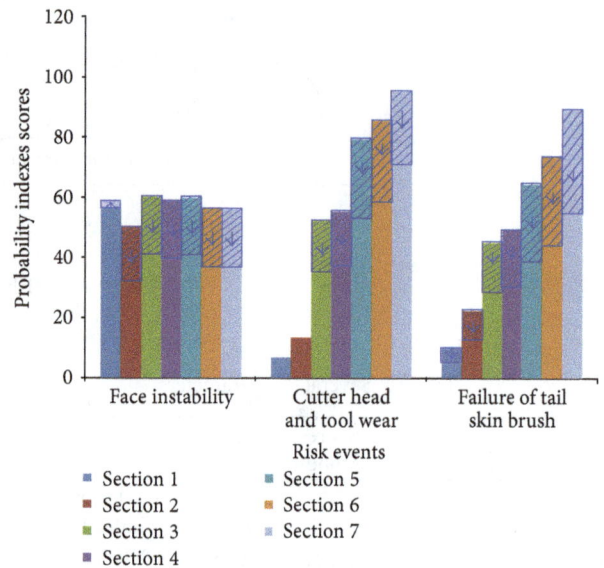

FIGURE 9: Risk probability index in the preliminary risk assessment and construction process.

6.3. Summary of Results and Analysis. Based on the three key parameters, risk assessment can be finalized after establishing a scoring rule. The accidents and the types of risks in each section must be provided.

The accidents and risk scores in each section are shown in Figure 8. In the figure, the risks which are represented by risk number are shown in Table 5. These data are considered as the preliminary risk assessment results. They do not consider the influence of construction. The construction indexes are difficult to evaluate before the construction period because the construction index is more dependent on the attitude of the construction personnel (whether or not the construction personnel are active and conscientious). The results in Figure 8 reflect the influence of the basic and design indexes. The risk assessment is conducted to inform the construction personnel what risks they should pay more attention to and what measures to take. The risk events that have high scores are face instability, tool wear, and shield tail seal failure. We suggest that construction personnel pay special attention to these three risk events in the shield-driving process and prepare risk control measures.

During the construction period, the score of the construction indexes is obtained according to the measures that the construction personnel have taken and their attitude. The risk probability indexes in the preliminary risk assessment and the construction process (including the construction indexes) are shown in Figure 9. The shadow in Figure 9 represents the influence of the construction index. The arrowheads indicate the increase or decrease in the index score. An upward arrowhead means the score increases, whereas a downward arrowhead means the score is reduced. The probability index almost decreases because several measures have been taken to reduce the risk and more attention has been given to the risk events. However, for the face instability risk event

in Section 1, the index increases because when tunneling begins, the operators of the shield machine need to adapt to the performance of the machine itself. They adjust and optimize the shield-tunneling parameters by monitoring the data until it achieves the optimum result. During this period, the construction increases the probability index.

Construction information is provided to verify the validity of this study's risk assessment model. Throughout the construction period, the face instability risk event does not happen. This finding is consistent with our risk assessment results. For the aforementioned risk event, the scores in most of the sections are around 40, which indicate that the probability is between "rarely" and "occasionally." Only the first section has the score of about 60, which means that the event is close to "possible" but the probability is still "occasionally." The frontal soil deformation of the tunneling face reflects the probability of the risk. The deformation values in the different sections are shown in Figure 10. At the initial point of 100 m, the range of the deformation values is between −30 mm and 10 mm. The deformation values at the other points are around −20 mm to 10 mm, and most of them are around −10 mm to 5 mm. The above analysis shows that the construction indexes affect the probability of the risk event. The changing trend is consistent with the result of the risk assessment.

The values of segment uplifting are also provided to verify the assessment model for the risk events, as shown in Figure 11. The segment-uplifting values are within the range of −10 mm to 30 mm, which is acceptable according to the code for the construction and acceptance of the shield-tunneling method [20].

Based on the monitoring data, risk assessment can be applied in subway construction. This assessment indicates which risks the construction personnel are not willing to accept, those they can accept, and those they ignore.

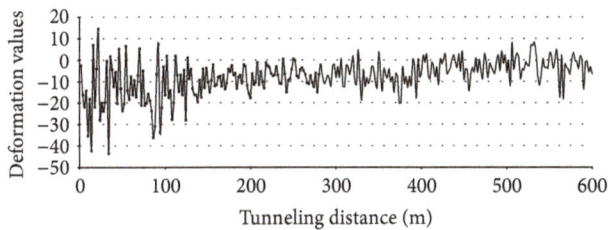

FIGURE 10: The frontal soil deformation values of the tunneling face.

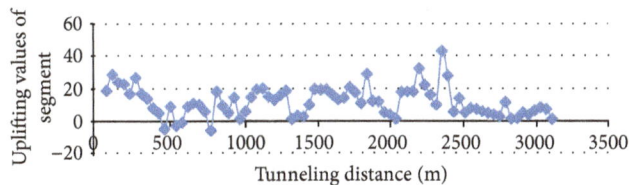

FIGURE 11: Values of segment uplifting.

7. Conclusion and Discussion

The purpose of this paper is to establish a qualitative-quantitative risk assessment model. The risk events change as the external environment changes; thus, the probability of the risk events happening also changes. If the value of the change is not large, adjusting the risk score by using the qualitative analysis method becomes difficult. With regard to general qualitative risk assessment, this study's model should easily adapt to the dynamic changes of the risk. With regard to general quantitative risk assessment, this model does not need to obtain the "real" probability, a process which costs much money. The probability index can be obtained by substituting the data in the scoring model with the data obtained from an existing document or information. Our assessment model can be compared to a thermometer and the risk value to temperature. The thermometer indicates what clothes people should wear; similarly, through our assessment model, construction personnel are informed of the safety situation and also the risks involved. Thus, the construction personnel would know how to respond in case an accident occurs.

An improved index method is established to address the above purpose. Four kinds of indexes are considered in the improved method, namely, (1) basic, (2) design, (3) construction, and (4) consequence indexes. Indexes (1), (2), and (3) constitute the probability index, and the calculation model of which is provided based on three accident occurrence modes. Considering the correlation between the basic and the design indexes, the coefficient matrix $K_{m,n}$ is introduced. The basic index model is then finalized. Different risk events have different occurrence mechanisms and risk factors. Thus, no single formula can express the relationship between the risk index and the known parameters. In this paper, the risk assessment for the river-crossing tunnel construction of Nanjing subway line 10 was utilized as an example, and face instability as a risk event was discussed in detail.

This risk assessment model can be applied to the assessment of disastrous accidents in the design and construction stages. In the design stage, the results of risk assessment can be obtained without considering the construction indexes. The results of risk assessment can be an important reference in the selection of design schemes. The construction indexes are considered in the construction stage. Based on the current construction conditions, a dynamic risk assessment was carried out for dynamic security control during the period of construction.

This risk assessment model has been verified by the Nanjing subway engineering. When the risk occurrence mechanism is not completely clear, the analysis of the risk events still relies on previous engineering experience, which is very crucial for the reliability of the index model. John Hudson, a famous British geotechnical engineering expert, talked about the importance of "collective memory" in the International Top-Level Forum on Engineering Science and Technology Development Strategy-Safe Construction and Risk Management of Major Underground Engineering [21]. He suggested that previous engineering data and experience should be sorted to establish a big shared database. The authors think that the establishment of this database is very important for the ongoing development of risk assessment methods.

Acknowledgment

This project was supported by the Science Fund for Creative Research Groups of the National Natural Science Foundation of China (no. 51021001).

References

[1] Q. H. Qian, "Continuous citifying and utilizing urban underground space," World Sci-Tech R and D, vol. 20, no. 3, pp. 4–9, 2003 (Chinese).

[2] Q. Qian and X. Rong, "State, issues and relevant recommendations for security risk management of china's underground engineering," Chinese Journal of Rock Mechanics and Engineering, vol. 27, no. 4, pp. 649–655, 2008 (Chinese).

[3] S. D. Eskesen, P. Tengborg, J. Kampmann, and T. Holst Veicherts, "Guidelines for tunnelling risk management," Tunnelling and Underground Space Technology, vol. 19, no. 3, pp. 217–237, 2004.

[4] H. H. Choi, H. N. Cho, and J. W. Seo, "Risk assessment methodology for underground construction projects," Journal of Construction Engineering and Management, vol. 130, no. 2, pp. 258–272, 2004.

[5] H. W. Huang, "State-of-art of the research on risk management in construction of tunnel and underground works," Proceedings of Risk Management on Metro and Underground Projects, pp. 16–26, 2005 (Chinese).

[6] Ministry of Housing and Urban-Rural Development of China, Code for Risk Management of Underground Works in Urban Rail Transit, Guangming Daily Publishing House, Peking, China, 2011.

[7] M. L. Lu, "Review on risk assessment methods for tunnelling and underground projects," Journal of Engineering Geology, vol. 14, no. 04, pp. 0462–0469, 2006 (Chinese).

[8] M. H. Faber and M. G. Stewart, "Risk assessment for civil engineering facilities: critical overview and discussion," *Reliability Engineering and System Safety*, vol. 80, no. 2, pp. 173–184, 2003.

[9] W. Kent Muhlbauer, *Pipeline Risk Management Manual Ideas, Techniques, and Resources*, Gulf Professional Publishing, Houston, Tex, USA, 3rd edition, 2004.

[10] G. X. Zhou, "An accident happened in Guangzhou metro construction, and caused one death," 2004 http://news.sohu.com/2004/03/19/78/news219497815.html.

[11] X. G. Luo, "collapse accident happened in construction of 10 line of Beijing subway," 2007, http://news.xinhuanet.com/politics/2007-05/08/content_6071419.htm.

[12] Nanjing Morning News, "Collapse Leads to Natural Gas Explosion in Nanjing Subway Tunnel," 2007, http://news.sina.com.cn/c/p/2007-02-06/031812237328.html.

[13] X. C. Wang, "Hangzhou subway collapse accident is recognized as a major liability accident," 2010, http://news.xinhuanet.com/society/2010-02/09/content_12961265.htm.

[14] F. Pellet, F. Descoeudres, and P. Egger, "The effect of water seepage forces on the face stability of an experimental microtunnel," *Canadian Geotechnical Journal*, vol. 30, no. 2, pp. 363–369, 1993.

[15] W. Broere, "Influence of excess pore pressures on tunnel face stability," in *(Re)Claiming the Underground Space*, J. Saveur, Ed., pp. 759–765, International Tunnelling Association, Swets and Zeitlinger, 2003.

[16] P. D. Buhan, A. Cuvillier, L. Dormieux, and S. Maghous, "Face stability of shallow circular tunnels driven under the water table: a numerical analysis," *International Journal For Numerical and Analytical Methods in Geomechanics*, vol. 23, pp. 79–95, 1999.

[17] Y. Li, "Stability analysis of large slurry shield-driven tunnel in soft clay," *Tunnelling and Underground Space Technology*, vol. 10, pp. 1–10, 2008 (Chinese).

[18] JSCE, *Tunnel Standard Code and Commentary*, China Building Industry Press, Peking, China, 2011.

[19] C. B. Chapman and S. Ward, "Project risk management processes," in *Techniques and Insights*, John Wiley & Sons, Chichester, UK, 1997.

[20] Ministry of Housing and Urban-Rural Development of China, *Code for Construction and Acceptance of Shield Tunnelling Method*, China Building Industry Press, Peking, China, 2008.

[21] A. John Hudson, "Design methodology for the safety of underground rock engineering," in *Proceedings of the International Top-Level Forum on Engineering Science and Technology Development Strategy-Safe Construction and Risk Management of Major Underground Engineering*, Wuhan, China, 2012.

6

A Synchronized Multipoint Vision-Based System for Displacement Measurement of Civil Infrastructures

Hoai-Nam Ho,[1] Jong-Han Lee,[2] Young-Soo Park,[1] and Jong-Jae Lee[1]

[1] Department of Civil and Environmental Engineering, Sejong University, Seoul 143-747, Republic of Korea
[2] Research and Engineering Division, POSCO E&C, Incheon 406-732, Republic of Korea

Correspondence should be addressed to Jong-Jae Lee, jongjae@sejong.ac.kr

Academic Editors: D. Choudhury and Q. Q. Liang

This study presents an advanced multipoint vision-based system for dynamic displacement measurement of civil infrastructures. The proposed system consists of commercial camcorders, frame grabbers, low-cost PCs, and a wireless LAN access point. The images of target panels attached to a structure are captured by camcorders and streamed into the PC via frame grabbers. Then the displacements of targets are calculated using image processing techniques with premeasured calibration parameters. This system can simultaneously support two camcorders at the subsystem level for dynamic real-time displacement measurement. The data of each subsystem including system time are wirelessly transferred from the subsystem PCs to master PC and vice versa. Furthermore, synchronization process is implemented to ensure the time synchronization between the master PC and subsystem PCs. Several shaking table tests were conducted to verify the effectiveness of the proposed system, and the results showed very good agreement with those from a conventional sensor with an error of less than 2%.

1. Introduction

Large-scale civil structures including bridges and buildings are exposed to various loads such as traffic loads and/or natural disasters (e.g., earthquakes, typhoons, cyclones, blizzards). Monitoring structural displacement under such dynamic loads plays an essential role in structural health monitoring. In fact, the direct measurement of structural displacement responses has been a challenge, especially for large-scale structures because traditional sensors such as the linear variable transformer (LVDT) require a stationary reference that is difficult to find a proper location in the field. In recent years, the global position system (GPS) [1–3] and the laser doppler vibrometer [4] have emerged as new noncontact measurement techniques, but their applications are still limited as a result of their high cost.

With increases in CPU capabilities, improvements in image capturing devices and the development of new post-processing image algorithms, vision-based displacement measurement is becoming one of the most common non-contact measurement techniques in civil engineering applications [5–9]. Compared with the other sensors, vision-based measurement provides several advantages: (1) it can provide direct measurements in both the time and the three-dimensional (3D) displacement; (2) it can measure displacement at multiple locations simultaneously in cost-effective manner; (3) it needs a less complicated and labor intensive setup.

Various vision-based systems that measure structural displacement have been developed. Wahbeh et al. [5] developed a high-fidelity video camera with a resolution of 520 lines and a digital zoom of 450 capabilities. The targets consisted of a 28 × 32 inch black steel sheet on which two high-resolution LEDs were mounted to measure the displacement of the Vincent Thomas Bridge, located in San Pedro, California. Then, vision-based systems for measuring the dynamic displacement of bridges in real time was introduced by the authors [6, 7] in 2006 and 2007, respectively. They attached target panels to a structure, captured moving targets by camcorders, and then calculated the amount of structural displacement by applying image processing techniques. In 2009, Fukuda et al. [8] presented a cost-effective (a term previously coined by Lee and Shinozuka [6]) vision-based displacement measurement applied to large-size civil engineering structures, such as bridges and buildings. They employed a TCP/IP (transmission control protocol/Internet

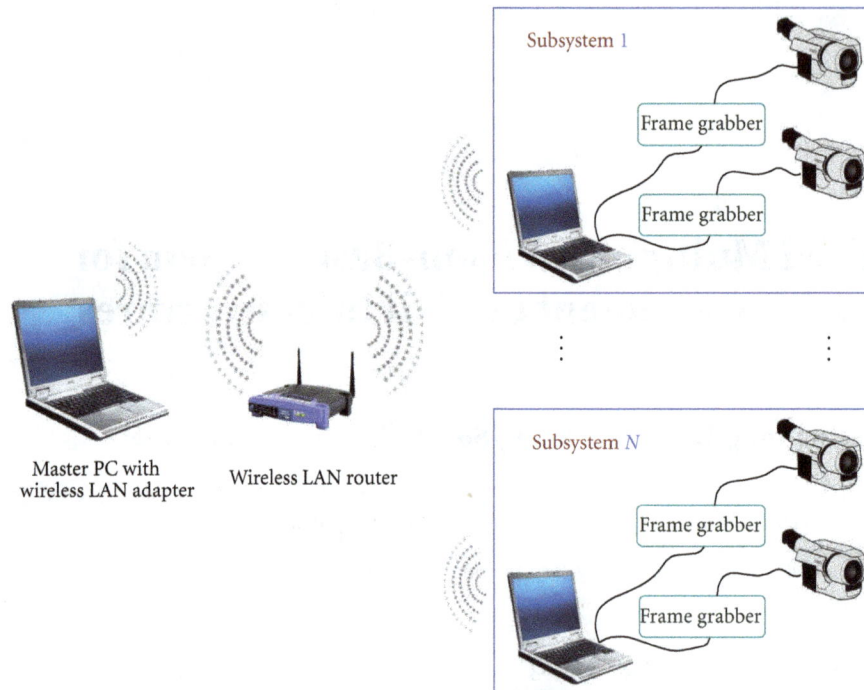

FIGURE 1: Schematic of the system.

protocol) for communications and carried out time synchronization for the time synchronization of the system. More recently, in 2011, Choi et al. [9] introduced a vision-based structural dynamic displacement system using an economical hand-held digital camera. A recorded video containing dynamic information of target panel attached directly to the civil structure was processed with image resizing method, and mm/pixel coefficient updating process, then the structure displacement was successfully determined by calculating the target position in each frame.

The existing vision-based systems for displacement measurement applied to civil engineering applications still have several limitations. There is little cost-effective system which can measure displacement at multiple locations simultaneously. The cost of frame grabbers which can support a few image capturing devices is still very expensive. Furthermore, it is not easy to increase the number of measurement locations more than the frame grabbers' capacity (usually less than 4). When using multiple frame grabbers, time synchronization for frame grabbers and high speed data transmission can be critical issues to be resolved.

The objective of this study is to introduce a synchronized multipoint vision-based system for dynamic real-time displacement measurement. Because of the huge amount of image data, real-time processing is an important issue to be resolved, particularly in the multipoint vision based system. Compared to the previous vision-based displacement measurement systems, the proposed system provides the following advantages: (1) it provides cost-effective multipoint measurement capabilities (i.e., it can support two camcorders at the subsystem level and utilize the commercial camcorders for real-time displacement measurement); (2) it

allows an increase in the number of measurement points in cost-effective manner; (3) it provides a user-friendly software interface. Furthermore, it uses TCP/IP to transfer the time synchronization and data connection of each subsystem via a network. To verify the efficiency, stability, and accuracy of this system, we conducted several shaking table tests.

2. Development of Advanced Vision-Based System

The schematic of the advanced vision-based system for real-time displacement measurement is shown in Figure 1. To measure structural displacement, target panels are attached to a desired location on the structure, and the images of the target panels are captured by camcorders at a remote distance referred to as a "fixed reference point" (the target panels can be marked directly on the structure without using the panels). Then images captured by camcorders are streamed into the slave PC via the frame grabber, and the displacement of the target is calculated by applying image processing techniques with premeasured calibration parameters [7].

2.1. Hardware. The hardware is composed of PCs, camcorders (including telescopic lenses that capture better image data from long distances), a wireless LAN router, and frame grabbers. In this system, the PCs can be commercial laptop computers with a minimum 2 GB of RAM. For camcorders, many kinds of available current commercial products can work well with this system. The wireless LAN router should comply with at least an 802.11 g wireless standard to ensure the stability of the data transaction between the master PC and the slave PCs.

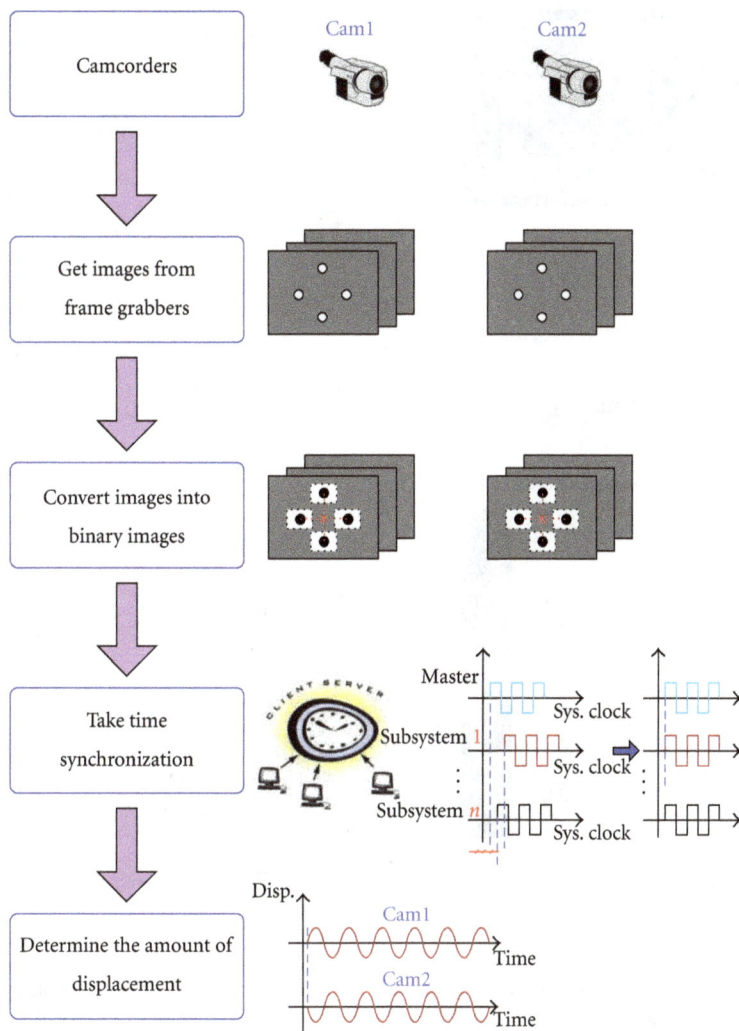

FIGURE 2: Flowchart of the software.

One hardware component that is particularly important is the frame grabber, which is basically considered as a *bridge* that connects the camcorders and the PCs. The frame grabbers help the PCs to adapt to the huge data flow from the single or multiple external camcorders. Subsequently, a low-quality frame grabber will result in poor performance and instability of the entire system. For our system, one commercial frame grabber called *myVision USB* [10] is selected based on the following factors.

(i) It is easy to move and connect, because the subsystems are designed to work outdoors with laptop PCs, the frame grabber has to be portable, small, and light weight.

(ii) It supports a minimum frame size of 640 × 480 pixels.

(iii) It has low power consumption.

(iv) It provides adequate frame quality at a reasonable cost.

2.2. Software. The software consists of two subprograms implemented by Visual C++ language. One is for the master PC and the other for the slave PCs. The general flowchart of the software program is depicted in Figure 2. Image signals are transferred from camcorders to frame grabbers, and then the real-time images are formed through implementation of the DiretShow libraries of the software. Since the start time of the camcorders can differ depending on the camcorder vendor and type, this system used frame gabbers to initialize the start time and send the frames immediately when required without waiting for the camcorders to start capturing. This connection method, which requires very little time for resetting up the system, can change camcorders that are out of order easily and conveniently, upgrade new camcorder models, and so on without changing or modifying the software.

The working window of the slave and master programs are shown in Figures 3 and 4, respectively. Initially, each region of interest (ROI) must be selected manually by clicking on the white dots of the target panels on the screen

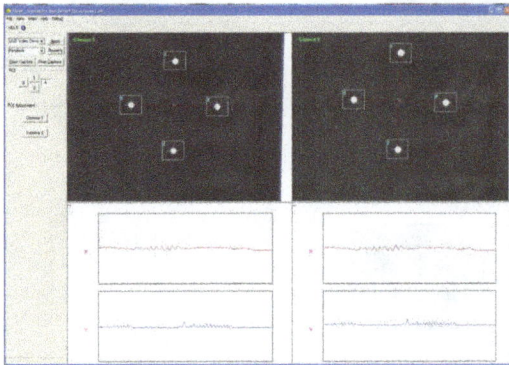

FIGURE 3: Slave program.

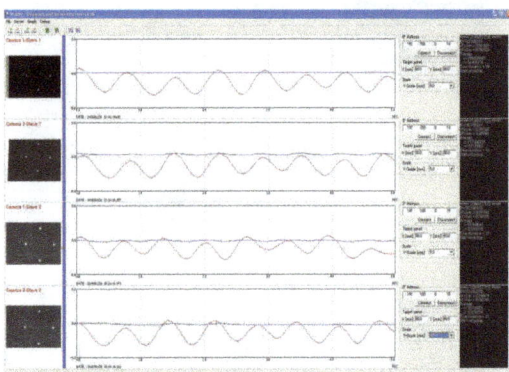

FIGURE 4: Master program.

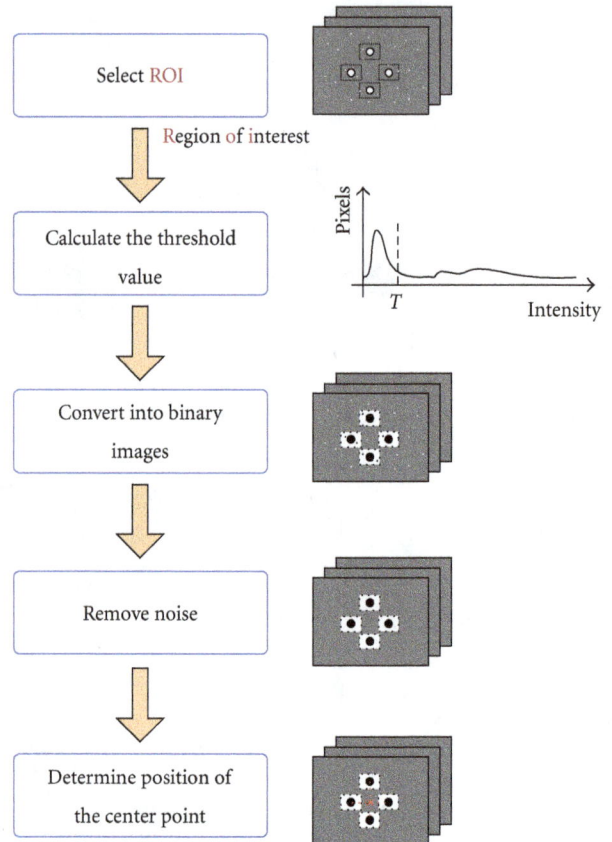

FIGURE 5: Flowchart of binary image conversion.

of the slave PCs. As shown in Figure 3, four ROIs should be defined in each slave screen of a subsystem. Theoretically, we can increase the number of cameras connected to the PCs, but because of computational restrictions the speed of current commercial PCs is not powerful enough to handle more than two camcorders with real-time processing. In addition, before measurements are taken, the IP address and the size of target panels must be accurately defined.

The master program exhibits variations in the wave forms of the target images of each subsystem defined in the slave program. Figure 4 exhibits the full-scale images (640×480) of two camcorders and the wave forms of horizontal (X) and vertical (Y) displacements in red and blue lines, respectively.

2.2.1. Conversion of Images into Binary Images. To convert grayscale images into binary images, we should begin by determining a suitable threshold value that concisely recognizes the position of target points. Therefore, an adaptive threshold technique using Otsu's method [11] is implemented in this system. The process of binary image conversion using the adaptive threshold value is summarized in Figure 5. Figure 6 illustrates the significant difference between applying the adaptive threshold algorithm and not applying it. As shown in Figure 6(a), some target points (white spots) can be missed when using an unreasonable threshold value, whereas there is little chance of missing

targets when deploying the adaptive threshold algorithm, shown in Figure 6(b). Figure 7 shows some typical target panel recognitions in different situations using adaptive threshold technique.

2.2.2. System Calibration for Time Synchronization. Time calibration process needs to be carried out to maintain time synchronization between the master PC and the two slave PCs, and this important process is clearly shown in Figure 8. The master PC measures the time lag due to the wireless communication and differences in internal time clocks between the master and slave PCs. The master PC first sends the same size of dummy time data to all the slave PCs. When the data reach the slave PCs, they immediately return the received dummy data to the master PC. Then the master PC measures the time gap between the sending and receiving time to calculate the time delay between all the subsystems. Subsequently, the master PC sends the internal clock time and the time delay of the slave PCs to each subsystem. Finally, the slave PCs adjust the internal clock according to the time data received from the master PC. Time is calibrated every 60 s, so the synchronization of time between the master PC and slave PCs remains consistent.

The actual structural displacement can be calculated using trigonometric transformation matrix and scaling factors. The transformation matrix is calculated from the positions of the detected white spots on the target and scaling

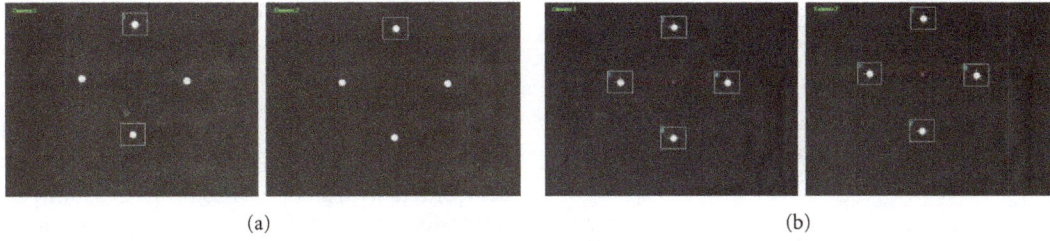

FIGURE 6: Target recognition. (a) The same threshold value for four spots without adaptive threshold technique and (b) each threshold value for each spot using adaptive threshold technique.

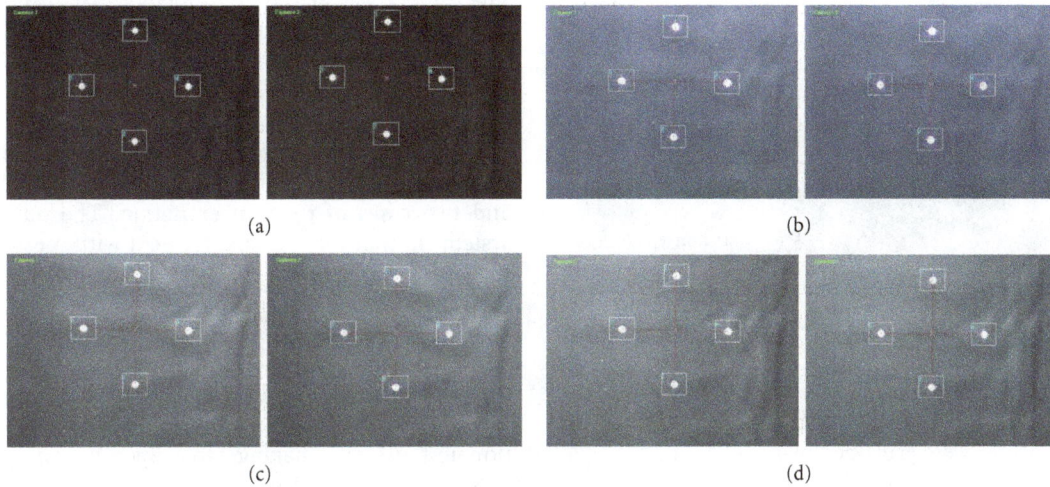

FIGURE 7: Target recognition in various situations with adaptive threshold technique: (a) dark condition, (b) bright light, (c) bright light in the upper right corner, and (d) bright light in the upper left corner.

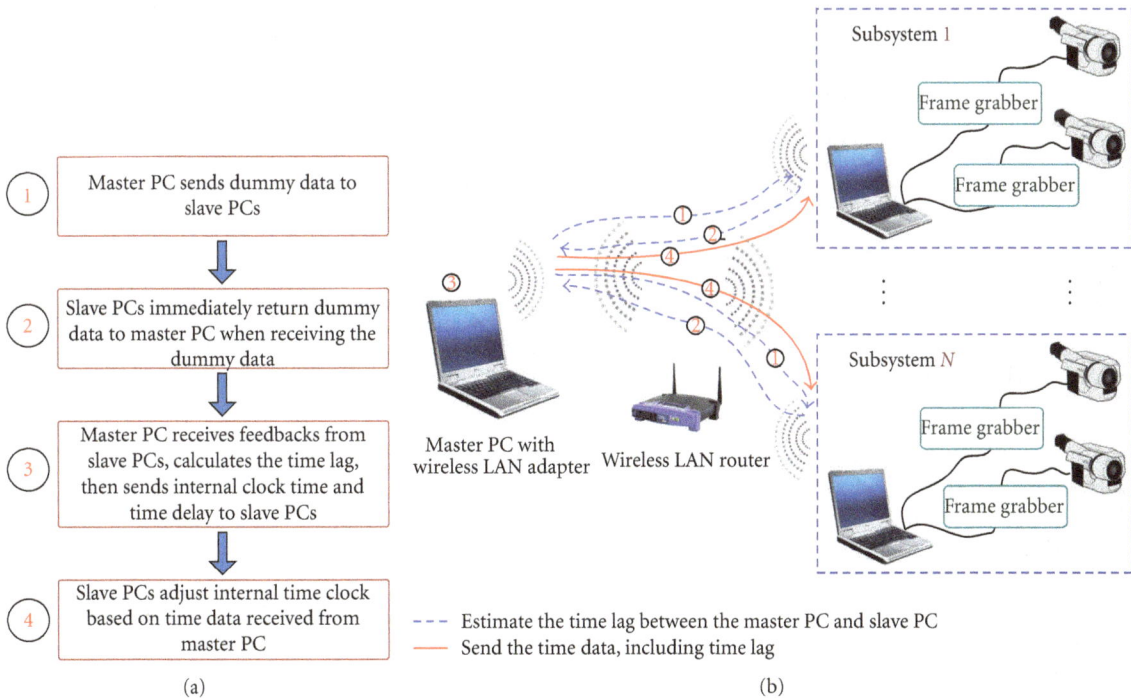

FIGURE 8: Time synchronization process: (a) time synchronization algorithm and (b) diagram of time synchronization process.

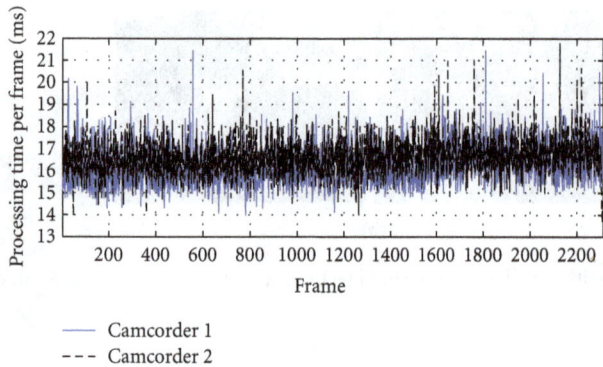

FIGURE 9: Processing time per frame.

FIGURE 10: Shaking table and target size.

factors relate the pixel information to the actual geometric information based on the predetermined geometry of the target. More details were explained by Lee et al. [7].

The proposed vision-based system can support two cameras at the subsystem level with 30 frames per second (fps), so it can be appropriate to track the motion of civil structures with the maximum frequency of less than 15 Hz (Nyquist frequency). In reality, civil structures usually have a natural frequency of lower than 4-5 Hz [8], thus the proposed system can be applicable to measuring the dynamic displacement of large-scale structures. However, for the purpose of real-time processing, the processing time per frame should be less than 33.3 milliseconds (ms). To verify the performance of the proposed system, processing time per frame at a subsystem PC was checked as shown in Figure 9. A total of 2300 frames were captured by camcorders and processed at the subsystem. The averaged and maximum processing time per frame are 17 ms and 21.5 ms, respectively, which is fast enough for real-time processing.

3. Experimental Verifications

Several laboratory tests were conducted to verify the proposed multipoint vision-based system and the time synchronization algorithm. To ensure good data transaction between the slave subsystems and the master PC, all the PCs and their components should be stable without viruses and spywares before conducting the verification tests.

3.1. Shaking Table Test. In order to evaluate the performance and the stability of the proposed synchronized multipoint vision-based system, several shaking table tests were performed. Figure 10 shows the shaking table and the target

used in the experiments. The results of this system were compared with those measured using LVDT.

The testing system consisted of two laptops, one Lenovo-R61 (Intel Core Duo 2.4 MHz, 2 GB of RAM), and one Acer-Asprire 5580 (Intel Core 2-Duo 1.66 MHz, 3 GB of RAM), Lenovo was used as the slave PC. Two JVC GZ-MS120 camcorders with an optical zooming capability of 40 times, a resolution of 640 × 480 pixels, and a frame rate of 30 fps were used. In addition, two *myVision USB* frame grabbers, two telescopic lenses, and one wireless LAN router complying with the 802.11 g wireless standard were implemented to transfer data between the master PC and slave PCs. The target size is 15 mm in vertical and horizontal directions. The camera was placed at 16 meters apart from the target. The number of pixels between the uppermost white spot and the lowermost white spot was 284, thus the physical resolution was 0.053 mm/pixel. The schematic of the testing configuration is given in Figure 11.

Figure 12 shows the results of laboratory tests using a shaking table with 2 Hz and 4 Hz frequencies of excitation and two cases of random excitation. The outputs of this system showed very good agreement with measurements of the LVDT with the variation in the maximum values of less than 2% in all cases.

3.2. Time Synchronization Test. Another shaking test verified the time synchronization algorithm used in this study. Figure 13 shows the configuration of the time synchronization test using a shaking table. In this test, three laptop PCs, one master and two slaves, were used, and the typical specifications as follows:

   (i) Lenovo R61, Intel Core Duo 2.4 MHz, 160 of HDD, 2 GB of RAM,

   (ii) Acer Asprire 5580, Intel Core 2-Duo 1.66 MHz, 160 GB of HDD, 3 GB of RAM,

   (iii) Lenovo X201 Tablet, Intel Core i7-640LM 2.13 MHz, 320 GB of HDD, 4 GB of RAM.

Lenovo X201 Tablet was used as the master PC. Two subsystems simultaneously tracked the same target panel at a distance of 16 meters. All the subsystems were connected to master PC via a wireless LAN access point placed 10 meters away from all the PCs. The distance between the systems was determined considering the performance of telescopic lenses and camcorders. The software, including the time synchronization algorithm, was installed on both the master and the slave PCs.

The system time of the slave PCs were synchronized based on the internal time clock of the master PC. For verifying the accuracy of the salve PCs' internal time clocks, each slave PC generated voltage signals through its serial port based on its internal clock. An oscilloscope connected to the serial ports of the slave PCs monitored the voltage signals generated by the slave PCs, and the time lag between the two slave PCs was determined by referring to the voltage signals. The time lag of the subsystems were checked every 60 s and found a time delay of 1.54 ms initially after the time synchronization process. The time delay increased with

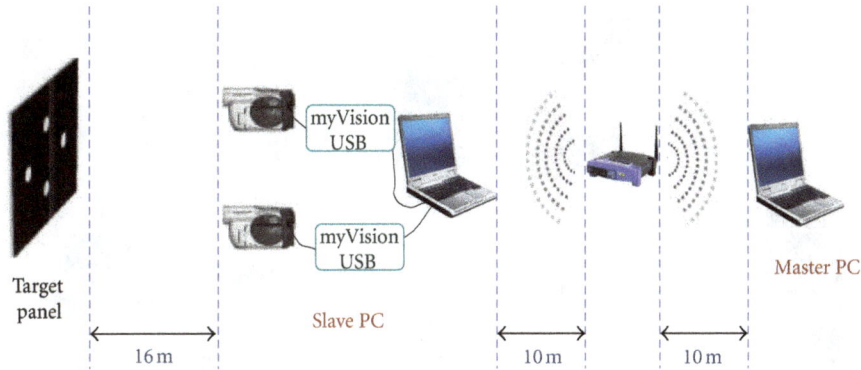

FIGURE 11: Experimental location setup.

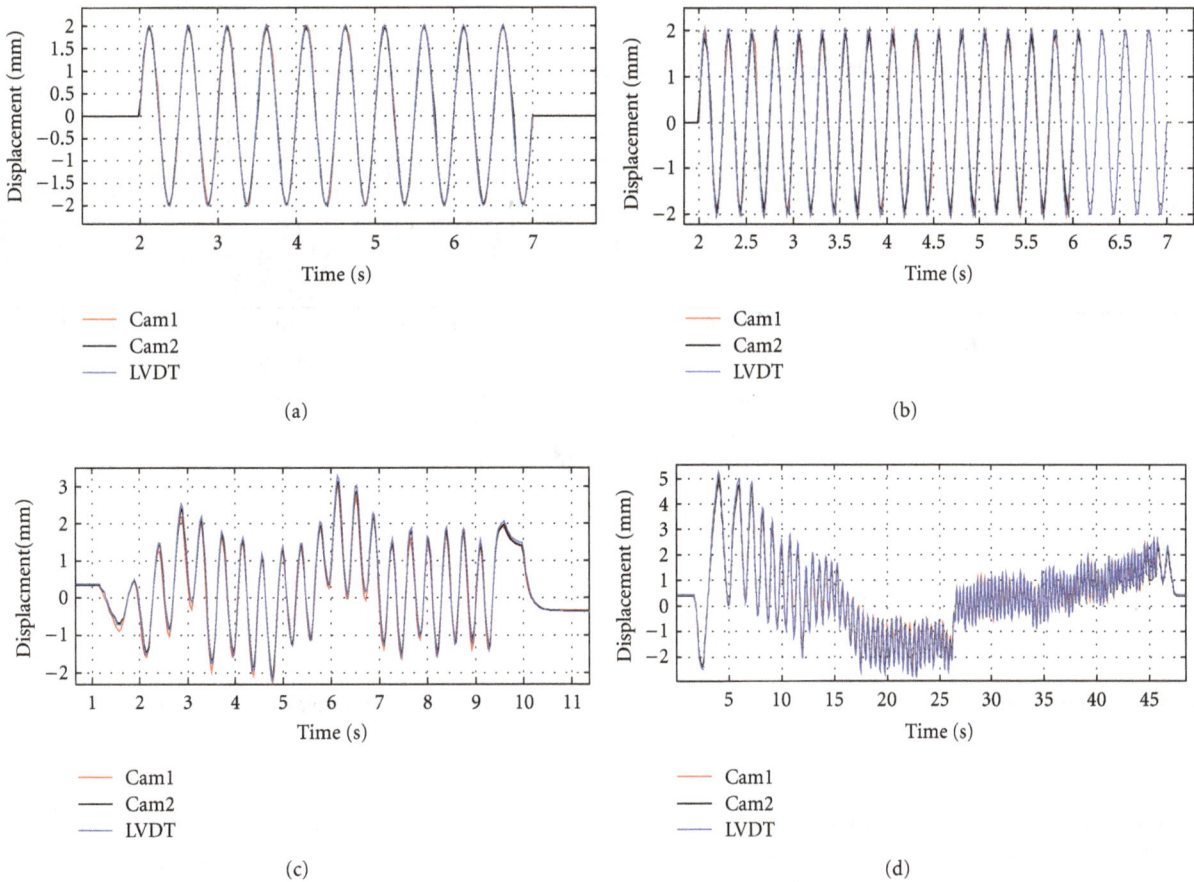

FIGURE 12: Results of shaking table test: (a) sine wave 2 Hz, (b) sine wave 4 Hz, (c) random case 1, and (d) random case 2.

time due to differences in the accuracy of the internal time clocks of the slave PCs, as shown in Figure 14(a). In this experiment, the time lag increased 1.66 ms an average of every 60 s. Thus, to avoid increasing in the time lag, the time synchronization process was performed every 60 s. Figure 14(b) compares the displacements measured from the LVDT with those obtained from the system with the periodical time synchronization process. With sinusoidal excitation of 1 Hz, the displacements obtained from the

system using different Laptop configurations were in good agreement with the LVDT measurements.

4. Conclusions

In this study, an advanced synchronized multipoint vision-based system using an image processing technique has been successfully developed for a real-time dynamic displacement measurement. To evaluate the efficiency, stability, and

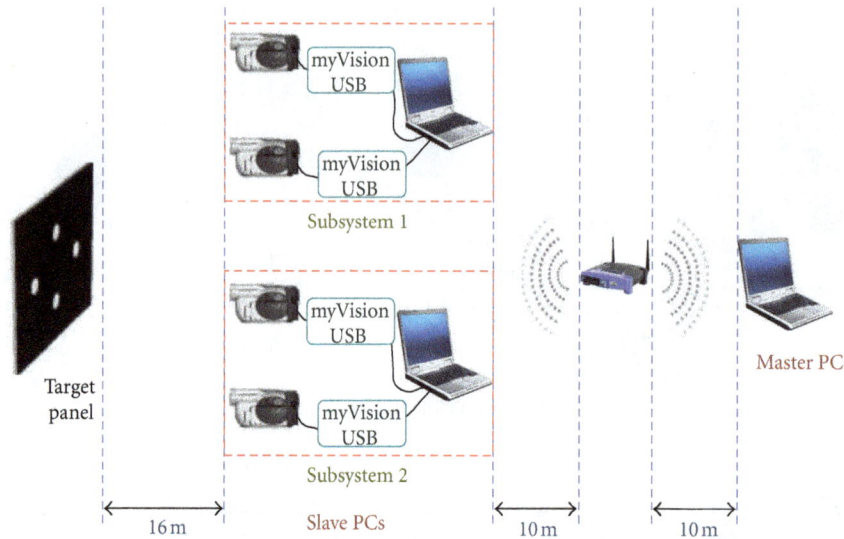

FIGURE 13: Experimental setup for the time synchronization test.

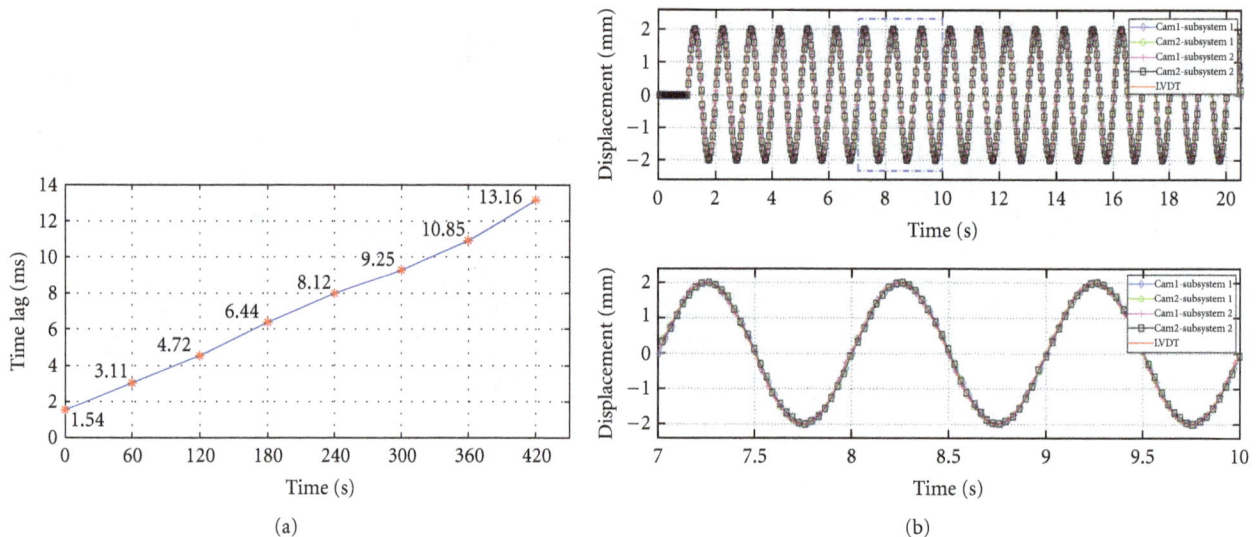

FIGURE 14: Result of the time synchronization test: (a) time lag between the subsystems and (b) displacement with an excitation frequency of 1 Hz.

accuracy of the system, several laboratory tests using the shaking table were conducted. Compared with the data measured from the LVDT, the outputs obtained from the proposed system showed very good agreements, and the variation in the maximum values between the LVDT measurement and system output was less than 2%.

The results of the laboratory tests showed that the proposed system could effectively be applied to large-scale civil infrastructures with low natural frequencies for real-time displacement monitoring and measurement. This system offers the following advantages over current displacement measurement systems for civil engineering.

(i) It simultaneously supports two camcorders at the subsystem level.

(ii) It is easy to install, operate, and maintain.

(iii) It can be set up quickly and configured at a low cost.

(iv) It is robust under complicated on-site light conditions.

(v) It can adjust region of interest (ROI) to fit the current targets.

(vi) It provides easy-to-expand measurement points at the subsystem level with a time synchronization process.

(vii) It can support remote control and data transfer via an internet connection using TCP/IP.

(viii) It can measure 2D relative motions easily and cost-effectively at two different locations using a single system supporting two cameras.

(ix) It provides a user-friendly software interface.

In conclusion, the proposed system can be a promising and cost-effective alternative to measure displacement at multiple locations for large civil structures.

Acknowledgments

This work was supported by the Korean Ministry of Land, Transportation Maritime Affairs (MLTM) through Core Research Project 4 of the Super Long Span Bridge R&D Center (08 CTIP-E01). This work was also sponsored by research project "Development of an Integrated Design Solution based on Codes and Field Data for the Advancement of the Plant Industry" (no. 10040909) funded by the Korean Government Ministry of Knowledge Economy and Korea Evaluation Institute of Industrial Technology (KEIT). Their supports are greatly appreciated.

References

[1] M. Çelebi, "GPS in dynamic monitoring of long-period structures," *Soil Dynamics and Earthquake Engineering*, vol. 20, no. 5–8, pp. 477–483, 2000.

[2] S. I. Nakamura, "GPS measurement of wind-induced suspension bridge girder displacements," *Journal of Structural Engineering*, vol. 126, no. 12, pp. 1413–1419, 2000.

[3] A. Knecht and L. Manetti, "Using GPS in structural health monitoring," in *Smart Structures and Materials*, vol. 4328 of *Proceedings of the SPIE*, pp. 122–129, Newport Beach, Calif, USA, March 2001.

[4] H. H. Nassif, M. Gindy, and J. Davis, "Comparison of laser Doppler vibrometer with contact sensors for monitoring bridge deflection and vibration," *NDT and E International*, vol. 38, no. 3, pp. 213–218, 2005.

[5] A. M. Wahbeh, J. P. Caffrey, and S. F. Masri, "A vision-based approach for the direct measurement of displacements in vibrating systems," *Smart Materials and Structures*, vol. 12, no. 5, pp. 785–794, 2003.

[6] J. J. Lee and M. Shinozuka, "A vision-based system for remote sensing of bridge displacement," *NDT and E International*, vol. 39, no. 5, pp. 425–431, 2006.

[7] J. J. Lee, Y. Fukuda, M. Shinozuka, S. Cho, and C. B. Yun, "Development and application of a vision-based displacement measurement system for structural health monitoring of civil structures," *Smart Structures and Systems*, vol. 3, no. 3, pp. 373–384, 2007.

[8] Y. Fukuda, M. Q. Feng, and M. Shinozuka, "Cost-effective vision-based system for monitoring dynamic response of civil engineering structures," *Structural Control and Health Monitoring*, vol. 17, no. 8, pp. 918–936, 2010.

[9] H. S. Choi, J. H. Cheung, S. H. Kim, and J. H. Ahn, "Structural dynamic displacement vision system using digital image processing," *NDT and E International*, vol. 44, no. 7, pp. 597–608, 2011.

[10] Withrobot Lab, 2012, http://www.withrobot.com.

[11] The OpenCV Reference Manual—Release 2.3., 2011, http://public.cranfield.ac.uk/c5354/teaching/dip/opencv/manual/.

7

Mixed Transportation Network Design under a Sustainable Development Perspective

Jin Qin,[1] Ling-lin Ni,[2] and Feng Shi[1]

[1] School of Traffic and Transportation Engineering, Central South University, Changsha 410075, China
[2] Dongfang College, Zhejiang University of Finance and Economics, Hangzhou 310012, China

Correspondence should be addressed to Jin Qin; csu qinjin@hotmail.com

Academic Editors: B. Uy and J. S. Zhang

A mixed transportation network design problem considering sustainable development was studied in this paper. Based on the discretization of continuous link-grade decision variables, a bilevel programming model was proposed to describe the problem, in which sustainability factors, including vehicle exhaust emissions, land-use scale, link load, and financial budget, are considered. The objective of the model is to minimize the total amount of resources exploited under the premise of meeting all the construction goals. A heuristic algorithm, which combined the simulated annealing and path-based gradient projection algorithm, was developed to solve the model. The numerical example shows that the transportation network optimized with the method above not only significantly alleviates the congestion on the link, but also reduces vehicle exhaust emissions within the network by up to 41.56%.

1. Introduction

Sustainable development, which is generally defined as "development that meets the needs of the present without compromising the ability of future generations to meet their own needs" [1], has in recent years become a widely appreciated concept with strong political and popular support. There are many components to this notion, such as economy, society, resources, and the environment. In particular, followed with the improvement of urbanization, the conflicts between transportation and the environment are becoming more and more serious. Therefore sustainable transport, which is defined as "satisfying current transport and mobility needs without compromising the ability of future generations to meet these needs" [2], has in recent decades become a very important goal in the field of transportation engineering.

The transportation network is one of the most critical infrastructures in urban structures and plays an essential role in meeting normal city operation. It is regularly improved to cope with the ever-growing demands in travel and problems of congestion. In order to optimize some given performance measures (e.g., total travel time or generalized cost) of the network, a crucial decision in the improvement of the network is the allocation of the limited resources to the capacity expansion of the existing links and/or the addition of new candidate links. Such decision problem is always referred to as a transportation network design problem (TNDP).

Morlok firstly proposed the quantitative TNDP in 1973, and since then it has been widely studied by researchers over the last 40 years. The TNDP is always formulated as a mathematical programming model with traffic equilibrium assignment, which can be categorized into three classes: continuous transportation network design problem (CTNDP), discrete transportation network design problem (DTNDP), and mixed transportation network design problem (MTNDP). CTNDP determines the optimal continuous expansion of capacity to existing links [3–8]. DTNDP selects the optimal link additions from a set of candidate links [9–12]. MNDP is a mixture of CTNDP and DTNDP, which simultaneously incorporates both discrete and continuous decision variables [13, 14]. Obviously, MTNDP is more sensible than others in the description of the practical transportation design problem. Nevertheless, MTNDP deals with discrete variables and continuous variables at the same time, so it is very difficult to solve.

Although there is much literature on transportation network design, we can easily find that their common objective is to minimize related economic costs, and goals related to sustainable development are always not taken into account. Guided by this transportation network design theory, the development of transportation engineering has brought about serious consequences regarding the ecological environment in recent decades. Given this situation it is very important to develop a study on the methods and theories of sustainable transportation network design.

This paper mainly studies MTNDP in the consideration of sustainable development, in which the sustainable factors, such as car exhaust emissions or land used and link load are involved. The problem is described as a bilevel programming model and its solution method is proposed.

The remainder of this paper is organized as follows. Hypothesis and notations are given in Section 2. Section 3 formulates the bilevel programming model for the problem. Section 4 designs a solution method based on the annealing algorithm and path-based gradient projection method. In Section 5, numerical experiments are conducted to test the proposed approach. The final section concludes the paper and briefly discusses future research directions.

2. Notations

The complexity of the MTNDP lies mainly within the setting of link-state decision variables, because a single variable cannot describe the continuous link capacity and discrete link connection state at the same time.

Nonetheless, we know that in reality it is meaningless to subtly improve the link capacity. The improvement of a link always comprises of adding new lanes or significantly increasing its capacity; namely, in practice the link capacity expansion is not continuous but within a limited status. Thus the link status can be divided into different discrete grade variables according to the link capacity, which could discretize the mixed problem and lay a foundation for the mathematical model formulation.

For example, according to the Road Commission of China Association for Engineering Construction Standardization [15] the link states are usually divided into 5 grades: high-speed link, 1st-level link, 2nd-level link, 3rd-level link, and 4th-level link. Each link grade represents different lanes, different unit construction costs, different design speeds, and different capacities. On this occasion it is possible to discretize the mixed problem by using the link grade decision variables.

It can be seen from Table 1 that we can set the link grade decision variable u with 10 integers, from 0–9 (in which 0 indicates that the link is not built) respectively representing the different link grades, meanwhile corresponding to the different link capacities and number of lanes, as shown in Table 1. So the link states in the mixed transportation network can be uniformly described by this link grade decision variable. The numbers of the decision variables need not to be restricted to the range of 0 to 9, and in practice can be set properly according to the needs.

In formulating the optimization model, the following notation is used:

A_1: set of existing links in the network,

A_2: set of candidate links to be added,

A_3: set of links that are determined to be added,

A: set of all links in the network, $A = A_1 \cup A_3$,

R: set of original nodes,

S: set of destination nodes,

r: origin node index, $r \in R$,

s: destination node index, $s \in S$,

rs: origin-destination (OD) pair index,

a: link index, $a \in A$,

K_{rs}: set of the paths connecting OD pair rs,

u_a: state variable of link $a \in A$; we assume that there are $n + 1$ grade states from 0 to n in the network, so $u_a \in \{0, 1, 2, \ldots, n\}$, in which $u_a = 0$ means link a will not be built,

u_a^0: the initial grade of link $a \in A$, $u_a^0 \in \{0, 1, 2, \ldots, n\}$,

l_a: length of link $a \in A$ (in kilometers),

q_{rs}: travel demand between OD pair rs,

x_a: total traffic flow on link $a \in A$,

c_a: capacity of link $a \in A$ is determined by the link grade u_a,

v_a: design speed of link $a \in A$ (kilometers/hour) is determined by the link grade u_a,

t_a^0: free-flow travel time on the link $a \in A$ (in minutes). And if $u_a = 0$, $t_a^0 = +\infty$,

t_a: travel time function of link $a \in A$ (in minutes) we use the BPR function as $t_a = t_a^0 \cdot (1 + \alpha(x_a/c_a)^\beta)$ (α, β are all given positive parameters),

$R_{\max}$: maximum values of allowed link load degree for all links,

$p_a(x_a)$: CO emission function of vehicles on the link $a \in A$,

$I_a(u_a)$: construction cost function per unit length of link $a \in A$ when its link grade is u_a,

E: upper limit value of land amount which can be used

$e(u_a, u_a^0)$: land-use scale function of the link $a \in A$, namely the land-use scale of the link a is a function on the initial state u_a^0 and final state u_a,

$I_a(u_a, u_a^0)$: the construction cost function per unit length of the link $a \in A$, that is to say, the construction cost of link a is a function of its initial state u_a^0 and final state u_a,

f_k^{rs}: traffic flow on path k connecting OD pair rs,

$\delta_{a,k}^{rs}$: path/link incidence variables, if path k connecting OD pair rs passes link $a \in A$, then $\delta_{a,k}^{rs} = 1$, otherwise $\delta_{a,k}^{rs} = 0$.

TABLE 1: Decision variables of link grade and its corresponding information in China.

Link status variable u	0	1	2	3	4	5	6	7	8	9
Link grade	No-way	4th-level	3rd-level	2nd-level	2nd-level	1st-level	1st-level	Highway	Highway	Highway
Lane	0	2	2	2	4	4	6	4	6	8
Design speed	0	40	60	80	100	100	100	120	120	120
Capacity	0	1500	3000	4000	6000	15000	20000	30000	45000	60000
Construction cost (tens of thousands of Yuan/km)	0	80	200	500	800	1500	3000	4000	6000	7000

3. Model Formulation

The TNDP consists in seeking a transportation network configuration that minimizes some objectives, subject to the equilibrium constraint.

3.1. Sustainable Development in Transportation Network Design.
From the perspective of sustainable transport, we should take the sustainability factors into consideration in the design of the transportation network. Moreover, according to the definition of sustainable transport we could conclude that the sustainability factors in the transportation network should mainly focus on the pollution caused by vehicle exhaust emissions, resources utilization, and link load.

3.1.1. Vehicle Exhaust Emissions.
Vehicles continuously generate hazardous exhaust gases when they are running, which is the major cause of air pollution. In order to protect the ecological environment and human health, reducing vehicle exhaust emissions has been an ongoing endeavor of many governmental authorities over the past few decades.

Vehicle exhaust emissions contain a range of pollutants, such as carbon monoxide (CO), nitric oxide (NO), and nitrogen dioxide (NO_2). Because it is almost solely emitted by vehicles, CO is considered as the indicator for the level of atmospheric pollution generated by vehicular traffic in the transportation network [16]. Thus if we can reduce one pollutant from vehicle emissions, other pollutants in the emissions will also be reduced. In this regard we only calculate CO as a control subject in this paper.

Obviously, vehicle exhaust emissions on the link depend on the link traffic flow. Yin and Lawphongpanich [17] proposed the following function to estimate vehicular CO emissions:

$$p_a(x_a) = 0.2038 \cdot t_a \cdot e^{0.7962(l_a/t_a)}, \tag{1}$$

where l_a is the length of link $a \in \mathbf{A}$ (in kilometers) and $t_a(x_a)$ is the travel time for link $a \in \mathbf{A}$ (in minutes), and $p_a(x_a)$ is the vehicular CO emissions on link a (in g/hour).

So the total CO emissions in the network are:

$$\sum_{a \in A} p_a(x_a) x_a = 0.2038 \times \sum_{a \in A} t_a \cdot x_a \cdot e^{0.7962(l_a/t_a)}. \tag{2}$$

From (2) it can be found that the total CO emission is associated with the vehicle travel time and the average travel speed on the link, so the CO emission in the network is difficult to determine. We could regard it as an environmental cost of the transportation network system, which should be minimized in the optimization objectives of the TNDP.

3.1.2. Link Load.
The service level of the transportation network perceived by travelers is generally the average travel speed or travel time. Obviously, the higher the average speed of the vehicles, the shorter the travel time, and the greater the perceived level of service.

It is easy to know that the average speed, $\bar{v}_a$, on the link, a, should be

$$\bar{v}_a = \frac{l_a}{t_a} = \frac{l_a}{t_a^0 \cdot \left(1 + \alpha(x_a/c_a)^\beta\right)} = \frac{l_a}{t_a^0} \cdot \frac{1}{1 + \alpha(x_a/c_a)^\beta}. \tag{3}$$

Because l_a, t_a^0, α, and β are all the parameters given, the average travel speed on the link, a, only depends on the value of x_a/c_a, which is usually known as the link load or congestion degree. Thus, to reach a certain service level, as well as to reserve some link capacity for future development, it is necessary to impose restrictions on the maximum link load.

According to the definition, $R_{\max}$ is the maximum load allowed for all links in the transportation network, then the link load constraint can be expressed as

$$\frac{x_a}{c_a} \leq R_{\max}. \tag{4}$$

3.1.3. Land Resource Utilization.
It is well known that urban land resources are limited and precious, which leads to the fact that they cannot be extravagantly used for traffic. A certain amount of land resource needs to be taken up when upgrading existing links or adding new links. To ensure sustainable development we should use as little land as possible on the premise that the expected goal is achieved. Therefore we should set an upper limit value for the amount of land used in the network design.

Obviously the amount of used land for the improvement of a link is relevant to the initial and final states of the link. According to the definition, the land-use scale function of link a is $e(u_a, u_a^0)$, thus the land-use scale constraint can be expressed as

$$\sum_{a \in A} e\left(u_a, u_a^0\right) \leq E, \tag{5}$$

where E is a given upper limit for the amount of land available in the network design.

3.1.4. Construction Funds Utilization. That construction fund is also a limited resource. Therefore, with the premise of achieving the expected construction goal, the total construction cost should be as low as possible. Thus, the total construction cost $\sum_{a \in A} l_a \cdot I(u_a, u_a^0)$ should be minimized in the problem, which is also to be minimized in the model.

3.2. Model Formulation. Similar to a conventional TNDP, the mixed sustainable TNDP can be formulated as a bilevel programming model, in which the upper-level model is the network structure design model and the lower-level model is the user equilibrium assignment model.

Mathematically, the formulation of the following bi-level programming model is

$$\min C = \sum_{a \in A} \left[\theta \cdot p_a \left(x_a \right) \cdot x_a + l_a \cdot I_a \left(u_a, u_a^0 \right) \right],$$

$$\text{s.t.} \quad R_{\min} \leq \frac{x_a}{c_a} \leq R_{\max},$$

$$\sum_{a \in A} e \left(u_a, u_a^0 \right) \leq E, \tag{6}$$

$$u_a \geq u_a^0,$$

$$u_a \in \{0, 1, 2, \ldots, n\},$$

where x_a is implicitly defined by the lower-level model:

$$\min T = \sum_{a \in A} \int_0^{x_a} t_a \left(x_a, c_a \left(u_a \right) \right) d_x,$$

$$\text{s.t.} \quad \sum_{k \in K_{rs}} f_k^{rs} = q_{rs}, \quad \forall s \in \mathbf{S}, \ \forall r \in \mathbf{R},$$

$$x_a = \sum_{r \in R} \sum_{s \in S} \sum_{k \in K_{rs}} f_k^{rs} \delta_{a,k}^{rs}, \quad \forall a \in \mathbf{A}, \tag{7}$$

$$f_k^{rs} \geq 0, \quad \forall s \in \mathbf{S}, \ \forall r \in \mathbf{R}, \ \forall k \in K_{rs}.$$

In the upper-level model, the optimization objective is to minimize the total investment costs and vehicle emissions of CO at the same time, where parameter θ represents the unit conversion factor. Since the existing links generally will not be closed or degraded in the network improvement, it is possible to set the link grade restraint as

$$u_a \geq u_a^0. \tag{8}$$

4. Method of Solution

4.1. Simulated Annealing Algorithm. As mentioned previously, the MTNDP is NP-hard, and this nature makes heuristics the natural choice for it. In this paper we adopted the simulated annealing (SA) algorithm as the basic method to solve the model.

SA is a probabilistic heuristic for global optimization problems for finding a good approximation to the global optimum of a given objective function in the search space. The strength of SA is illustrated by the breadth of studies

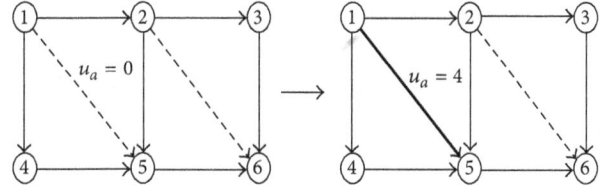

FIGURE 1: Neighboring function based on one link level being transformed.

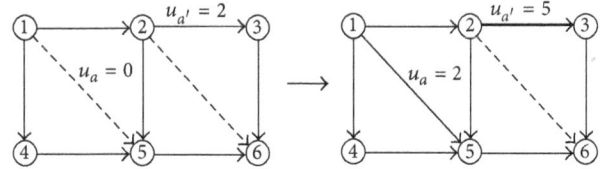

FIGURE 2: Neighboring function based on two link levels being transformed.

found in areas such as logistics network design, computer network design, and machine scheduling, in which SA has been proven as an effective tool for approximating globally optimal solutions to many NP-hard problems.

Although SA studies abound, the area of transportation network design has not been addressed in the literature. It is from this point that the current study embarks.

The neighborhood function plays a crucial role in the performance of SA. In order to improve the efficiency of local search we adopt neighborhood operations by simultaneously adjusting the grades of 1 or 2 links randomly in the neighborhood function. Figures 1 and 2 describe these two neighborhood operations respectively, which randomly select 1 or 2 links from the current network to conduct a link grade transformation operation (the solid lines represent the existing links, while the dashed lines represent the candidate links to be added). Note that the link grades can only be transformed into higher states.

In order to describe the procedure of the SA, $\mathbf{S}, \mathbf{S}', \overline{\mathbf{S}}$ are used to represent the different solutions to the model, $C(\mathbf{S})$ is used to represent the objective function value of the solution $\mathbf{S}$ in the upper-level model. Then the detailed steps of the SA algorithm can be described as follows.

Step 1 (initialization). Set the initial and final temperature values as T_0 and t_f, respectively. The cooling rate $\alpha(0 < \alpha < 1)$ is specified along with the maximum number of iterations N at each temperature value. Take the current network configuration as the initial solution $\mathbf{S}$ and the global optimal solution $\overline{\mathbf{S}}$. Define link set $\overline{\mathbf{A}} = \phi$. The iteration counter $i = 0$.

Step 2 (equilibrium assignment). Conduct equilibrium assignment in the initial network and obtain the equilibrium link flows. Check all link load constraints in the upper-level model. If link a violates the constraint then $\overline{\mathbf{A}} = \overline{\mathbf{A}} \cup a$, and the penalty cost in the objective function should be imposed, that is $C(\mathbf{S}) = C(\mathbf{S}) + n \sum_{a \in \overline{\mathbf{A}}} l_a I(u_a, 0)$ (where n is the given positive integer). Set the global optimal solution $\overline{\mathbf{S}} = \mathbf{S}$ and $C(\overline{\mathbf{S}}) = C(\mathbf{S})$.

Step 3 (generate a feasible neighboring solution). Generate a neighboring solution $\mathbf{S}'$ by altering the current solution $\mathbf{S}$ according to the given neighborhood function. Then check the feasibility of solution $\mathbf{S}'$. If $\mathbf{S}'$ is infeasible, it is necessary to repeat Step 3 to regenerate the neighboring solution; otherwise calculate $C(\mathbf{S}')$ and proceed to Step 4.

Step 4 (equilibrium assignment in a new network configuration). Conduct equilibrium assignment in the new network configuration (neighboring solution $\mathbf{S}'$). Get the new equilibrium link flows. Reset $\overline{A} = \phi$ and check the link load constraints in the upper model. If link a violates the link load constraints, then $\overline{A} = \overline{A} \cup a$ and $C(\mathbf{S}') = C(\mathbf{S}') + n \sum_{a \in \overline{A}} l_a I(u_a, 0)$.

Step 5 (evaluate the current solution with the neighboring solution). If $C(\mathbf{S}') \leq C(\overline{\mathbf{S}})$, $\overline{\mathbf{S}} = \mathbf{S}$ as well as $C(\overline{\mathbf{S}}) = C(\mathbf{S})$. If $C(\mathbf{S}') \leq C(\mathbf{S})$, then set $\mathbf{S} = \mathbf{S}'$, $C(\mathbf{S}) = C(\mathbf{S}')$, proceed to Step 7, otherwise proceed to Step 6.

Step 6 (examine the metropolis condition). The probability at which the relatively inferior neighboring solution should be accepted is $P(\mathbf{S}') = \exp(-(C(\mathbf{S}') - C(\mathbf{S}))/T_i)$, where T_i is the current temperature. A random number ρ is then generated from the interval $(0, 1)$. If $\rho < P(\mathbf{S}')$, then $\mathbf{S} = \mathbf{S}'$ and $C(\mathbf{S}) = C(\mathbf{S}')$.

Step 7 (increment counters). $i = i + 1$. If $i \leq N$, return to Step 3, otherwise proceed to Step 8.

Step 8 (adjust the temperature). Adjust the temperature value by the cooling rate. Mathematically this is $T_{i+1} = \alpha \cdot T_i$, where T_i is the temperature used to compute the acceptance probability at iteration i.

Step 9 (convergence check). If $T_{i+1} < t_f$, stop and output the optimal solution $\overline{\mathbf{S}}$ and the set $\overline{A}$. Otherwise reset $i = 1$, and return to Step 4.

In the above algorithm, the global optimal solution $\overline{\mathbf{S}}$ is saved in Step 5, which is the best solution found so far. And only if the neighborhood solution $\mathbf{S}'$ is better than the optimal solution $\overline{\mathbf{S}}$, do we replace the optimal solution with the neighborhood solution.

It should be noted that if and only if $\overline{A} = \phi$ after the algorithm terminates, the model has the optimal solution. Otherwise it can be deemed that the initial problem has no feasible solution, so then we should relax the link load constraints and/or land-use scale constraints appropriately and recalculate to obtain the optimal solution.

The method of solution for the lower-level traffic equilibrium model can be divided into two types, which are link-based and path-based. Compared to the link-based algorithm, the path-based algorithm can provide more traffic flow information with a higher computational efficiency [18, 19]. So we adopt the path-based gradient projection (pGP)

algorithm to solve the low-level model. The following is a step-by-step description of the procedure for the pGP algorithm [18].

Step 1 (initialization). Generate an initial path for each OD pair.

Step 1.1. Set $x_a(0) = 0$, $t_a = t_a[x_a(0)]$, for all a and $K_{rs}(0) = \phi$.

Step 1.2. Set iteration counter $n = 1$.

Step 1.3. Solve the shortest problem and get $\overline{k}_{rs}(n)$. Update the path set:

$$K_{rs}(n) = \overline{k}_{rs}(n) \cup K_{rs}(n-1), \quad \forall r, s. \quad (9)$$

Step 1.4. Perform "All or No" assignment: $f_{\overline{k}_{rs}}^{rs} = q_{rs}$, for all r, s.

Step 1.5. Calculate the flows on roads according to the flow on the path:

$$x_a(n) = \sum_{r \in R} \sum_{s \in S} \sum_{k \in K_{rs}(n)} f_k^{rs}(n) \delta_{ka}^{rs}, \quad \forall a. \quad (10)$$

Step 2 (column generation). Generate the shortest path based on the current link travel times and augment the set of generated path if it is new.

Step 2.1. Increment iteration counter $n = n + 1$.

Step 2.2. Update link travel times: $t_a(n) = t_a(x_a(n-1))$, for all a.

Step 2.3. Solve the shortest problem: $\overline{k}_{rs}(n)$, for all r, s.

Step 2.4. Augment path $\overline{k}_{rs}(n)$ into path set $K_{rs}(n-1)$ if it is not already existed. If $\overline{k}_{rs}(n) \notin K_{rs}(n-1)$, then $K_{rs}(n) = \overline{k}_{rs}(n) \cup K_{rs}(n-1)$, for all r, s. Otherwise, if $\overline{k}_{rs}(n) \in K_{rs}(n-1)$, then tag the shortest path in the path set $K_{rs}(n-1)$ as $\overline{k}_{rs}(n)$ and set $K_{rs}(n) = K_{rs}(n-1)$.

Step 3 (equilibration). Solve the path-formulated traffic assignment problem over the restricted set of paths generated thus far.

Step 3.1. Calculate first derivative path costs $d_k^{rs}(n)$, $d_{\overline{k}_{rs}}^{rs}(n)$ and second derivative path costs $s_k^{rs}(n)$, where

$$s_k^{rs}(n) = \sum_{a \in A} t_a'(n) \left(\delta_{ka}^{rs} - \delta_{\overline{k}_{rs}(n)a}^{rs} \right)^2,$$

$$\forall k \in K_{rs}(n), \ \forall k \neq \overline{k}_{rs}(n), \ \forall r, s,$$

$$d_k^{rs}(n) = \sum_{a \in A} t_a(n) \delta_{ka}^{rs}, \quad \forall r, s,$$

$$d_{\overline{k}_{rs}}^{rs}(n) = \sum_{a \in A} t_a(n) \delta_{\overline{k}_{rs}(n)a}^{rs}, \quad \forall k \in K_{rs}(n),$$

$$\forall k \neq \overline{k}_{rs}(n), \quad \forall r, s.$$

$$(11)$$

Step 3.2. Update the flows on the nonshortest paths:

$$f_k^{rs}(n+1)$$

$$= \max\left\{\left[f_k^{rs}(n) - \alpha(n)(s_k^{rs}(n))^{-1}\left(d_k^{rs}(n) - d_{\overline{k}_{rs}(n)}^{rs}(n)\right)\right], 0\right\},$$

$$\forall k \in K_{rs}(n), \forall k \neq \overline{k}_{rs}(n), \forall r, s,$$

$$(12)$$

where $\alpha(n)$ is the stepsize and it could be set as $\alpha(n) = $ function$(n) = (n+1)^{-1}$.

Step 3.3. If $f_k^{rs}(n+1) = 0$, then drop the path k: $K_{rs}(n) = K_{rs}(n) \setminus k$.

Step 3.4. Update the flows on the shortest paths:

$$f_{\overline{k}_{rs}(n)}^{rs}(n+1) = q_{rs} - \sum_{k \in K_{rs}, k \neq \overline{k}_{rs}(n)} f_k^{rs}(n+1), \quad \forall r, s. \quad (13)$$

Step 3.5. Update link flows:

$$x_a(n+1) = \sum_{r \in R} \sum_{s \in S} \sum_{k \in K_{rs}(n)} f_k^{rs}(n+1) \delta_{ka}^{rs}, \quad \forall a. \quad (14)$$

Step 4 (termination). Terminate the algorithm if it satisfies the stopping criterion.

Step 4.1. If $\max_{r,s} \sum_{k \in K_{rs}, k \neq \overline{k}_{rs}(n)} f_k^{rs}(n)(d_k^{rs}(n) - d_{\overline{k}_{rs}}^{rs}(n))(q_{rs} d_k^{rs}(n))^{-1} \leq \varepsilon$, then terminate, output the link flows; otherwise, proceed to Step 2.

5. Numerical Examples

In this section a transportation network, illustrated in Figure 3, is used to show the numerical results of the proposed model and solution method, which consists of 12 nodes and 5 OD pairs: $q_{1,11} = 4900$, $q_{1,12} = 10000$, $q_{2,8} = 8000$, $q_{2,12} = 9000$, and $q_{5,12} = 8000$. The 17 solid lines represent the existing links and the 6 dashed lines represent the candidate links. The length and initial grade of each link are given as the first and second values in parentheses, respectively. The available link grade s and its related information are shown in Table 1.

As mentioned previously, the construction cost of the link is determined by its initial and final link grade. We could regard the link construction cost function $I_a(u_a, u_a^0)$ as

$$I_a\left(u_a, u_a^0\right) = I_a\left(u_a\right) - I_a\left(u_a^0\right). \quad (15)$$

TABLE 2: Comparison of OD traveling times.

OD	Travel time in initial network (hour)	Travel time in optimized network (hour)	Gap (%)
(1, 11)	11.690	2.513	78.50
(1, 12)	12.664	3.018	76.17
(2, 8)	9.973	1.934	80.61
(2, 12)	11.720	2.354	79.92
(5, 12)	4.677	2.608	44.24

From Table 1 we could conclude that if the link grade u_a is determined, the free-flow speed on the link, as the designed speed v_a, can also be found (kilometers/hour), and thus the free-flow time on link a is

$$t_a^0 = \frac{l_a}{v_a}. \quad (16)$$

For the travel time function, $t_a = t_a^0 \cdot (1 + \alpha(x_a/c_a)^\beta)$, the parameters are uniformly set as $\alpha = 0.15$, $\beta = 4$, thus the travel time on link a is

$$t_a = t_a^0 \cdot \left(1 + \alpha\left(\frac{x_a}{c_a}\right)\right)^\beta = \frac{l_a}{v_a} \cdot \left(1 + 0.15\left(\frac{x_a}{c_a}\right)\right)^4. \quad (17)$$

The land-use amount function can be set specifically as

$$e\left(u_a, u_a^0\right) = 8\left(u_a - u_a^0\right) + 30\left(L\left(u_a\right) - L\left(u_a^0\right)\right), \quad (18)$$

where $L(u_a)$ refers to the number of lanes owned by link a at the link grade of u_a.

For the link load constraint, set $R_{\max} = 0.85$ for all links. The upper limit value of land amount $E = 1000$, the unit matching coefficient is $\theta = 100$.

The heuristic parameters are fixed as follows: cooling rate $\alpha = 0.7$, and the maximum number of iterations at each temperature value $N = 10|R||S|$.

For these mentioned models and solution algorithms, the results can be achieved with the help of Visual C# 2005 programming. A personal computer with Intel DUO Core 2.0 Ghz CPU, 4 G RAM and Windows XP Professional operating system was used for the test. The optimal solution for the problem can be obtained within about 4 seconds.

The travel times for each OD pair in the initial network and optimized network are listed in Table 2. Compared with the initial network, the OD travel times in the optimized network fell by 44.24–80.61%. The reconstruction of the transportation network significantly improves its performance.

Figure 4 shows the information on the grades and loads of various links under the flow equilibrium state before and after optimization, where the information marked above the line is the initial link grade and load, and the information below the line is the corresponding information after optimization. From the figure we can determine that most link loads before optimization exceed the allowable range. The congestion is obvious in the links, especially in links such as (1,5), (2,3),

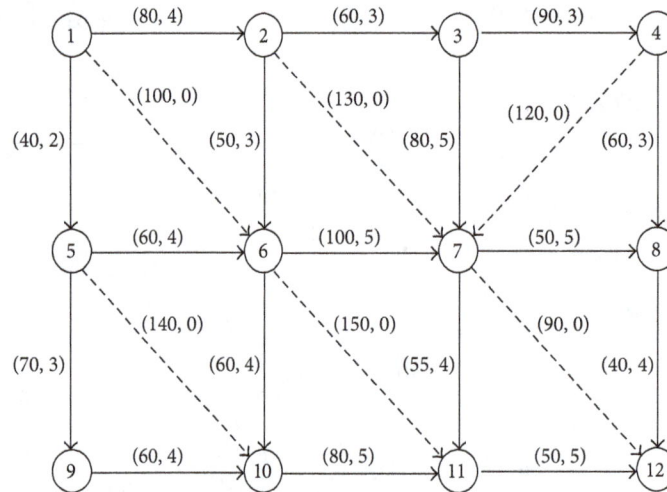

FIGURE 3: 12-node transportation network.

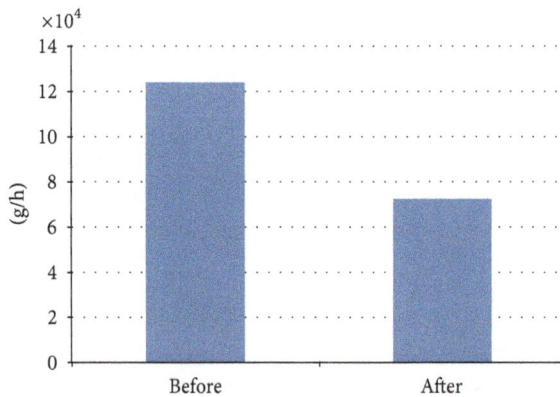

FIGURE 4: CO emission amount in network before and after optimization.

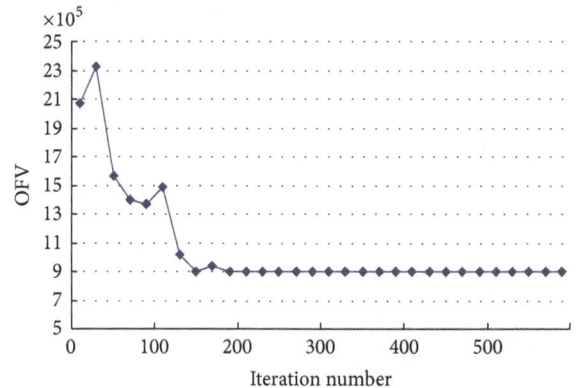

FIGURE 5: Objective value of upper model versus iteration number of SA.

(2,6), (5,9), and (6,10). These links have less flexibility and low elasticity for demand, which is not conducive to sustainable development of the transportation network. However, in the optimized network links (1,5), (2,3), (2,6), (5,6), (6,7), and (6,10) are upgraded, whilst at the same time constructing new links (1,6), (2,7), and (7,12). The loads of links with flows are in the range (0.29, 0.84), which all meet the related constraints. Thus, the new links play a good role in diverting the flows on the congestion links in the network and effectively improve the gridlock of the transportation network.

The total investment of the network improvement project is 192.8 million Yuan, and the land-use amount is 862. The total CO emission in the transportation network drops by 41.56%, reduced from 123,729.00 g/hour before optimization to 72,301.74 g/hour after optimization. It shows a significant improvement in environmental protection in the transportation network.

In addition it can be seen from Figure 5 (OFV is the objective function value of the upper-level model) that that the SA-pGP algorithm determines the optimal solution after about 200 iterations. The algorithm convergence speed and

convergence effect are really good, which also indicates that the SA algorithm proposed in this paper is effective. Figure 6 shows that the objective function of the lower model decreases quickly with the number of iterations in the computation process of the GP algorithm (OFV is the objective function value of the lower-level model) and converges to an optimal solution by the 25–30th iteration.

6. Conclusions

This paper studied the sustainable MTNDP. Link-grade decision variables were used to discretize the mixed problem, based on which the bi-level programming model is developed. The upper model is built to minimize the CO emissions of vehicles on the links and also the total investment, in terms of land-use scale and link load constraints, the lower one is the deterministic traffic assignment model. A heuristic algorithm is proposed to solve the model, in which the SA algorithm and the path-based GP algorithm are used to solve the upper and lower models, respectively.

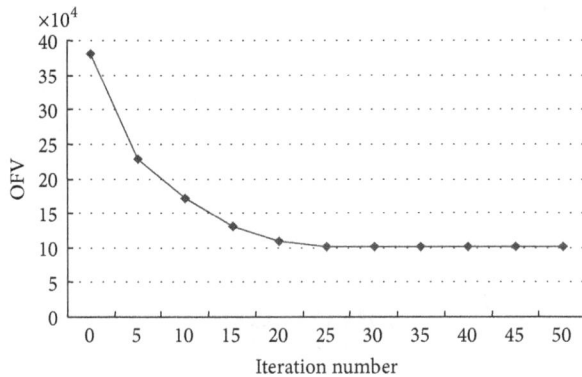

FIGURE 6: Objective value of lower model versus iteration number of GP.

The numerical examples demonstrated that the optimization model and the solution method are effective, in which the SA is not only more efficient and robust with regard to the calculations, but it also rapidly converges to the optimal solution. The path-based GP algorithm calculates quickly and converges steadily. Compared to the initial transportation network, the network optimization significantly improved congestion and pollution in the network. Thereby, the bi-level programming model of the sustainable mixed transportation network design and its solution algorithm can be used to solve real transportation network design problems, and it could provide some scientific evidence for decision making during transportation network planning when considering sustainable development.

Acknowledgment

The work was supported by the National Natural Science Foundation of China (Grant no. 71101155).

References

[1] World Commission on Environment and Development, *Our Common Future*, Oxford University Press, Oxford, UK, 1987.

[2] W. R. Black, "Sustainable transportation: a US perspective," *Journal of Transport Geography*, vol. 4, no. 3, pp. 151–159, 1996.

[3] M. Abdulaal and L. J. LeBlanc, "Continuous equilibrium network design models," *Transportation Research B*, vol. 13, no. 1, pp. 19–32, 1979.

[4] J. X. Ban, H. X. Liu, M. C. Ferris, and B. Ran, "A general MPCC model and its solution algorithm for continuous network design problem," *Mathematical and Computer Modelling*, vol. 43, no. 5-6, pp. 493–505, 2006.

[5] S. W. Chiou, "Bilevel programming for the continuous transport network design problem," *Transportation Research B*, vol. 39, no. 4, pp. 361–383, 2005.

[6] H. Sun, Z. Y. Gao, and J. C. Long, "The robust model of continuous transportation network design problem with demand uncertainty," *Journal of Transportation Systems Engineering and Information Technology*, vol. 11, no. 2, pp. 70–76, 2011.

[7] T. Xu, H. Wei, and Z. D. Wang, "Study on continuous network design problem using simulated annealing and genetic algorithm," *Expert Systems with Applications*, vol. 36, no. 2, pp. 2735–2741, 2009.

[8] G. Zhang and J. Lu, "Genetic algorithm for continuous network design problem," *Journal of Transportation Systems Engineering and Information Technology*, vol. 7, no. 1, pp. 101–105, 2007.

[9] D. E. Boyce and B. N. Janson, "A discrete transportation network design problem with combined trip distribution and assignment," *Transportation Research B*, vol. 14, no. 1-2, pp. 147–154, 1980.

[10] H. Farvaresh and M. M. Sepehri, "A single-level mixed integer linear formulation for a bi-level discrete network design problem," *Transportation Research E*, vol. 47, no. 5, pp. 623–640, 2011.

[11] Z. Gao, J. Wu, and H. Sun, "Solution algorithm for the bi-level discrete network design problem," *Transportation Research B*, vol. 39, no. 6, pp. 479–495, 2005.

[12] H. Poorzahedy and M. A. Turnquist, "Approximate algorithms for the discrete network design problem," *Transportation Research B*, vol. 16, no. 1, pp. 45–55, 1982.

[13] Y. Sun, R. Song, S. W. He, and Q. Chen, "Mixed transportation network design based on immune clone annealing algorithm," *Journal of Transportation Systems Engineering and Information Technology*, vol. 9, no. 3, pp. 103–108, 2009.

[14] P. Luathep, A. Sumalee, W. H. K. Lam, Z. C. Li, and H. K. Lo, "Global optimization method for mixed transportation network design problem: a mixed-integer linear programming approach," *Transportation Research B*, vol. 45, no. 5, pp. 808–827, 2011.

[15] *Road Commission of China Association for Engineering Construction Standardization*, The People Jiaotong Press, Beijing, China, 2004.

[16] A. Alexopoulos, D. Assimacopoulos, and E. Mitsoulis, "Model for traffic emissions estimation," *Atmospheric Environment B*, vol. 27, no. 4, pp. 435–446, 1993.

[17] Y. Yin and S. Lawphongpanich, "Internalizing emission externality on road networks," *Transportation Research D*, vol. 11, no. 4, pp. 292–301, 2006.

[18] A. Chen, D. H. Lee, and R. Jayakrishnan, "Computational study of state-of-the-art path-based traffic assignment algorithms," *Mathematics and Computers in Simulation*, vol. 59, no. 6, pp. 509–518, 2002.

[19] D. H. Lee, Y. Nie, A. Chen, and Y. C. Leow, "Link- and path-based traffic assignment algorithms computational and statistical study," *Transportation Research Record*, vol. 1783, pp. 80–88, 2002.

8

A Novel Evaluation Method for Building Construction Project Based on Integrated Information Entropy with Reliability Theory

Xiao-ping Bai and Xi-wei Zhang

School of Management, Xi'an University of Architecture and Technology, Xi'an Shanxi 710055, China

Correspondence should be addressed to Xiao-ping Bai; xxppbai@163.com

Academic Editors: D. Choudhury and Z. Guan

Selecting construction schemes of the building engineering project is a complex multiobjective optimization decision process, in which many indexes need to be selected to find the optimum scheme. Aiming at this problem, this paper selects cost, progress, quality, and safety as the four first-order evaluation indexes, uses the quantitative method for the cost index, uses integrated qualitative and quantitative methodologies for progress, quality, and safety indexes, and integrates engineering economics, reliability theories, and information entropy theory to present a new evaluation method for building construction project. Combined with a practical case, this paper also presents detailed computing processes and steps, including selecting all order indexes, establishing the index matrix, computing score values of all order indexes, computing the synthesis score, sorting all selected schemes, and making analysis and decision. Presented method can offer valuable references for risk computing of building construction projects.

1. Introduction

The evaluation decision of building construction schemes is a complex multiobjective and multifactor problem, selecting a reasonable goal structure system of evaluation schemes and the optimal scheme is very important in the building construction decision process [1].

Until now, there have been many references studying the optimization decision problem of building construction schemes, and many concepts about it have been set up. Among them, some research results only aim at the small range special field; for example, some researchers study pit bracing construction scheme decision problem [2]. Analytical Hierarchy Process (AHP) is a common method; however, its convictive power is not strong enough because it lacks quantitative data analysis [3–5]. In [6], authors integrated value engineering principle with technical and economic factors to make evaluating and decision making of construction schemes, its advantage is more considering evaluation factors included in construction schemes, but the selection of evaluation values is fuzzy [6]. In addition, grey correlation

[7], minimum variance [8], fuzzy decision, projection pursuit, and other methods are also used in decision-making of project schemes [9, 10].

The optimization decision-making evaluation of construction schemes is a multiobjective process; many targets should be analyzed. Some references select cost, progress, quality, and reliability as evaluation targets, such as in [11, 12], analysis of time-cost-quality tradeoff optimization in construction project management is presented [11, 12], but on the whole there is a shortage of systematic deep discussion, and qualitative study is dominant.

Reliability method is rarely used in the decision making of construction schemes, which is usually simply mentioned in three elements of the project in management references. In [13], reliability method is applied in the evaluation of the construction procedure [13].

In multiobjective optimization decision-making evaluation of projects, the relatively important degree of each evaluation index usually should be considered. The most direct and simple method expressing the important degree of each evaluation target is to give each target relevant weight. The

Table 1: The detailed composition of building construction cost.

The total building construction cost
Direct cost
Direct labor cost (C_1)
Basic wage (C_{11})
Wage allowance (C_{12})
Auxiliary wage (C_{13})
Welfare expense (C_{14})
Labor protection expense (C_{15})
Direct material cost (C_2)
Material initial cost (C_{21})
Material transportation miscellaneous cost (C_{22})
Transportation loss cost (C_{23})
Purchasing and storage cost (C_{24})
Material inspection and testing cost (C_{25})
Direct mechanical cost (C_3)
Mechanical depreciation cost (C_{31})
Repair cost (C_{32})
Installing/removing and external transportation cost (C_{33})
Mechanical labor cost (C_{34})
Fuel power cost (C_{35})
Road toll and vehicle and vessel subsidy expenses (C_{36})
Direct measure cost (C_4)
Environmental protection expense (C_{41})
Civilized construction expense (C_{42})
Safe construction expense (C_{43})
Temporary construction expense (C_{44})
Night construction expense (C_{45})
Large-scale mechanical equipment inside and out expense (C_{46})
Concrete formwork expense (C_{47})
Scaffold expense (C_{48})
Equipment protection expense (C_{49})
Dewatering expense (C_{410})
Indirect cost (C_5)
Construction stipulated expense (C_{51})
Enterprise administration expense (C_{52})

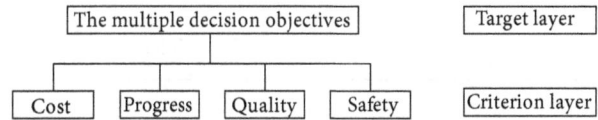

Figure 1: The multiobjective evaluation chart of building project construction schemes.

numerical example shows that the optimal coordination saves more than 50% of waste in system costs, compared to the worst-case scenario [16].

Making use of many existed studying results, this paper integrates engineering economics, risk and reliability theories, and information entropy theory to present a set of detailed engineering management decision methods of building construction projects combined with the concrete example. Presented detailed methods and steps can offer the reference for engineering management decision of building construction projects.

On the basis of summarizing and absorbing some existed references, this paper selected engineering cost, progress, quality, and safety as first-order criterion indexes, shown in Figure 1. For every first-order criterion index, further extended analysis calculation was done.

2. Calculating Cost of Building Construction Schemes

Combined with practical engineering experiences, the building construction cost of engineering projects includes direct cost and indirect cost. The direct cost also includes direct labor cost (C_1), direct material cost (C_2), direct mechanical cost (C_3), and direct measure cost (C_4). The indirect cost (C_5) includes building construction stipulated expense and enterprise administration expense. The detail composition of building construction cost can be expressed by Table 1.

The total building construction cost can be calculated by

$$C = C_1 + C_2 + C_3 + C_4 + C_5. \tag{1}$$

3. Calculating the Progress Score Value of Building Construction Schemes Combined with Reliability Theory

Combined with practical engineering experiences, this paper divided the whole building construction project engineering into 10 first-order progress segments; moreover, the first-order progress segment is divided into detailed second-order progress segments, as shown in Table 2. The progress score value of every second-order progress segment can be given out directly by domain experts.

The total progress score value can be calculated by

$$R = \sum_{i=1}^{10} R_i, \tag{2}$$

where R_i is the sum of progress score value of the i first-order progress segment.

entropy is a very ideal criterion to be applied for evaluating different decision-making processes. Applying the entropy principle to determine the weights of evaluation indexes has the scientific and the accuracy nature. In 1991, two Chinese scholars GU Changyao and QIU Wanhua firstly defined complex entropy and apply it in decision analysis. In 1994, QIU Wanhua also presents group decision-making complex entropy model [14].

In [15], the authors presented the evolution of concepts, an overview of research and applications pertaining to reliability in construction production, and the use of reserves, robust itineraries, and contingency of time and cost. It describes areas of management advisory systems in relation to the cycle of risk analysis [15].

In [16], a biobjective genetic algorithm was employed to solve the multiperiod network optimization problem, and a

TABLE 2: The detailed segment composition of building construction progress.

First-order progress influence factor segments	Construction progress	
	Second-order progress influence factor segments	Progress score value
Site leveling progress (T_1)	Scene investigation progress (T_{11})	R_{11}
	Removing obstacle progress (T_{12})	R_{12}
	Calibrating range, setting benchmarks, and grid progress (T_{13})	R_{13}
	Measuring elevation progress (T_{14})	R_{14}
	Calculating earthwork cut and fill engineering quantities progress (T_{15})	R_{15}
	Land leveling progress (T_{16})	R_{16}
	Field compaction progress (T_{17})	R_{17}
Ground foundation progress (T_2)	Earth excavation progress (T_{21})	R_{21}
	Cushion placement progress (T_{22})	R_{22}
	Foundation placement progress (T_{23})	R_{23}
	Foundation wall placement progress (T_{24})	R_{24}
	Base pillar, ground ring beam progress (T_{25})	R_{25}
	Backfilling earthwork progress (T_{26})	R_{26}
Body engineering progress (T_3)	Reinforcement engineering progress (T_{31})	R_{31}
	Formwork engineering progress (T_{32})	R_{32}
	Concrete engineering progress (T_{33})	R_{33}
	Masonry engineering progress (T_{34})	R_{34}
Roof engineering progress (T_4)	Clean up, leveling progress (T_{41})	R_{41}
	Processing insulation layer progress (T_{42})	R_{42}
	Waterproofing and drainage progress (T_{43})	R_{43}
	Setting qualified seaming progress (T_{44})	R_{44}
	Ventilation, exhaust progress (T_{45})	R_{45}
Decoration and fitment progress (T_5)	Groundwater seepage test of bathroom and kitchen progress (T_{51})	R_{51}
	Laying chisel trough and circuit pipeline progress (T_{52})	R_{52}
	Sealing and burying trunk progress (T_{53})	R_{53}
	Decoration and fitment progress of public wall (T_{54})	R_{54}
	Installing doors and windows progress (T_{55})	R_{55}
Water, electricity, and HVAC progress (T_6)	Water supply and sewerage progress (T_{61})	R_{61}
	Laying circuit progress (T_{62})	R_{62}
	Installing heating progress (T_{63})	R_{63}
	Water, heating, and electrical test progress (T_{64})	R_{64}
Equipment installation progress (T_7)	Installing elevator progress (T_{71})	R_{71}
	Installing fire-fighting equipment progress (T_{72})	R_{72}
	Installing emergency device progress (T_{73})	R_{73}
Goods transportation progress (T_8)	Goods transportation scheme 1 progress (T_{81})	R_{81}
	Goods transportation scheme 2 progress (T_{82})	R_{82}
	Goods transportation scheme 3 progress (T_{83})	R_{83}
Environmental factor progress (T_9)	Natural environment factor progress (T_{91})	R_{91}
	Social environment factor progress (T_{92})	R_{92}
Other factors progress (T_{10})	Construction design drawings change progress (T_{101})	R_{101}
	Construction machinery, materials changes progress (T_{102})	R_{102}
	Progress affected by weak supervision (T_{103})	R_{103}
	Progress affected by supply failure of the municipal system (T_{104})	R_{104}
	Progress affected by great political and social activities (T_{105})	R_{105}

TABLE 3: The detailed quality influence factor segment composition of the building construction schemes.

| First-order influence quality factor segments | Engineering quality | |
	Second-order influence quality factor segments	Quality score value
Site leveling quality segment (Q_1)	Machinery performance influence quality segment (Q_{11})	A_{11}
	Measuring tool influence quality segment (Q_{12})	A_{12}
	Climate influence quality segment (Q_{13})	A_{13}
	Worker influence quality segment (Q_{14})	A_{14}
Ground and foundation quality segment (Q_2)	Equipment performance influence quality segment (Q_{21})	A_{21}
	Materials reliability influence quality segment (Q_{22})	A_{22}
	Personnel quality influence quality segment (Q_{23})	A_{23}
	Drawings reliability influence quality segment (Q_{24})	A_{24}
	Construction plans influence quality segment (Q_{25})	A_{25}
	Construction technology influence quality segment (Q_{26})	A_{26}
	Construction environment influence quality segment (Q_{27})	A_{27}
Main body engineering quality segment (Q_3)	Construction machinery influence quality segment (Q_{31})	A_{31}
	Construction technology influence quality segment (Q_{32})	A_{32}
	Manager's quality influence quality segment (Q_{33})	A_{33}
	Construction technique maturation influence quality segment (Q_{34})	A_{34}
	Construction materials influence quality segment (Q_{35})	A_{35}
	Construction worker technology influence quality segment (Q_{36})	A_{36}
	Construction drawing reliability influence quality segment (Q_{37})	A_{37}
Decoration and fitment engineering quality segment (Q_4)	Decoration and fitment material influence quality segment (Q_{41})	A_{41}
	Construction worker influence quality segment (Q_{42})	A_{42}
	Constructor influence quality segment (Q_{43})	A_{43}
	Construction equipment influence quality segment (Q_{44})	A_{44}
Building roof quality segment (Q_5)	Waterproof and thermal insulation material reliability influence quality segment (Q_{51})	A_{51}
	Construction technology and procedure influence quality segment (Q_{52})	A_{52}
	Working environment influence quality segment (Q_{53})	A_{53}
Building water supply and drainage, heating quality segment (Q_6)	Water supply and drainage, heating ventilating pipe influence quality segment (Q_{61})	A_{61}
	Constructor technology level influence quality segment (Q_{62})	A_{62}
	Construction technology influence quality segment (Q_{63})	A_{63}
Building equipment and installation quality segment (Q_7)	Equipment reliability influence quality segment (Q_{71})	A_{71}
	Professional technology constructor influence quality segment (Q_{71})	A_{72}

TABLE 4: The detailed composition of building construction safety influence factor.

Human safety factor (S_1)	Matter safety factor (S_2)	Environment safety factor (S_3)
Not wearing security protection apparatus (S_{11})	Not setting safety protection (S_{21})	Construction natural conditions (S_{31})
Unsafe costume (S_{12})	Incorrect safety protection and dice marking (S_{22})	Narrow construction work surface (S_{32})
Wrong mechanical operation (S_{13})	Mechanical equipment being in the unsafe state (S_{23})	Disorderly construction yard (S_{33})
Ignoring safety warning (S_{14})	Mechanical equipment being in nonnormal state (S_{24})	Artificial lighting in the night is lacking (S_{34})
Using unsafe equipment (S_{15})	The material stacked in the unsafe state (S_{25})	Wrong operation process design (S_{35})
Body or spirit reason (S_{16})	Having harmful material (S_{26})	Bad ventilation (S_{36})
Rash advance operation (S_{17})		Improper protection for traffic line (S_{37})
Being in unsafe site (S_{18})		
Error handling of dangerous goods (S_{19})		

TABLE 5: Taking value of three risk degree indexes.

Possibility		Probability		Result	
Caused accident possibility	Taking the value of M	Caused accident probability	Taking the value of P	The result caused by accident	Taking the value of F
Sustainably occur	10	Very large possible occurrence	10	Extreme large accident, shutdown, and rectification	10
Often occur	8	Large possible occurrence	8	Very large accident, causing death	9
Occur for many times	6	Possible occurrence	6	Severe accident, having severe injury	7
Occur in a few times	5	Occur for once a while	5	Accident, having slight injury	5
Occur for very few times	3	Rare occurrence	3	Small accident, having minor injures	3
Basically not occur	1	Basically impossible	1	Very small accident, no injury	1

TABLE 6: The calculated cost value of 4 construction schemes (unit: ten thousand yuan).

	C_1	C_2	C_3	C_4	C_5
Scheme 1	89.70	110.85	69.80	13.52	22.71
Scheme 2	76.80	121.90	73.75	13.62	22.89
Scheme 3	81.45	115.37	60.15	12.85	21.59
Scheme 4	79.45	118.69	70.23	13.42	22.54

For calculating R_i, the authors make use of related knowledge in reliability theory. Considering that the progress relations of various second-order progress segments among goods transportation progress (T_8) are parallel, so R_8 can be calculated by

$$R_8 = 1 - \left(1 - R_{81}\right)\left(1 - R_{82}\right)\left(1 - R_{83}\right). \tag{3}$$

For other R_i except for R_8, the progress relation of various second-order progress segments are a series, so R_i except for R_8 can be calculated by

$$R_i = \sum_{j=1}^{n_i} R_{ij}, \tag{4}$$

where n_i is the number of second-order progress segments among the i first-order progress segment.

4. Calculating Quality Score Value of Building Construction Schemes Combined with Reliability Theory

The authors divide the whole building construction project engineering into 7 first-order influence quality factor segments. Every first-order influence quality factor segment is divided into detailed second-order segments, as shown in Table 3.

For calculating the quality score value of every second-order influence quality factor segment, this paper divides them into two types; one type can be calculated by related reliability method, including collecting failure data, putting forward hypotheses by the frequency histogram, estimating

parameters, and testing hypothesis. The other type can be calculated by the expert evaluation method.

Combined with practical engineering knowledge, the quality relations of various second-order quality segments included in every first-order quality segment are a series, so the quality score value of every first-order quality segment can be calculated by

$$A_i = \sum_{j=1}^{n_i} A_{ij}, \tag{5}$$

where n_i is the number of second-order quality segments among the i first-order quality segment.

The total quality score value can be calculated by

$$A = \sum_{i=1}^{7} A_i. \tag{6}$$

5. Calculating Safety Score Value of Building Construction Schemes

Factors affecting building construction safety mainly include direct factor and indirect factor. Direct factors include human factor, matter factor, and environment factor; indirect factors include management factor, and it is caused by three direct factors. This paper further analyzes three direct factors; the detailed composition of building construction safety influence factor is shown in Table 4.

In this paper, the safety score value can be calculated by

$$W_i = M_i \times P_i \times F_i, \tag{7}$$

where M expresses the possibility risk degree index caused by unsafe factors in i safety factor, P expresses the probability risk degree index caused by i unsafe factor, and F expresses the produced result risk degree index after the accident in i safety factor. The value of three indexes can be obtained by experts according to Table 5.

The total quality score value can be calculated by

$$W = \sum_{i=1}^{3} W_i. \tag{8}$$

6. Case Studies and Synthesis Computational Method Based on Integrated Information Entropy with Reliability Theory

6.1. Case Analysis. Taking a building construction engineering project, for example, it is located in the third ring road east section of a city outskirt, has convenient traffic environment. There are residential buildings on the east, west, and north sides of this project; a Greenbelt Park is located in the south side of it. The total land area is 15 acres, the plot ratio is 2.1, and it is planned to be completed in one stage. The building engineering construction will begin on March 1, 2013; the planned construction period is 12 months.

The construction scheme 1 is described as follows. The expected period of engineering construction is 12 months. The month construction completed rate of this scheme, respectively, is 8%, 10%, 11%, 10 %, 8%, 8%, 10%, 11%, 8%, 6%, 6%, and 4%. The month construction progresses in winter and summer is slower than other months because of the effects of the natural environment.

In construction scheme 2, the expected period of engineering construction is 11 months. The month construction completed rate of this scheme, respectively, is 9%, 11%, 11%, 11%, 9%, 9%, 12%, 11%, 8%, 8%, and 7%.

In construction scheme 3, the expected period of engineering construction is 11 months. The month construction completed rate of this scheme, respectively, is 9%, 10%, 11%, 10%, 9%, 8%, 10%, 10%, 8%, 7%, 6%, and 2%.

In construction scheme 4, the expected period of engineering construction is 12 months. The month construction completed rate of this scheme, respectively, is 7%, 9%, 9%, 9%, 8%, 8%, 10%, 10%, 9%, 8%, 7%, and 6%.

According to Table 1 and formula (1) to calculate, respectively, the cost of 4 construction schemes, the result is shown in Table 6.

According to Table 2, formula (2), (3), and (4) to calculate, respectively, progress score value of 4 building construction schemes, the result is shown in Table 7.

According to Table 3, formula (5) and (6) to calculate, respectively, quality score value of 4 building construction schemes, the result is shown in Table 8.

According to Tables 4 and 5, and formula (7) and (8) to calculate, respectively, safety score value of 4 building construction schemes, the result is shown in Table 9.

According to above detailed calculating methods and steps, the result is shown as a calculated total value of 4 indexes including cost, progress, quality, and safety of 4 building construction schemes, as shown in Table 10.

6.2. Detailed Computing Steps of Entropy Weight. Regarding a multiobjective decision making problem that has m selected schemes and n evaluation indexes, detailed computing steps of entropy weight are as follows.

(1) Establishe evaluation index matrix including each evaluation index and corresponding evaluation value:

$$A = (a_{ij})_{m \times n}. \tag{9}$$

(2) Standardize evaluation index matrix.

TABLE 7: The calculated progress score value of 4 building construction schemes.

	Scheme 1	Scheme 2	Scheme 3	Scheme 4
T_{11}	0.98	0.96	0.98	0.97
T_{12}	0.92	0.93	0.94	0.95
T_{13}	0.97	0.95	0.86	0.95
T_{14}	0.87	0.86	0.92	0.98
T_{15}	0.99	0.87	0.95	0.95
T_{16}	0.96	0.91	0.89	0.88
T_{17}	0.89	0.97	0.97	0.93
T_{21}	0.95	0.94	0.97	0.94
T_{22}	0.92	0.98	0.86	0.89
T_{23}	0.93	0.91	0.94	0.93
T_{24}	0.99	0.98	0.89	0.92
T_{25}	0.98	0.92	0.99	0.89
T_{26}	0.88	0.97	0.97	0.99
T_{31}	0.98	0.87	0.95	0.91
T_{32}	0.94	0.95	0.92	0.98
T_{33}	0.97	0.88	0.92	0.87
T_{34}	0.91	0.86	0.94	0.87
T_{41}	0.89	0.91	0.91	0.92
T_{42}	0.92	0.87	0.96	0.98
T_{43}	0.89	0.89	0.88	0.96
T_{44}	0.92	0.94	0.91	0.87
T_{45}	0.97	0.85	0.90	0.96
T_{51}	0.88	0.98	0.94	0.87
T_{52}	0.85	0.91	0.87	0.91
T_{53}	0.98	0.94	0.86	0.92
T_{54}	0.89	0.95	0.86	0.86
T_{55}	0.99	0.91	0.91	0.92
T_{61}	0.86	0.88	0.87	0.96
T_{62}	0.98	0.96	0.89	0.98
T_{63}	0.92	0.88	0.90	0.92
T_{64}	0.91	0.91	0.90	0.95
T_{71}	0.90	0.93	0.88	0.95
T_{72}	0.88	0.95	0.91	0.94
T_{73}	0.92	0.86	0.90	0.85
T_{81}	0.89	0.91	0.91	0.85
T_{82}	0.92	0.88	0.94	0.88
T_{83}	0.88	0.90	0.86	0.96
T_{91}	0.86	0.93	0.96	0.99
T_{92}	0.89	0.95	0.95	0.86
T_{101}	0.89	0.87	0.96	0.88
T_{102}	0.85	0.85	0.89	0.98
T_{103}	0.93	0.94	0.98	0.91
T_{104}	0.94	0.97	0.91	0.89
T_{105}	0.98	0.88	0.93	0.86

For the index that is "the bigger, the better," the standardized value r_{ij} of the evaluation index can be calculated by

$$r_{ij} = \frac{a_{ij} - \min a_{ij}}{\max a_{ij} - \min a_{ij}}. \tag{10}$$

TABLE 8: The calculated quality score value of 4 building construction schemes.

	Scheme 1	Scheme 2	Scheme 3	Scheme 4
Q_{11}	0.97	0.92	0.96	0.90
Q_{12}	0.93	0.91	0.98	0.93
Q_{13}	0.99	0.91	0.97	0.96
Q_{14}	0.92	0.94	0.98	0.96
Q_{21}	0.93	0.97	0.97	0.98
Q_{22}	0.95	0.93	0.94	0.99
Q_{23}	0.94	0.99	0.91	0.93
Q_{24}	0.91	0.97	0.96	0.92
Q_{25}	0.91	0.94	0.96	0.96
Q_{26}	0.94	0.96	0.96	0.98
Q_{27}	0.92	0.97	0.93	0.90
Q_{31}	0.98	0.97	0.97	0.91
Q_{32}	0.91	0.98	0.92	0.92
Q_{33}	0.95	0.90	0.97	0.93
Q_{34}	0.92	0.92	0.91	0.97
Q_{35}	0.95	0.91	0.91	0.97
Q_{36}	0.91	0.96	0.96	0.90
Q_{37}	0.95	0.93	0.98	0.95
Q_{41}	0.94	0.98	0.98	0.92
Q_{42}	0.92	0.97	0.97	0.93
Q_{43}	0.92	0.94	0.98	0.95
Q_{44}	0.93	0.94	0.93	0.96
Q_{51}	0.90	0.92	0.91	0.94
Q_{52}	0.96	0.96	0.95	0.91
Q_{53}	0.94	0.98	0.96	0.96
Q_{61}	0.97	0.94	0.91	0.93
Q_{62}	0.90	0.94	0.92	0.97
Q_{63}	0.96	0.98	0.96	0.94
Q_{71}	0.92	0.92	0.93	0.98
Q_{72}	0.90	0.93	0.92	0.94

TABLE 9: The calculated safety score value of 4 building construction schemes.

Schemes	Scheme 1			Scheme 2			Scheme 3			Scheme 4		
Risk degree indexes	M	P	F	M	P	F	M	P	F	M	P	F
S_{11}	5	8	3	8	5	5	6	5	3	1	5	3
S_{12}	8	3	2	4	1	7	5	8	2	5	7	2
S_{13}	6	1	5	8	6	3	4	4	3	2	8	3
S_{14}	8	1	1	7	6	4	8	4	2	9	2	3
S_{15}	2	6	3	6	3	9	3	7	6	6	8	5
S_{16}	5	4	5	4	3	3	1	5	7	7	3	1
S_{17}	5	6	7	5	3	3	2	5	2	1	5	8
S_{18}	5	5	9	7	6	4	4	7	7	5	5	5
S_{19}	5	7	7	8	3	8	7	4	5	1	8	6
S_{21}	6	5	4	5	2	3	5	6	5	3	7	9
S_{22}	8	6	2	2	9	1	1	3	1	5	3	8
S_{23}	8	9	4	9	4	7	4	1	9	5	8	4
S_{24}	8	6	5	2	1	5	4	9	4	9	1	9
S_{25}	2	8	4	8	7	2	8	8	4	8	1	3
S_{26}	6	6	1	9	7	2	3	7	6	8	6	3
S_{31}	5	8	3	8	5	4	2	6	8	3	4	9
S_{32}	5	6	3	8	3	7	6	7	2	9	7	7
S_{33}	5	8	1	6	2	1	5	4	9	2	8	9
S_{34}	1	3	8	7	2	8	6	8	2	2	1	5
S_{35}	8	3	2	2	3	6	5	6	9	7	4	9
S_{36}	1	5	9	8	3	2	1	2	5	7	7	8
S_{37}	6	7	7	3	1	2	5	5	8	8	5	4

For the index that is "the smaller, the better," the standardized value r_{ij} of the evaluation index can be calculated by

$$r_{ij} = \frac{\max a_{ij} - a_{ij}}{\max a_{ij} - \min a_{ij}}. \quad (11)$$

So the standardized decision-making matrix $(R = (r_{ij})_{m \times n}, r_{ij} \in [0,1])$ can be obtained.

(3) Calculate the entropy value of each evaluation index:

$$H_j = -k \sum_{i=1}^{m} f_{ij} \ln f_{ij} \quad (j = 1, \ldots, n)$$

$$\times \left(f_{ij} = \frac{r_{ij}}{\left(\sum_{i=1}^{m} r_{ij}\right)}, k = \frac{1}{(\ln m)} \right). \quad (12)$$

(4) Calculate the comprehensive feudatory degree Z_i of each evaluation object:

$$\omega_j = \frac{1 - H_j}{n - \sum_{j=1}^{n} H_j}, \quad (13)$$

$$Z_i = (r_{ij})_{m \times n} \omega^T. \quad (14)$$

TABLE 10: The calculated total value of 4 indexes including cost, progress, quality, and safety of 4 building construction schemes.

	Cost	Progress	Quality	Safety
Scheme 1	306.58	0.192	0.233	2527.00
Scheme 2	308.96	0.173	0.291	2233.00
Scheme 3	291.40	0.203	0.368	2450.00
Scheme 4	304.33	0.172	0.260	2886.00

6.3. Calculations Combined with Case. Standardizes evaluation index matrix composed of value in Table 10. Cost and safety indexes are ones that is "the smaller, the better", making use of a formula (11) to calculate. Progress and quality indexes are ones that is "the bigger, the better", making use of a formula (10) to calculate. The standardization result is shown in Table 11".

Making use of formula (12) and (13) to calculate the entropy value and entropy weight of 4 schemes, the result is shown in Table 12.

TABLE 11: The standardization result of 4 building construction schemes based on 4 decision indexes.

	Cost	Progress	Quality	Safety
Scheme 1	0.136	0.645	0.001	0.55
Scheme 2	0.001	0.032	0.430	1.00
Scheme 3	1.000	1.000	1.000	0.67
Scheme 4	0.264	0.001	0.200	0.01

TABLE 12: The calculated entropy value and entropy weight of 4 schemes.

Indexes	Cost	Progress	Quality	Safety
Entropy value (H_i)	0.678	0.567	0.873	0.413
Entropy weighs (ω_i)	0.219	0.295	0.087	0.400

Calculates comprehensive feudatory degree Z_i of each evaluation scheme by formula (14) are shown as follows:

$$Z_i = \left(r_{ij}\right)_{m\times m}\omega^T$$

$$= \begin{bmatrix} 0.136 & 0.645 & 0.001 & 0.55 \\ 0.001 & 0.032 & 0.430 & 1.00 \\ 1.000 & 1.000 & 1.000 & 0.67 \\ 0.264 & 0.001 & 0.200 & 0.01 \end{bmatrix} \begin{bmatrix} 0.219 \\ 0.295 \\ 0.087 \\ 0.400 \end{bmatrix} \tag{15}$$

$$= [0.440, 0.447, 0.869, 0.080].$$

7. Conclusions

Based on the above calculation results, 4 building construction project schemes can be selected according to such sequence: Scheme 3 > Scheme 2 > Scheme 1 > Scheme 4. Generally, the optimization decision of building construction schemes is usually multiobjective optimization decision-making problem affected by many factors. This paper selects cost, progress, quality, safety as the four first-order evaluation indexes, and further deployment analyses of these indexes integrate engineering economics, risk and reliability theories, and information entropy theory to present a new evaluation optimization method for building construction projects based on integrated information entropy with the reliability theory combined with a case study. Presented detailed methods and steps can offer the reference for engineering management decision for the building construction projects.

Acknowledgments

This work was supported in part by NSFC (59874019), Shanxi Province Education Department Research Project (12JK0803), Shanxi Province Key Discipline Construction Special Fund Subsidized Project (E08001), Shanxi Province Higher Education Philosophical Social Science Key Research Base Construction Special Fund Subsidized Project (DA08046), and Shanxi Province Higher Education Philosophical Social Science Characteristic Discipline Construction Special Fund Subsidized Project (E08003, E08005).

References

[1] Y. Xu, Y. Wang, and B. Yao, "Construction project stakeholder collaboration group decision making based on entropy theory," *Chinese Journal of Management Science*, vol. 16, pp. 117–121, 2008.

[2] Y. Feng and K. Shi, "Optimum decision-making of deep foundation pit construction project based on the least variance priority method," *Building Science*, vol. 25, no. 1, pp. 12–15, 2009.

[3] B. Tian, *Management Science in Engineering Project*, Southwest Jiaotong University Press, 2009.

[4] Y. Chen and X. Peng, "Method of analytical hierarchy process making for decision on construction scheme," *Journal of Zhengzhou University of Light Industry (Natural Science)*, vol. 22, pp. 198–200, 2007.

[5] J. Chen, "On construction scheme selected based on value engineering," *Shanxi Architecture*, vol. 12, no. 36, p. 202, 2010.

[6] S. Gao and H. Du, "Study on comprehensive evaluation method about engineering construction scheme based on Grey Correlation Degree," *Coal Mine Engineering*, no. 1, pp. 37–39, 2003.

[7] Y. Feng, "Optimum decision-making of construction project based on the least variance priority method," *Mathematics in Practice and Theory*, vol. 36, no. 3, pp. 171–173, 2006.

[8] W. Oiu, *Management Decision and Application Entropy*, Mechanical Industry Press, 2001.

[9] J. Wang and E. Liu, "Analysis of time-cost-quality tradeoff optimization in construction project management," *Journal of Systems Engineering*, vol. 19, no. 2, pp. 148–150, 2004.

[10] J. Touboul, "Projection pursuit through relative entropy minimization," *Communications in Statistics*, vol. 40, no. 6, pp. 854–878, 2011.

[11] Q. Liu and Q. Yang, "The control of cost, duration, 'quality and safety in project management of construction,'" *Journal of Ningxia Institute of Technology (Natural Science)*, vol. 9, no. 1, pp. 31–33, 1997.

[12] N. Lu, Y. Shi, X. Gao, W. Li, and X. Liao, "Calculation method of construction working procedure," *Journal of Xi'an University of Architecture & Technology (Natural Science Edition)*, vol. 38, no. 3, pp. 311–315, 2006.

[13] W. Qiu, "An entropy model on group decision system," *Control and Decision*, vol. 10, no. 1, pp. 51–53, 1995.

[14] L. Ma and Q. Gao, "Analysis of organizational structure for human resource management department based on structure-entropy model," *Industrial Engineering Journal*, no. 4, pp. 86–90, 2010.

[15] Z. Turskis, M. Gajzler, and A. Dziadosz, "Reliability, risk management, and contingency of construction processes and projects," *Journal of Civil Engineering and Management*, vol. 18, no. 2, pp. 290–298, 2012.

[16] J. Oh, H. Kim, and D. Park, "Bi-objective network optimization for spatial and temporal coordination of multiple highway construction projects," *KSCE Journal of Civil Engineering*, vol. 15, no. 8, pp. 1449–1455, 2011.

9

Effect of Particle Shape on Mechanical Behaviors of Rocks: A Numerical Study Using Clumped Particle Model

Guan Rong,[1,2] Guang Liu,[1,3] Di Hou,[1] and Chuang-bing Zhou[1]

[1] State Key Laboratory of Water Resources and Hydropower Engineering Science, Wuhan University, Wuhan, Hubei 430072, China
[2] Earth Sciences Division, Lawrence Berkeley National Laboratory, Berkeley, CA 94720, USA
[3] Key Laboratory of Rock Mechanics in Hydraulic Structural Engineering, Ministry of Education, Wuhan University, Wuhan 430072, China

Correspondence should be addressed to Guan Rong; rg mail@163.com

Academic Editors: W. Chen and J. Mander

Since rocks are aggregates of mineral particles, the effect of mineral microstructure on macroscopic mechanical behaviors of rocks is inneglectable. Rock samples of four different particle shapes are established in this study based on clumped particle model, and a sphericity index is used to quantify particle shape. Model parameters for simulation in PFC are obtained by triaxial compression test of quartz sandstone, and simulation of triaxial compression test is then conducted on four rock samples with different particle shapes. It is seen from the results that stress thresholds of rock samples such as crack initiation stress, crack damage stress, and peak stress decrease with the increasing of the sphericity index. The increase of sphericity leads to a drop of elastic modulus and a rise in Poisson ratio, while the decreasing sphericity usually results in the increase of cohesion and internal friction angle. Based on volume change of rock samples during simulation of triaxial compression test, variation of dilation angle with plastic strain is also studied.

1. Introduction

Rocks are made from minerals, which are various in chemical composition and crystal morphology. Most Rocks in nature are composed of irregular mineral particles strongly bonded together [1]. The shapes of mineral particles are crucial for mechanical behaviors of rocks. Particle shape has been recognized to affect the mechanics characters of granular material, which has been revealed in several publications [2, 3]. The internal friction angle of different shaped particles was investigated through triaxial compression test by Shinohara et al. [4]. Dodds [5] expounded the influences of particle shape and stiffness effects on soil behavior. Liu et al. [6] quantified particle shapes by digital image method and expounded the correlation of mechanical performance and shape factors defined in their paper. Since various particle shapes in a material, it is very difficult to examine the influence of a specific particle shape. Johanson [7] used plastic pellets of different shapes (round, heart, and stars) coated with soft Tacky Wax to make samples, respectively. Although this method

can obtain consistent samples of distinct shape, it is not proper to produce rock or sand samples because of different performance between plastic pellets and these materials.

Many scholars studied mechanical behaviors of granular matter using numerical simulation method [8–10]. Compared with other continuum approaches, the particle mechanics approach based on discrete element method can reproduce the processes of fracture initiation and growth at quasimicro- or macroscopic levels and treat the problems relating to discontinuous large deformation with some simple assumptions and constitutive models [11, 12]. Particle flow codes (*PFC2D* and *PFC3D*) are the most widely used particle mechanics codes [1, 11]. Numerical tests by Kock showed that composition and texture of sediments were relevant to frictional strength and development of shear zone [13]. Furthermore, with the help of *PFC2D*, peak strength, internal friction angle, and thickness of shear zones were also proven to be connected with particle shape [14–16].

However, researches referring to this issue are still not wide and enough; a systematic study to this problem is of urgent demand for understanding mechanism of deformation and failure from quasi-microlevels. And past study results concentrated on materials such as sand or soil, but a few papers discussed how particle shapes in rock affect the mechanical behaviors of rocks [17]. Unlike loose particle materials sand and soil, rock particles are strongly bonded together. The cements in rocks cause different interaction mechanisms between rock particles. Consequently, Potyondy and Cundall [1] proposed a bonded-particle model for rock, where rock is represented by a dense packing of spherical particles that are bonded together. For different particle shapes, there is no a unified quantitative evaluation method due to their complexity. Some results described particle shape mainly based on the qualitative approach, lacking quantitative analysis of particle shapes.

Hence, the main purpose of this paper is to examine the influences of particle shape on the mechanical behaviors of rocks. From the result of quartz sandstone triaxial compression test and mineral particle shape in quartz sandstone, four representative particle shapes were created. We use sphericity index as the particle shape factor to characterize four representative particles. Then mechanical behaviors of four samples formed by four representative particles were studied, respectively.

2. Basic Theory of Particle Flow Method

A general particle flow model simulates the mechanical behavior of materials based on discrete element method. The Newton's laws of motion and the force-displacement law provide the fundamental relationship among force, displacement and particle motion.

Mechanical behaviors of materials are simulated in terms of the movement of each particle and the interparticle forces acting at each contact point. At each contact point, contact behaviors consist of stiffness, slip, and bond [18].

In the contact normal direction, the stiffness behavior provides the relation among the contact normal force component F^n, the total normal displacements U^n, and the contact normal stiffness (unit: Pa/m) K^n. In the contact tangential direction, the stiffness behavior relates the increment of shear force ΔF^s, and the increment of shear displacement ΔU^s, as follows:

$$F^n = K^n U^n,$$
$$\Delta F^s = -k^s \Delta U^s,$$

(1)

where K^n and k^s are the contact normal stiffness and tangent stiffness (unit: Pa/m), respectively. In this paper, the linear contact models are adopted, where K^n and k^s are independent of displacement. The slip behavior relates slip condition, given by

$$F^s_{\max} = \mu |F^n|,$$

(2)

where $F^s_{\max}$ is maximum allowable shear contact force and μ is the friction coefficient. At every contact point, if $|F^s| > F^s_{\max}$,

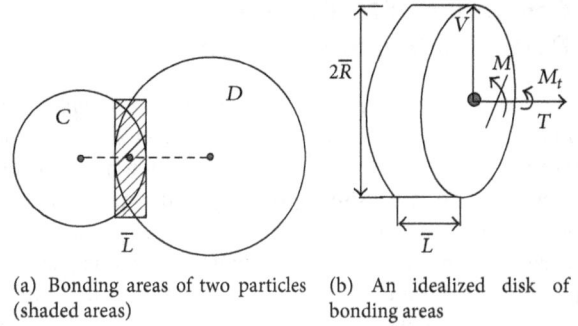

(a) Bonding areas of two particles (shaded areas) (b) An idealized disk of bonding areas

FIGURE 1: Schematic diagram of parallel bond.

then slip is allowed to occur. And during the next calculation cycle, the magnitude of F^s is set to $F^s_{\max}$.

The bond behavior acts as a kind of glue joining the two particles. *PFC3D* can simulate not only the loose particle materials such as sand and soil but also the rock-like materials where particles are strongly bonded together [1]. A contact-bond model and a parallel-bonded model are two widely used bond behaviors. Compared with the contact-bonded model, the parallel-bonded model is more suitable for rock, as the parallel-bonded model can transmit both forces and moments at contact point between particles, and this better meets the interaction mechanism of particles. And the stiffness in the parallel-bonded model consists of the contact stiffness and bonding stiffness. In the parallel-bonded model, the bond breakage will cause bonding stiffness inactive, which results in stiffness reduction. In this respect, the performance of stiffness reduction is consistent with experiments where the rock-like materials may failure in either tension or shearing with an associated reduction in stiffness [12, 19]. So the parallel-bonded model is used in this paper.

In a 3D model, the bonding areas of two particles with parallel bond are equivalent to a disk (Figure 1(a)). T, V, M, and M_t are tensile force, shear force, bending moment, and twisting moment acted on the bonding cross-section, respectively. The maximum normal stresses $\sigma_{\max}$ and shear stresses $\tau_{\max}$ carried by the bonding material can be written as [18]

$$\sigma_{\max} = \frac{T}{A} + \frac{|M|}{I}\overline{R},$$
$$\tau_{\max} = \frac{V}{A} + \frac{|M_t|}{J}\overline{R},$$

(3)

where A and I are the area and moment of inertia, respectively, of the parallel-bond cross-section. And J is the polar moment of inertia of the parallel-bond cross section. R is the radius of the disk (Figure 1(b)). The parallel-bond breaks when either normal stresses or shear stresses exceed corresponding maximum stresses.

3. Generation of Models and Determination of Parameters

3.1. Generation of Representative Particles and Samples. The mineral particles formed by different morphological structures of crystals in rocks are irregular and often have no fixed shapes. In a numerical simulation of granular material, some representative particles substitute for the real particles [20]. According to the characteristic of quartz sandstone particles, four kinds of representative particles were built to simulate complex particle shapes in quartz sandstone, as shown in Figure 2(b). A common feature of these representative particles adopted in this paper is that they are all made up of a big spherical particle and several small spherical particles. The big spherical particle acts as the main body of representative particles, and the several small spherical particles act as the rugged edges on the surface of rock particles. Figure 2(b) shows the relationship of size between the big spherical particle and the small ones. The samples in Figure 2(c) were formed by four representative particles respectively, and they were used to research mechanical behaviors of different particle shapes. The sample 5 (Figure 2(a)) was created by combinations of the representative particles from 1#–4# to simulate real quartz sandstone.

The representative particle 1# is a ball, which can be created directly in PFC3D. The representative particles 2#–4# can be generated using clumped particles. The clumped particles model is widely used model to generate non-spherical particles [15, 16]. The clumped particles are as a whole consisting of two or more balls in *PFC3D*. The clumped particles can create a group of cement particles that behave as a single rigid body, and these clumped particles may overlap as a deformable body that will not break apart regardless of the forces acting upon it [18].

Sample 1 was first created; then the balls in sample 1 were replaced with representative particles 2#–4#, which produced samples 2–4, respectively. The replacements obey the following three principles [18]. (1) Replace ball with clumped particle, each of which has the same volume as the ball that it replaces. The porosity of sample remains the same after replacement. (2) Each clumped particle is oriented randomly by rotating them about the axes by a random angle. (3) Volume-based centroid of the clumped particle coincides with centroid of the replaced ball.

Considering the generation approach of samples 2–4, the number of balls in sample 4 is four times greater than in sample 1. Although balls in a clumped particle is treated as a single rigid body in calculation cycle, the replacement process mentioned previously will cost too much time if too many balls are generated in the sample 1. To improve calculation speed, we suggest not using too many balls in sample 1.

3.2. Quantitative Description of Particle Shapes. Shape factors to quantify particle shapes are discussed in detail within a number of literatures [6, 21]. Roundness and aspect ratios are very commonly used shape factors [3, 9]. To two-dimensional particles, Kong and Peng [14] suggested to determine shape

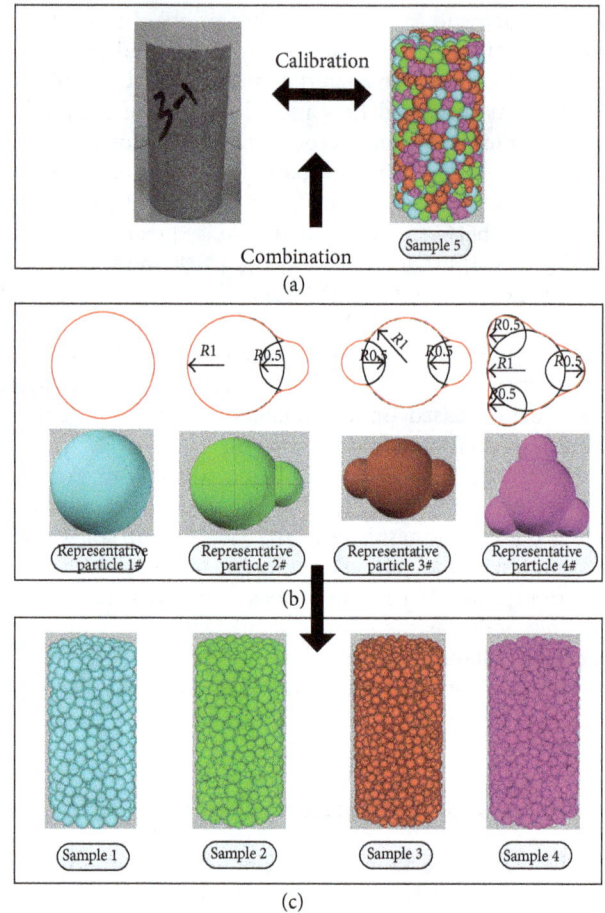

FIGURE 2: The representative particles and samples.

factors by (4) after analyzing the mechanical behaviors from mesoscopic levels:

$$F = \alpha F_1 + \beta F_2,$$
$$\alpha + \beta = 1, \tag{4}$$

where F is shape parameter, F_1 and F_2 are two parameters related to roundness of particle and roughness of surface, respectively, with α and β being corresponding weighting coefficients.

The Fourier analysis technique also an important method to characterize particle shapes. The digitized particle outline is described by Fourier analysis which is originally based on the x-, y-coordinate detection method for contour curves [4, 22].

In this paper, we discuss the mechanical response of three-dimensional particles. Combined with previous shape factors method, considering the feasibility and utility of method, we define sphericity as a shape factors for particles

$$S = \frac{S_s}{S_p}, \tag{5}$$

where S_s is surface area of a sphere whose volume is the same with the particle. S_p is the surface area of particle. S is

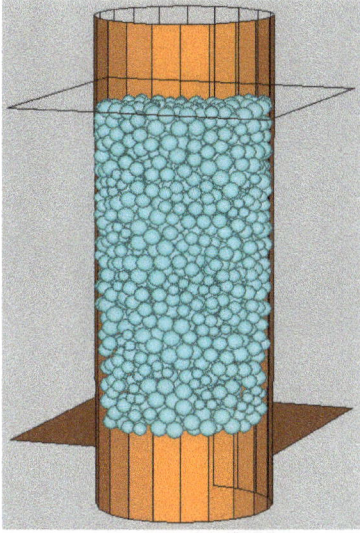

FIGURE 3: Sample and loading plane.

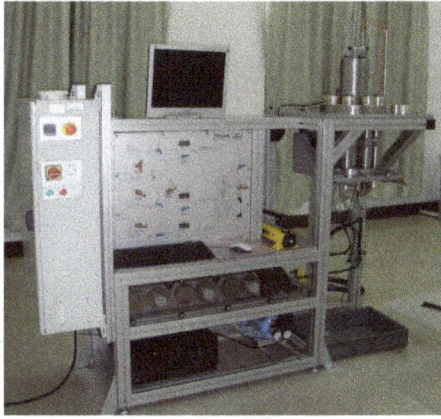

FIGURE 4: Triaxial coupling test instrument.

TABLE 1: Shapes factor of representative particles.

Number	1#	2#	3#	4#
Sphericity	1.000	0.959	0.924	0.892

which load or unload the sample by moving along the axis of the cylinder. The side face of cylinder is a servo plane, which keeps the confining pressure constant by expansion or contraction.

3.3. Numerical Calibration of Microscopic Parameters. Since some microscopic parameters in *PFC* models cannot be obtained directly from laboratory experiments, numerical calibration is required. The microscopic parameters are adjusted to simulate stress-strain curve of quartz sandstone. The quartz sandstone specimens (Figure 4) come from Luojia Mountain, Wuhan, China. For a standard test specimen, test specimen shall be right circular cylinders with a height-to-diameter ratio of 2.0 and a diameter preferable not less than 50 mm. The quartz sandstone specimens used for laboratory test are cylinders having diameter of 50 mm and height of 100 mm. Density of the specimens are 2.65 g/cm^3. The quartz sandstone specimens are greyish white. The quartz sandstone specimens are composed of 95% quartz clasts, 5% feldspars, and other minerals. Particles sizes are between 0.25–0.50 mm. And the cements are mainly siliceous; the specimens are clastic texture.

The triaxial test instrument is researched and developed by the University of Lille and is produced by the company of Top Industria in France (Figure 5). In laboratory experiment, confining pressure is 8 MPa. The quartz sandstone specimens are loaded until destruction, a rupture plane can be seen from Figure 4. Meanwhile, the numerical sample 5 (Figure 6) is used to simulate triaxial compression test in PFC. Figure 6(a) is sample 5 before the numerical test, and Figure 6(b) shows the distribution of microcracks at the peak stress. The microcracks are showed in white. When a bond breaks either for tension or shearing failure, a new microcrack forms in PFC. From Figure 6(b), we can see the microcracks are concentrated in the diagonal, top, and bottom of the sample. In theory, the microcracks should mainly distribute in the maximum shear stress plane. But in this numerical test, the loads from load planes transmit to the sample only by the contact point of particles and load planes which results in stress concentration easily at the interfaces of sample and load planes. And the bond strength is not homogeneous the standard deviations of bond strengths is showed in Table 2.

In the process of numerical calibration, when the microscopic parameters of sample 5 correspond with the parameters of quartz sandstone, the stress-strain curves both experiment and numerical simulation should be similar (Figure 6).

Figure 7 shows the comparison of experiment and numerical simulation. As can be seen in Figure 7, two curves come close before stress get to peak strength. There are slight differences in postpeak region. The main reason is small stiffness of test machine and slow servo response speed of test machine, which cause the large intervals and uneven

spherity of particle. Equation (5) can be expressed by the volume and surface area of particle, where V_p is the particle volume

$$S = \frac{4\pi}{S_p} \left(\frac{3V_p}{4\pi} \right)^{2/3} \qquad (6)$$

In order to obtain the volume and surface area of representative particles, four representative particles are created in *AutoCAD*, and their volumes and surface areas are easy to get by *AutoCAD*. Then the sphericities of four representative particles are counted by (4). The result is shown in Table 1. As is shown in Table 1, actually, sphericity expresses the degree of similarity between the sphere and particle. The closer the shape of a particle comes to a sphere, the nearer the sphericity approximates to 1.

All the samples are cylinders with height of 80 mm and diameter of 40 mm. The minimal radius of the balls is 2.5 mm, and the ratio of the maximum radius to the minimal radius is 1.5. Figure 3 shows the location of load planes and a sample. The upper plane and lower plane are two load planes,

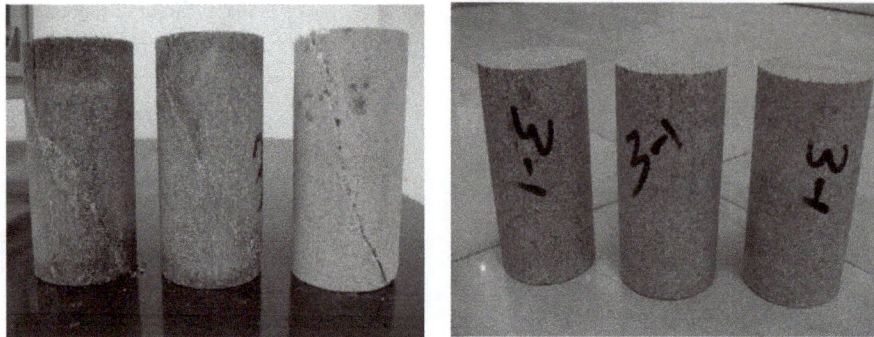

Figure 5: Intact and cracked quartz sandstone samples.

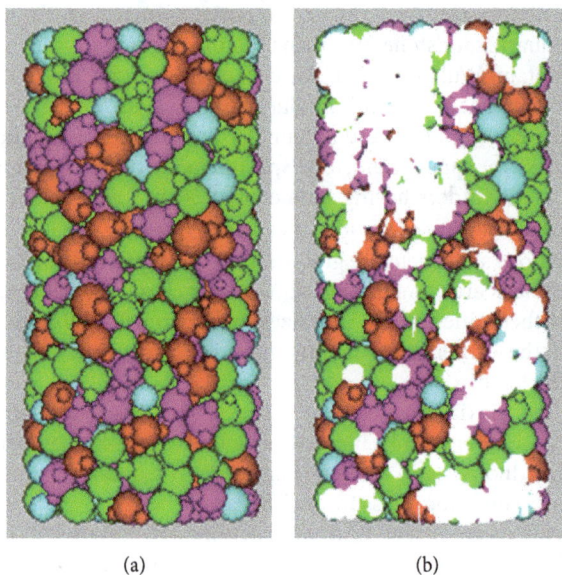

(a)   (b)

Figure 6: The samples before and during the numerical test.

Table 2: Microscopic parameters of *PFC3D* model.

Name	Value
Minimum ball radius, $R_{\min}$ (mm)	2.5
Ball radius ratio, $R_{\max}/R_{\min}$	1.5
Ball density (Kg/m^3)	2600
Ball-ball contact Young's modulus (GPa)	25
Young's modulus of parallel bond (GPa)	20
Ball stiffness ratio, k_n/k_s	2.5
Parallel bond stiffness ratio	2.5
Particle friction coefficient	1.0
Parallel bond normal strength, mean (MPa)	130
Parallel bond normal strength, std. dev (MPa)	10
Parallel bond shear strength, mean (MPa)	130
Parallel bond shear strength, std. dev (MPa)	10

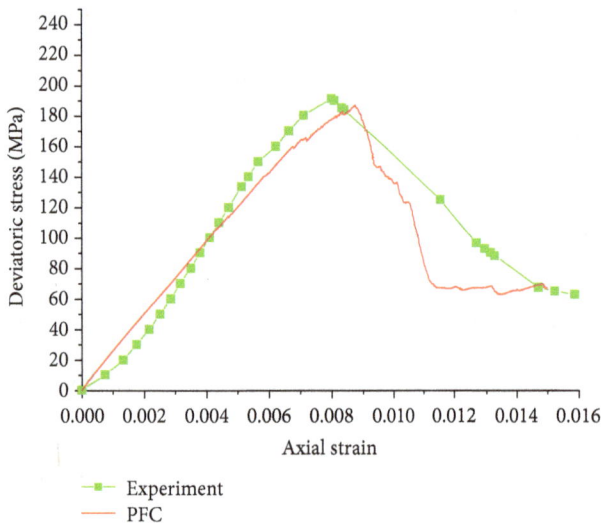

Figure 7: The comparison between numerical test and experiment.

characteristics of measuring point. So it is reasonable to exist slight differences.

The agreement between numerical simulation and laboratory results is acceptable (Figure 7), which shows microscopic parameters adopted in the numerical simulation are suitable for the quartz sandstone. The corresponding model parameters are listed in Table 2.

4. Results Analysis of Numerical Test

4.1. The Influences of Particle Shapes on Strength. To study the relation of particle shapes and mechanical behaviors, numerical triaxial compression test of samples 1–4 (Figure 2) was made under different confining pressures with model parameters listed in Table 2.

Stress-strain curves with confining pressures 2–20 MPa are shown in Figures 8(a)–8(d). The sphericity of particles is noted in parentheses. As can be seen in Figure 8, particle shapes affect stress-strain curve considerably. This impact manifests mainly in the relation between peak strength and particle shapes. Specifically, peak strength decreases with the increasing of sphericity of particles under the same confining pressure. The modulus of elasticity of samples varies with particle shapes, which can be seen from the slope

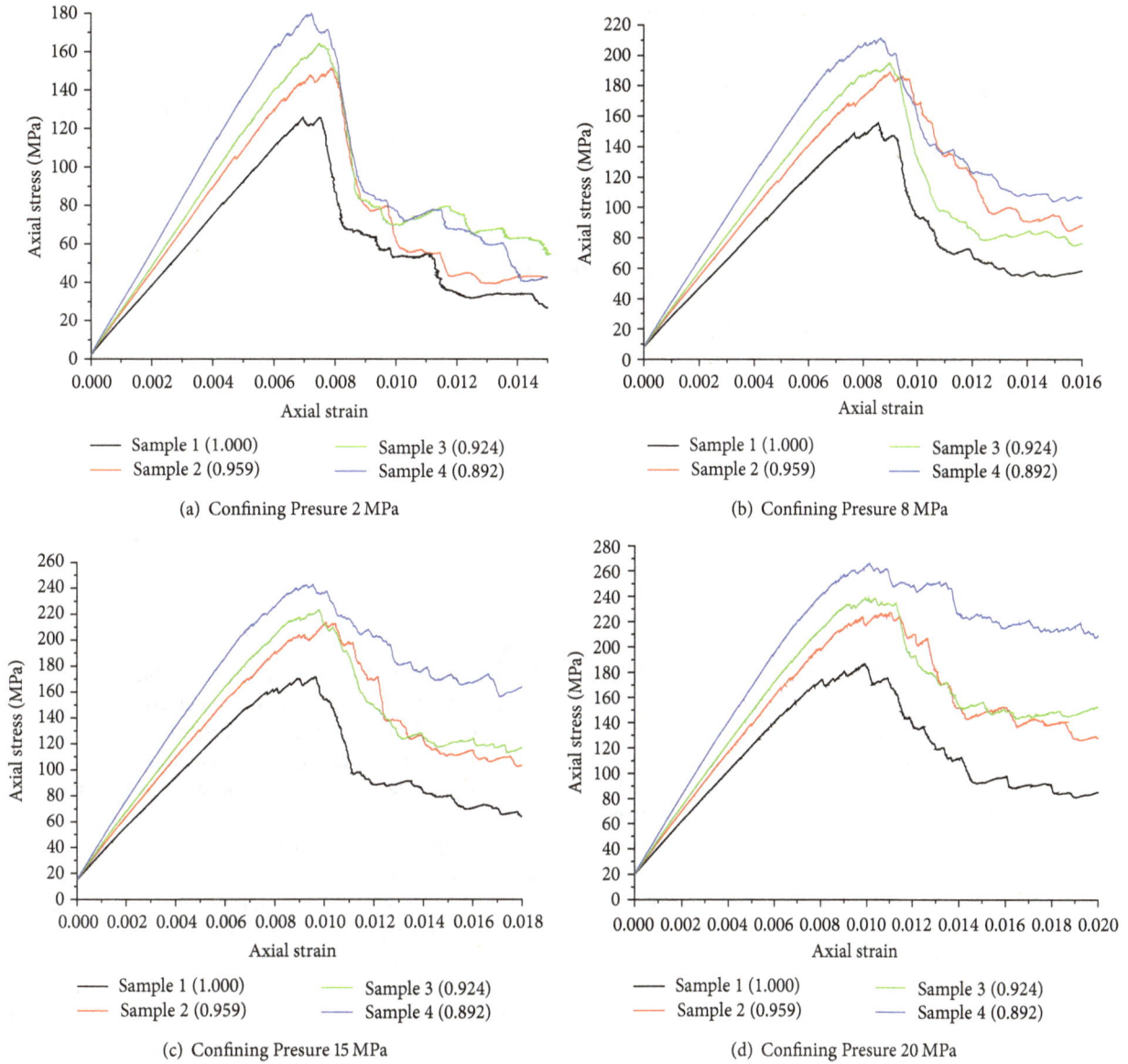

(a) Confining Presure 2 MPa

(b) Confining Presure 8 MPa

(c) Confining Presure 15 MPa

(d) Confining Presure 20 MPa

FIGURE 8: Stress-strain curves of four samples.

of curves before peak point. The sample with smaller particles sphericity has a higher modulus of elasticity.

Comparing stress-strain curves of different confining pressure (Figures 8(a)–8(d)), we found the influences of particle shapes on residual strength depend on confining pressure. With a high confining pressure and a small sphericity of particles, stress drop at peak point is inconspicuous. Under such condition, the residual strength is close to peak strength. As can be seen from Figure 8(d), when confining pressure maintains a high value (for example 20 MPa), stress drop from peak point is related to sphericity, the smaller the sphericity; the smaller the stress drop. So the residual strength of different samples under high confining pressure varies greatly. While with a low confining pressure, there are little differences about residual strength among different samples.

At mesoscale level, the results can be explained as follow. Before stress go to peak point, particles bond and interlock together. The smaller the particles sphericity, the greater the degree of interlocking is, correspondingly, the greater the overall peak strength. In the post-peak stage, bonds of particles crack. Particles with big sphericity approaching the spheroid, slip and roll easily, which leads to rapid stress drop. While for small sphericity particles, the effect of interlocking, and friction is stronger, which makes samples remain a higher residual strength in the postpeak stage.

4.2. The Influences of Particle Shapes on Crack Initiation Stress and Crack Damage Stress. In the failure process of rock, crack initiation stress and crack damage stress are two important indicators. The crack initiation stress σ_{ci} marks crack initiation and stable propagation. When stress exceeds crack damage stress σ_{cd}, the crack growth is unstable, and σ_{cd} is also the beginning of rock dilation.

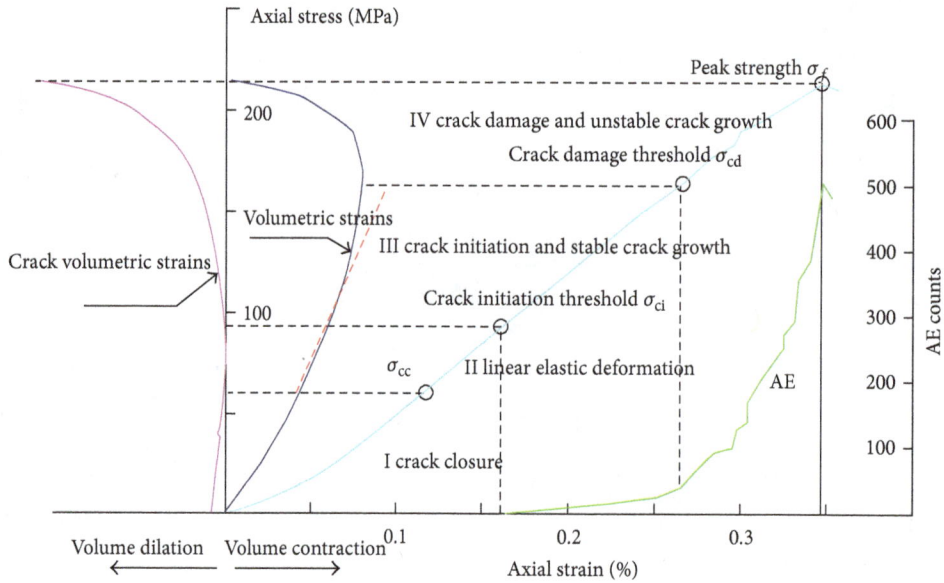

FIGURE 9: Schematic diagram of stress-strain curves of rocks.

According to the research results of progressive failure process by Martin [23], Eberhardt [24], Eberhardt et al. [25], and Diederichs et al. [26], there are three major ways to ascertain σ_{ci}, as is shown in Figure 9. (1) Acoustic Emission Test (AE). Eberhardt put out that σ_{ci} is the stresses when new AE counts first rise above background. (2) The stress-volumetric strain curves. When stresses get to σ_{ci}, the stress-volumetric strain curves deviate from the elastic line. (3) Crack volumetric strains. The crack volumetric strains deviate from zero with σ_{ci} coming. The crack volumetric strains are defined as (7) by Martin [23]:

$$\varepsilon_{cv} = \varepsilon_v - \frac{(1 - 2\nu)(\sigma_1 - \sigma_3)}{E}, \qquad (7)$$

where ε_{cv} and ε_v are the crack volumetric strains and volumetric strains, respectively, ν is poisson ratio, and E is elastic modulus.

The crack damage stresses σ_{cd} can be ascertained by AE or the stress-volumetric strain curves. When stress reaches σ_{cd}, AE counts curves show transition and AE counts increase rapidly. In addition, from the stress-volumetric strain curves we can observe volumetric strains rate is near zero and a transition point arises with the crack damage stresses, σ_{cd} coming, which are shown in Figure 9.

This paper determined the crack initiation stress and crack damage stress of rock samples by crack volumetric strains and the stress-volumetric strain curves. Relevant curves of four samples under confining pressure 15 MPa are shown in Figure 10. As mentioned before, the stress that crack volumetric strains deviate from zero is σ_{ci}, marked with green circle in Figure 10. And the σ_{cd} can be obtained from volumetric strain reversal, marked with red circle in Figure 8.

Further, Figure 11 shows how the crack initiation stress and crack damage stress vary with particles sphericity. As we can see from Figure 11, the crack initiation stress and crack

damage stress reduce with particles sphericity increasing. In other words, particle shapes affect crack initiation and unstable crack growth in the process of rock failure. With a smaller particles sphericity, rocks can support a higher load before appearing crack initiation and unstable crack growth.

4.3. The Influences of Particle Shapes on Cohesion and Internal Friction Angle. Cohesion c and internal friction angle ϕ are two important material parameters of the Mohr-Coulomb strength theory. The Mohr-Coulomb yield criterion is expressed by (8) using principal stresses. The Mohr-Coulomb yield criterion assumes the cohesion c and internal friction angle ϕ of rock are both constant, but this assumption is limitation in practical application. Since some scholars [27–29] found cohesion c and internal friction angle ϕ of rock were not constant in triaxial compression test,

$$\frac{1}{2}(\sigma_1 - \sigma_3) - \frac{1}{2}(\sigma_1 + \sigma_3)\sin\phi - c\cos\phi = 0, \qquad (8)$$

where σ_1 and σ_3 are maximum principal stress and minimum principal stress, respectively. Supposing rock samples yield at peak strength, and strength σ_p and confining pressure σ_c are substituted into (8), we get (9):

$$\sigma_p = \frac{2c\cos\phi}{1 - \sin\phi} + \frac{1 + \sin\phi}{1 - \sin\phi}\sigma_c. \qquad (9)$$

Equation (9) shows the relation between the peak strength and confining pressure under triaxial condition.

Based on numerical triaxial compression tests with the confining pressure of 2 MPa, 8 MPa, 15 MP, and 20 MPa, $\sigma_p - \sigma_c$ curve of sample 1 is displayed (Figure 12). As is shown by Figure 12, there is a good linear relation between σ_p and σ_c. Then $\sigma_p - \sigma_c$ curve can be obtained by beeline fitting, which is given in Figure 12. Comparing the linear equation

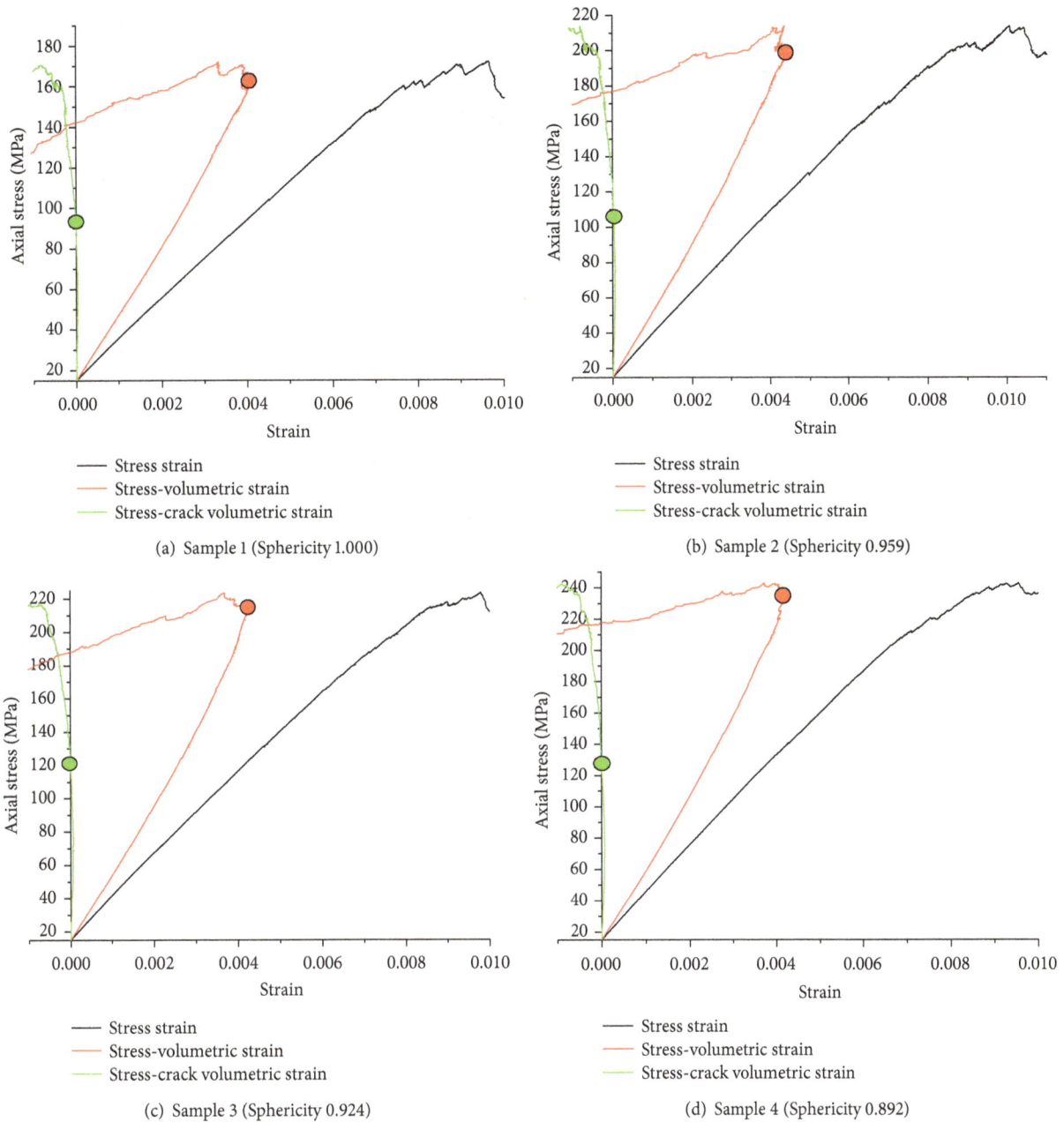

(a) Sample 1 (Sphericity 1.000)

(b) Sample 2 (Sphericity 0.959)

(c) Sample 3 (Sphericity 0.924)

(d) Sample 4 (Sphericity 0.892)

FIGURE 10: The crack initiation stress and crack damage stress of four rock samples.

in Figure 12 and (9), cohesion c and internal friction angle ϕ can be calculated by the gradient a and the constant b of linear equation (see (10)):

$$\phi = \arcsin\left(\frac{a-1}{a+1}\right),$$

$$c = \frac{b(1-\sin\phi)}{2\cos\phi}. \tag{10}$$

Along the same ways, cohesion c and internal friction angle ϕ of other samples are acquired. Figure 13 shows how

cohesion c and internal friction angle ϕ vary with particles sphericity. As can be found in Figure 13, basically, cohesion c and internal friction angle ϕ of samples reduce with particles sphericity rising. When particles sphericity is between 0.92 and 0.96, the variation of internal friction angle ϕ is gentle.

4.4. The Influences of Particle Shapes on Elastic Modulus and Poisson Ratio. Elastic modulus showed a downward trend as particles sphericity rise in Figure 8. To facilitate our analysis, suppose rocks samples are isotropic and compress in the

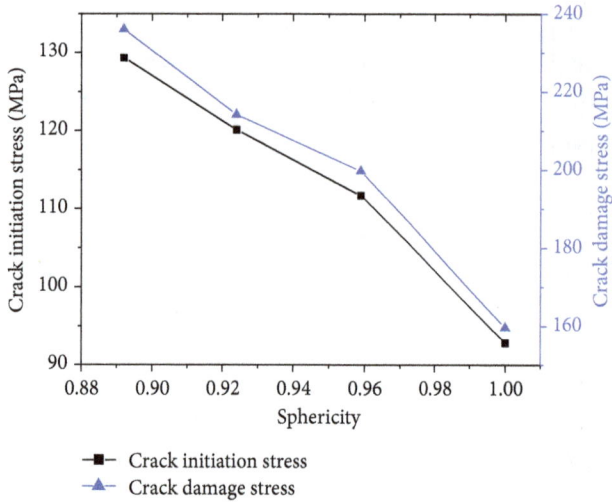

FIGURE 11: The relation among the crack initiation stress, crack damage stress, and sphericity.

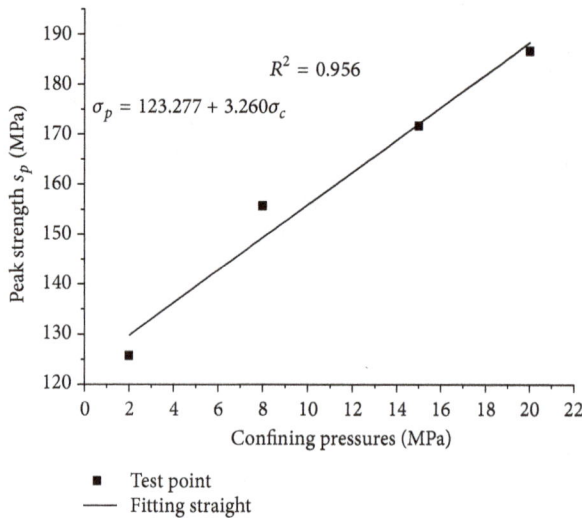

FIGURE 12: The peak strength versus confining pressure for sample 1.

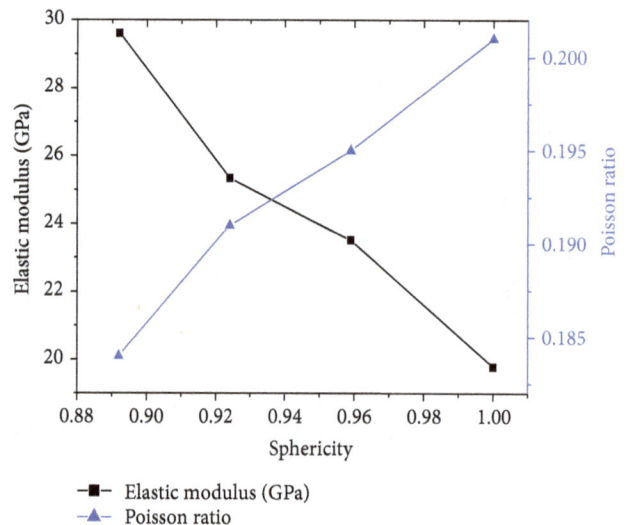

FIGURE 13: The variation of cohesion and internal friction angle with sphericity.

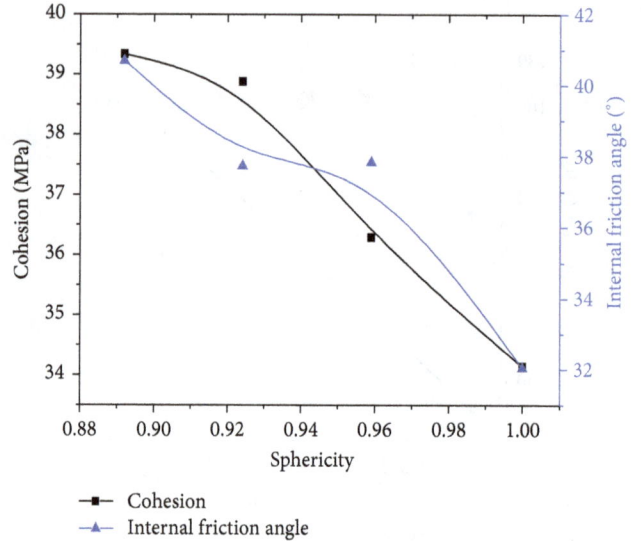

FIGURE 14: The variation of modulus of elasticity and Poisson ratio with sphericity.

y-direction. In numerical triaxial compression tests, elastic modulus E and Poisson ratio ν can be calculated as follows:

$$E = \frac{\Delta \sigma_y}{\Delta \varepsilon_y},$$

$$\nu = -\frac{\Delta \varepsilon_x + \Delta \varepsilon_z}{2\Delta \varepsilon_y} = -\frac{\Delta \varepsilon_V - \Delta \varepsilon_y}{2\Delta \varepsilon_y}, \qquad (11)$$

where $\Delta \varepsilon_x$, $\Delta \varepsilon_y$, and $\Delta \varepsilon_z$ are strain increments in x-, y- and z-directions, respectively. And $\Delta \varepsilon_V$ is volumetric strain increment, $\Delta \sigma_y$ is stress increment in y direction. Ii is obviously that E and ν change with load process. But, for simplicity sake, elastic modulus E and Poisson ratio ν are subject to the point of half peak stress.

Figure 14 shows how elastic modulus and Poisson ratio vary with particles sphericity under confining pressure

15 MPa. As particles sphericity increase, elastic modulus of samples drops and Poisson ratio rises. As can be seen from the magnitude of Poisson ratio, there are only small differences between Poisson ratio in spite of sphericity influence.

4.5. Dilation Effect of Samples.

In the process of numerical triaxial compression test, the samples are compressed first, and then they are dilated. In continuum mechanics, dilation angle is the widely used parameter to discribe the dilation effect. Dilation angle is not a constant during the deformation process of rock [30].

But, considering the rock materials may not obey Drucker's stability postulate [31], which recites that the work of the external agency on the displacement produced must

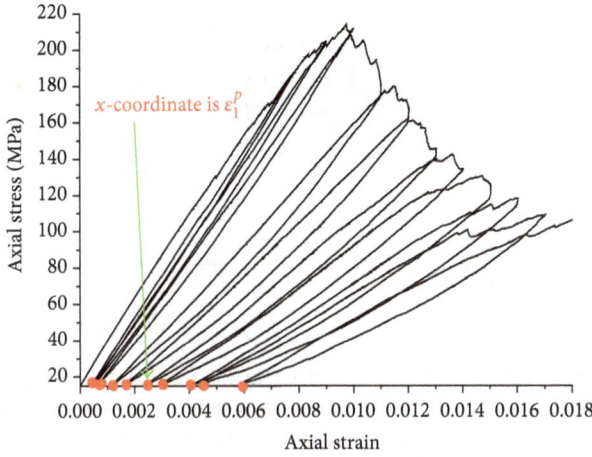

FIGURE 15: Loading and unloading of sample 2.

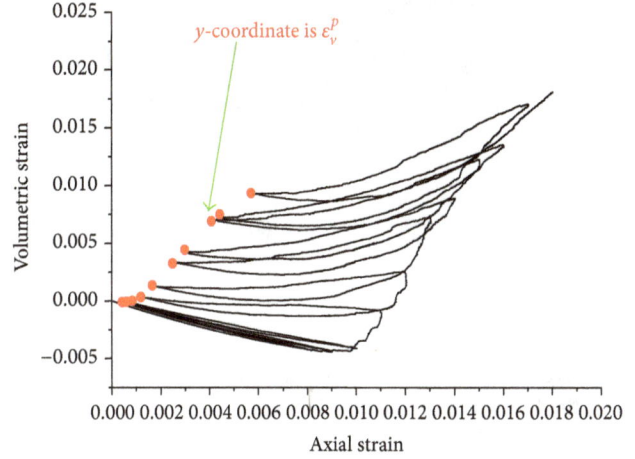

FIGURE 16: Volume strain curve of sample 2.

be positive or zero, the nonassociated flow rule should be adopted as follows:

$$\dot{\varepsilon}_{ij}^p = \lambda \frac{\partial g}{\partial \sigma_{ij}}, \qquad (12)$$

where λ is a plastic multiplier and g is plastic potential. One of the commonly used plastic potential assumptions states that [32]

$$g = \sigma_1 - K_\psi \left(\sigma_{ij}, \eta \right) \sigma_3,$$

$$K_\psi \left(\sigma_{ij}, \eta \right) = \frac{1 + \sin \psi \left(\sigma_{ij}, \eta \right)}{1 - \sin \psi \left(\sigma_{ij}, \eta \right)}, \qquad (13)$$

where ψ is dilation angle and σ_1, σ_3 are the maximum and minimum principal stress, respectively. σ_{ij} is the stress tensor and η is the plastic parameter. Here, η can be expressed as shear plastic strain, as follows:

$$\eta = \gamma^p = \varepsilon_1^p - \varepsilon_3^p. \qquad (14)$$

The dilation angle of rock can be obtained by (15) [33], where $\dot{\varepsilon}_v^p$ and $\dot{\varepsilon}_1^p$ are volumetric and axial plastic strain increments, respectively,

$$\sin \psi = \frac{\dot{\varepsilon}_v^p}{-2\dot{\varepsilon}_1^p + \dot{\varepsilon}_v^p}. \qquad (15)$$

Volumetric and axial plastic strain can be obtained by loading and unloading cycles [32, 34]. Since the dilation effect in prepeak stage is not obvious, we focus on evolvement laws of dilation angle in postpeak stage. Figure 15 shows the stress-strain curve from loading and unloading cycles under confining pressure 15 MPa. The numerical test process is as follows.

The confining pressure and axial stress of 15 MPa are applied on samples first, in which case samples work in elastic behavior. That is plastic strains still do not occur; this state is called original state. Then axial stresses increase to load

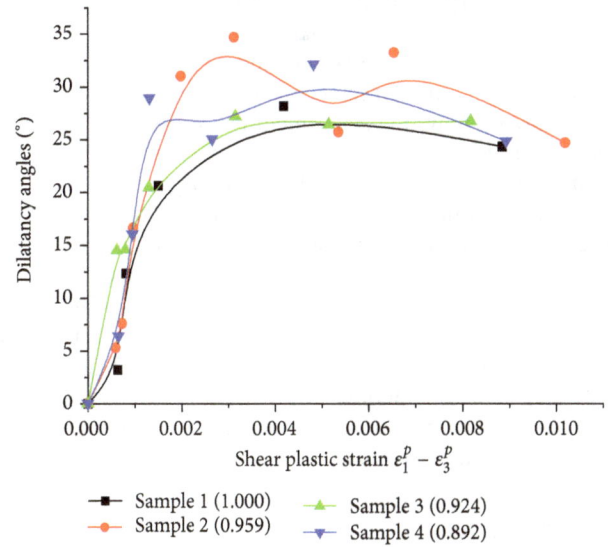

FIGURE 17: The relation between dilation angle and shear plastic strain.

samples and axial strain is recorded from the original state. When axial strains reach to 0.008, 0.009, 0.0100, 0.011, 0.012, 0.013, 0.014, 0.015, 0.016, and 0.017, unloading, respectively until samples return to the original state. The strains in original state are axial plastic strains ε_1^p, marked with red circle in Figure 15, and then load to the next unloading point, and so forth. In the process of loading and unloading cycle, volumetric strains are recorded, as it is shown in Figure 16. The volumetric strains ε_v^p are marked with red circle in Figure 16.

Equation (15) is conveniently written as incremental form, as follows:

$$\sin \psi = \frac{\Delta \varepsilon_v^p}{-2\Delta \varepsilon_1^p + \Delta \varepsilon_v^p}. \qquad (16)$$

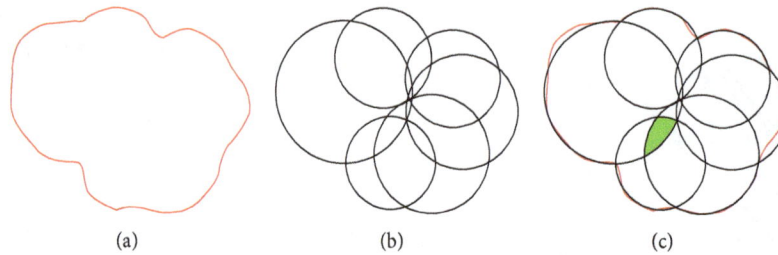

FIGURE 18: Simulating complex particle shape.

Given the deformation conditions of triaxial compression test, the following formula is deduced.

$$\varepsilon_2^p = \varepsilon_3^p,$$
$$\varepsilon_v^p = \varepsilon_1^p + 2\varepsilon_3^p. \tag{17}$$

Dilation angles can be calculated according to (16). The volumetric strains ε_v^p and axial plastic strain ε_1^p can be obtained from loading and unloading cycles in Figures 15 and 16. Shear plastic strain η can be calculated using (14) and (17).

Figure 17 shows the evolvement process of dilation angle with shear plastic strain. As can be seen from Figure 17, dilation angles increase rapidly in the beginning. Dilation angles reach a maximum around shear plastic strain 0.003. Then, there was no significant change on dilation angles. According to the research of Alejano and Alonso [32] and Zhao and Cai [34], dilation angles will start to fall when shear plastic strain is larger. As shown in Figure 17, although dilation angles of four samples have a little difference, there are no essential distinctions on trends of curves among four samples. That is, particle shapes only impact size of dilation angles rather than trends of curves.

5. Discussions

The real shape of mineral particles in rocks is quite complex, while the shape of representative particles used in this paper is simple. Theoretically, the method forming representative particles by clumped particles is also effective for complex particles (Figure 18). The red lines represent the planar contour for a complex particle, and the black lines represent the clumped particles in Figure 18. As is shown in Figure 18(c), the clumped particles can be a good approximation of the complex particle. That is to say, the real shapes of mineral particles in rocks can be simulated using clumped particles in PFC. Deserved to be mentioned, when the overlap parts in a clumped particle involve over two particles (the green part in Figure 18(c)), the clump volume automatically calculated in PFC is inaccurate [18]. In this case, tools such as AutoCAD can be use to compute the clump volume.

The sphericity is not the only shape factor to characterize particle shapes. As a matter of fact, the number and type of representative particles used in this paper are limited. To be more precise, for this kind of representative particles, which are formed by a big spherical particle and several small spherical particles, the sphericity is a relatively efficient shape factor. However, for other kinds of particles, such as elongated particles, aspect ratios and other shape factors may be more effective. It is verified in our study that there are good correlations between the main mechanical parameters of samples and the sphericity of particles. But, the sphericity does not significantly influence the residual strength and dilation effect. And, when particles shapes get more complex, the relation between the sphericity and the internal friction angle becomes unclear. The main reason for these problems may lie in the fact that the sphericity only expresses the degree of similarity between a sphere and particle. In other words, although the sphericity expresses much of morphological characters especially for simple particle shapes, some morphological characters of particle cannot be entirely reflected by the sphericity. So we need to take into account more parameters to further express particle shapes.

6. Conclusions

Our numerical experiments show that the mechanical behaviors of rock are influenced by their particle shapes; the primary conclusions are as follows.

(1) The sphericity index is an applicable shape factor to measure particles shapes. The sphericity describes the proximity of a particle to a sphere, which directly influences the interlocking ability of particles.

(2) The crack initiation stress, crack damage stress, and peak stress of rock are affected by particle shapes. Specifically, three stress indices decrease with increasing of sphericity. The increasing sphericity also leads to smaller elastic modulus and larger poisson ratio. And the decreasing sphericity causes particle interlocking of different degrees which restrains slip and rotation, consequently, cohesion and internal friction angle rise.

(3) To samples of different particle shapes, the evolvement process of dilation angle with shear plastic strain was researched. The result shows the trends of dilation angle changing with shear plastic strain are similar, but values of dilation angle have a little difference.

Acknowledgments

The research work presented in this paper is sponsored by the National Basic Research Program of China ("973" Program)

(Grant nos. 2011CB013501 and 2010CB732005), the National Natural Science Foundation of China (Grant no. 50979081) and Program for New Century Excellent Talents in University (Grant no. NCET-11-0406). The authors would like to thank their support.

References

[1] D. O. Potyondy and P. A. Cundall, "A bonded-particle model for rock," *International Journal of Rock Mechanics & Mining Sciences*, vol. 41, no. 8, pp. 1329–1364, 2004.

[2] J. M. Ting, L. Meachum, and J. D. Rowell, "Effect of particle shape on the strength and deformation mechanisms of ellipse-shaped granular assemblages," *Engineering Computations*, vol. 12, no. 2, pp. 99–108, 1995.

[3] A. Lin, "Roundness of clasts in pseudotachylytes and cataclastic rocks as an indicator of frictional melting," *Journal of Structural Geology*, vol. 21, no. 5, pp. 473–478, 1999.

[4] K. Shinohara, M. Oida, and B. Golman, "Effect of particle shape on angle of internal friction by triaxial compression test," *Powder Technology*, vol. 107, no. 1-2, pp. 131–136, 2000.

[5] J. Dodds, *Particle Shape and Stiffness-Effects on Soil Behavior*, Institute of Technology, Atlanta, Ga, USA, 2003.

[6] Q.-B. Liu, W. Xiang, M. Budhu, and D.-S. Cui, "Study of particle shape quantification and effect on mechanical property of sand," *Rock and Soil Mechanics*, vol. 32, no. 1, pp. 190–197, 2011.

[7] K. Johanson, "Effect of particle shape on unconfined yield strength," *Powder Technology*, vol. 194, no. 3, pp. 246–251, 2009.

[8] S. J. Antony and M. R. Kuhn, "Influence of particle shape on granular contact signatures and shear strength: new insights from simulations," *International Journal of Solids and Structures*, vol. 41, no. 21, pp. 5863–5870, 2004.

[9] J. Härtl and J. Y. Ooi, "Numerical investigation of particle shape and particle friction on limiting bulk friction in direct shear tests and comparison with experiments," *Powder Technology*, vol. 212, no. 1, pp. 231–239, 2011.

[10] D. Liu, T. T. Xie, G. Ma, and X. L. Chang, "Effect of particle shape on mechanical characters of rockfill in True triaxial numerical experiments," *Chinese Journal of Water Resources and Power*, vol. 29, no. 9, pp. 68–71, 2011.

[11] T. Koyama and L. Jing, "Effects of model scale and particle size on micro-mechanical properties and failure processes of rocks—a particle mechanics approach," *Engineering Analysis with Boundary Elements*, vol. 31, no. 5, pp. 458–472, 2007.

[12] J.-W. Park and J.-J. Song, "Numerical simulation of a direct shear test on a rock joint using a bonded-particle model," *International Journal of Rock Mechanics & Mining Sciences*, vol. 46, no. 8, pp. 1315–1328, 2009.

[13] I. Kock and K. Huhn, "Influence of particle shape on the frictional strength of sediments—a numerical case study," *Sedimentary Geology*, vol. 196, no. 1-4, pp. 217–233, 2007.

[14] L. Kong and R. Peng, "Particle flow simulation of influence of particle shape on mechanical properties of quasi-sands," *Chinese Journal of Rock Mechanics and Engineering*, vol. 30, no. 10, pp. 2112–2119, 2011.

[15] D.-D. Shi, J. Zhou, W.-B. Liu, and Y.-B. Deng, "Exploring macro- and micro-scale responses of sand in direct shear tests by numerical simulations using non-circular particles," *Chinese Journal of Geotechnical Engineering*, vol. 32, no. 10, pp. 1557–1565, 2010.

[16] D.-D. Shi, J. Zhou, W.-B. Liu, and M.-C. Jia, "Numerical simulation for behaviors of sand with non-circular particles under monotonic shear loading," *Chinese Journal of Geotechnical Engineering*, vol. 30, no. 9, pp. 1361–1366, 2008.

[17] S.-J. Xu, X.-T. Yin, and F.-N. Dang, "Mechanical characteristics of rock and soil affected by particle size of crystal and mineral," *Rock and Soil Mechanics*, vol. 30, no. 9, pp. 2581–2587, 2009.

[18] Itasca Consulting Group, Inc., *PFC3D (Particle Flow Code), Version 4.0*, Minneapolis, Minn, USA, 2008.

[19] N. Cho, C. D. Martin, and D. C. Sego, "A clumped particle model for rock," *International Journal of Rock Mechanics & Mining Sciences*, vol. 44, no. 7, pp. 997–1010, 2007.

[20] H. Abou-Chakra, J. Baxter, and U. Tüzün, "Three-dimensional particle shape descriptors for computer simulation of non-spherical particulate assemblies," *Advanced Powder Technology*, vol. 15, no. 1, pp. 63–77, 2004.

[21] X.-B. Tu and S.-J. Wang, "Particle shape descriptor in digital image analysis," *Chinese Journal of Geotechnical Engineering*, vol. 26, no. 5, p. 659, 2004.

[22] T. Shibata and K. Yamaguchi, "Shift x, y-coordinate detection of line figures and the extraction of particle shape information," *Powder Technology*, vol. 81, no. 2, pp. 111–115, 1994.

[23] C. D. Martin, *The Strength of Massive Lac du Bonnet Granite around Underground Opening*, Department of Civil & Geological Engineering, University of Manitoba, Winnipeg, Canada, 1993.

[24] E. Eberhardt, *Brittle Rock Fracture and Progressive Damage in Uniaxial Compression*, Department of Civil Engineering, University of Saskatchewan, Saskatoon, Canada, 1998.

[25] E. Eberhardt, D. Stead, and B. Stimpson, "Quantifying progressive pre-peak brittle fracture damage in rock during uniaxial compression," *International Journal of Rock Mechanics & Mining Sciences*, vol. 36, no. 3, pp. 361–380, 1999.

[26] M. S. Diederichs, P. K. Kaiser, and E. Eberhardt, "Damage initiation and propagation in hard rock during tunnelling and the influence of near-face stress rotation," *International Journal of Rock Mechanics & Mining Sciences*, vol. 41, no. 5, pp. 785–812, 2004.

[27] A. E. Schwartz, "Failure of rock in the triaxial shear test," in *Proceedings of the 6th US Rock Mechanics Symposium*, pp. 109–135, Rolla, Mo, USA, 1964.

[28] M. Singh and K. S. Rao, "Bearing capacity of shallow foundations in anisotropic non-hoek-brown rock masses," *Journal of Geotechnical and Geoenvironmental Engineering*, vol. 131, no. 8, pp. 1014–1023, 2005.

[29] V. Hajiabdolmajid, *Mobilization of Strength in Brittle Failure of Rock*, Department of Mining Engineering, Queen's University, Kingston, Canada, 2001.

[30] E. Detournay, "Elastoplastic model of a deep tunnel for a rock with variable dilatancy," *Rock Mechanics and Rock Engineering*, vol. 19, no. 2, pp. 99–108, 1986.

[31] D. C. Drucker, "A definition of stable inelastic material," *Journal of Applied Mechanics*, vol. 26, no. 1, pp. 101–106, 1959.

[32] L. R. Alejano and E. Alonso, "Considerations of the dilatancy angle in rocks and rock masses," *International Journal of Rock Mechanics & Mining Sciences*, vol. 42, no. 4, pp. 481–507, 2005.

[33] P. A. Vermeer and R. de Borst, "Non associated plasticity for soils, concrete and rock," *Heron*, vol. 29, no. 3, pp. 3–64, 1984.

[34] X. G. Zhao and M. Cai, "A mobilized dilation angle model for rocks," *International Journal of Rock Mechanics & Mining Sciences*, vol. 47, no. 3, pp. 368–384, 2010.

10

Effect of Metakaolin on Strength and Efflorescence Quantity of Cement-Based Composites

Tsai-Lung Weng,[1] Wei-Ting Lin,[2,3] and An Cheng[2]

[1] *Physics Division, Tatung University, 40 Zhongshan North Road, 3rd Section, Taipei 104, Taiwan*
[2] *Deptartment of Civil Engineering, National Ilan University, 1 Shen-Lung Road, Ilan 260, Taiwan*
[3] *Institute of Nuclear Energy Research, Atomic Energy Council, Executive Yuan, Taoyuan 325, Taiwan*

Correspondence should be addressed to Tsai-Lung Weng; wengabc@yahoo.com.tw

Academic Editors: D. Aggelis, S. Chen, M. Jha, and E. Lui

This study investigated the basic mechanical and microscopic properties of cement produced with metakaolin and quantified the production of residual white efflorescence. Cement mortar was produced at various replacement ratios of metakaolin (0, 5, 10, 15, 20, and 25% by weight of cement) and exposed to various environments. Compressive strength and efflorescence quantify (using Matrix Laboratory image analysis and the curettage method), scanning electron microscopy, and X-ray diffraction analysis were reported in this study. Specimens with metakaolin as a replacement for Portland cement present higher compressive strength and greater resistance to efflorescence; however, the addition of more than 20% metakaolin has a detrimental effect on strength and efflorescence. This may be explained by the microstructure and hydration products. The quantity of efflorescence determined using MATLAB image analysis is close to the result obtained using the curettage method. The results demonstrate the best effectiveness of replacing Portland cement with metakaolin at a 15% replacement ratio by weight.

1. Introduction

Efflorescence is a fine, white, powdery deposit of water-soluble salts left on the surface of concrete as the water evaporates. This deposit is detrimental to the durability of cementitious materials and a stubborn problem for researchers in the field of masonry and concrete [1]. Until recently, it was assumed that calcium hydroxide ($Ca(OH)_2$, CH) forming within cement-based composites is responsible for efflorescence; however, CH does not contribute sufficiently towards the soluble alkali sulfates required for efflorescence to occur. Alkali sulfates penetrate through pores within the composites toward the surface. Reducing the number and size of these pores restricts the movement of salts to the surface. One approach is consolidating grout through mechanical vibration to reduce voids in the grout while improving the bond between the steel and the masonry wall. Producing composites with a denser microstructure also reduces the porous nature of the material, making it difficult for salts to migrate [2, 3].

In recent years, supplementary cementitious materials (SCMs), such as fly ash, slag, and silica fume, have been used to replace a portion of the aggregate or cementitious material in cement-based composites. The aim has been to improve the mechanical properties by taking advantage of their extremely fine spherical particles [4–7]. The pozzolanic reaction of SCMs produces an additional binder, which increases the density of the microstructure, thereby reducing permeability. The problem of efflorescence can be greatly reduced by including SCMs in cement-based composites.

Metakaolin has been widely studied for its highly pozzolanic properties, suggesting that metakaolin could be used as an SCM. Unlike other SCMs that are secondary products or by-products, metakaolin is a primary product, obtained by calcining kaolin clay within a temperature range of 650 to 800°C [8, 9]. Metakaolin is increasingly being used to produce

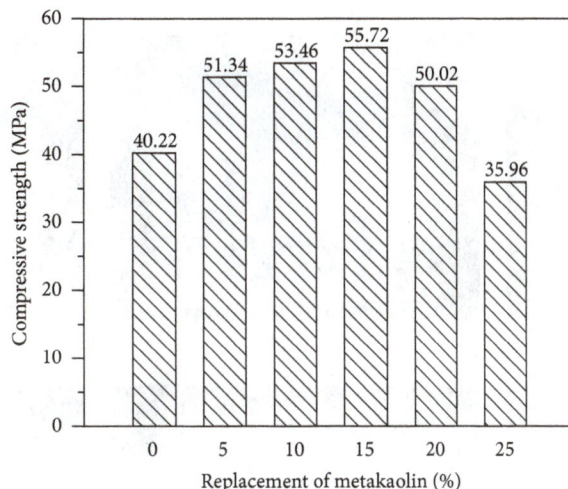

FIGURE 1: Histogram: compressive strength versus replacement of metakaolin (age = 28 days).

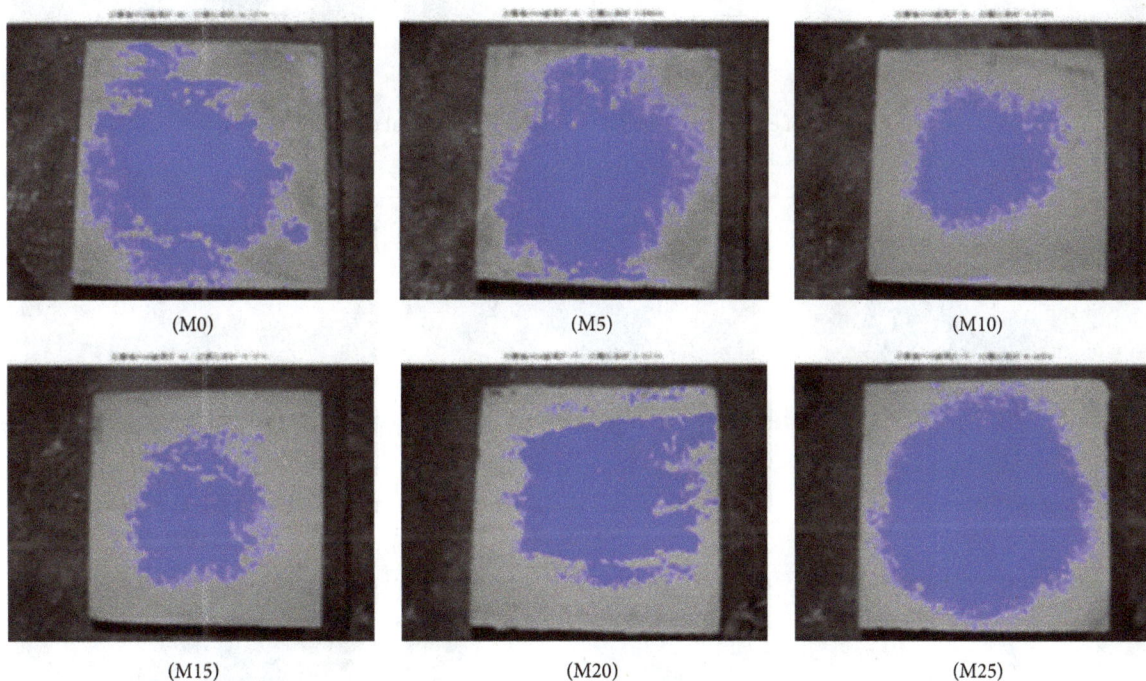

FIGURE 2: Quantity of efflorescence under normal environmental conditions (7 days).

materials with higher strength, denser microstructure, lower porosity, higher resistance to ions, and improved durability [10–12].

Very few researchers have addressed the problem of efflorescence in metakaolin cement-based composites. This study sought to determine the appropriate quantity of metakaolin required (as a replacement for cement) to reduce efflorescence. We employed specimens with various replacement ratios of metakaolin (0%, 5%, 10%, 15%, 20%, and 25%) at a water/cement (w/c) ratio of 0.5. The occurrence of white efflorescence was investigated under various curing environments, at the curing age of 3, 7, and 28 days.

2. Experimental Program

2.1. Materials and Specimens. We produced matrices of ASTM Type I Portland cement, silica sand, tap water, and metakaolin. The specific gravity and fineness modulus of the silica sand were 2.64 and 2.40, respectively. The physical and chemical properties of the metakaolin are presented in Tables 1 and 2.

Metakaolin was added as a replacement for cement at the following percentages: 0, 5, 10, 15, 20, and 25% of the weight of cement with the water/cementitious ratio (w/c) set to 0.50. The mixes were then exposed to the following environments:

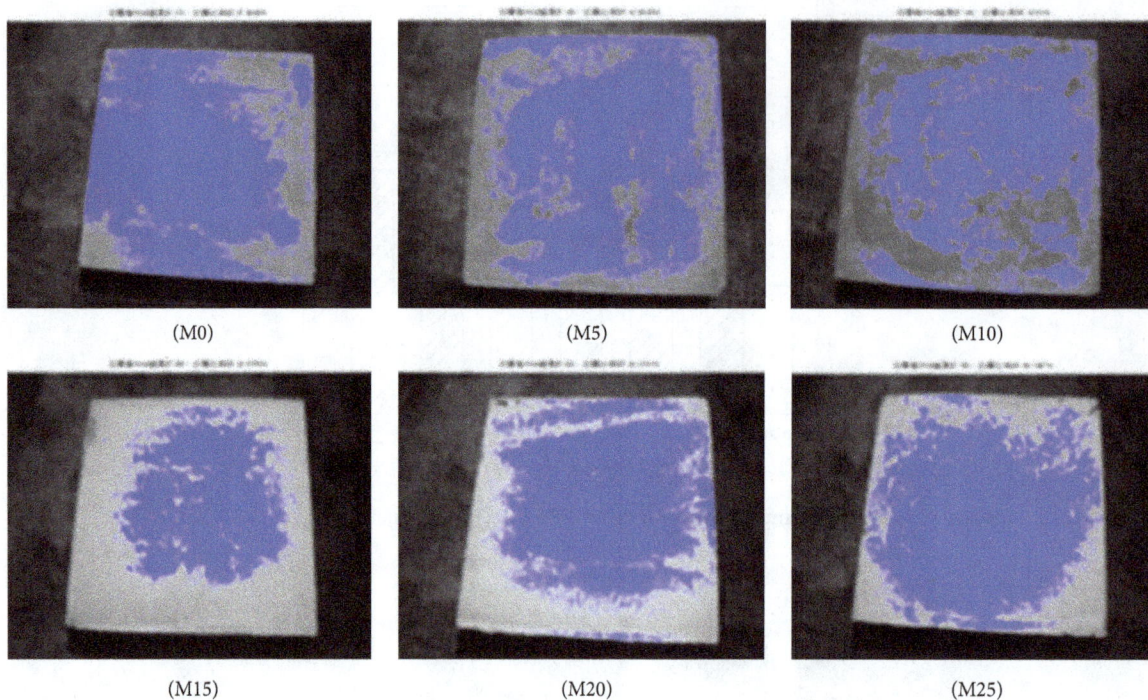

FIGURE 3: Quantity of efflorescence under normal environmental conditions (56 days).

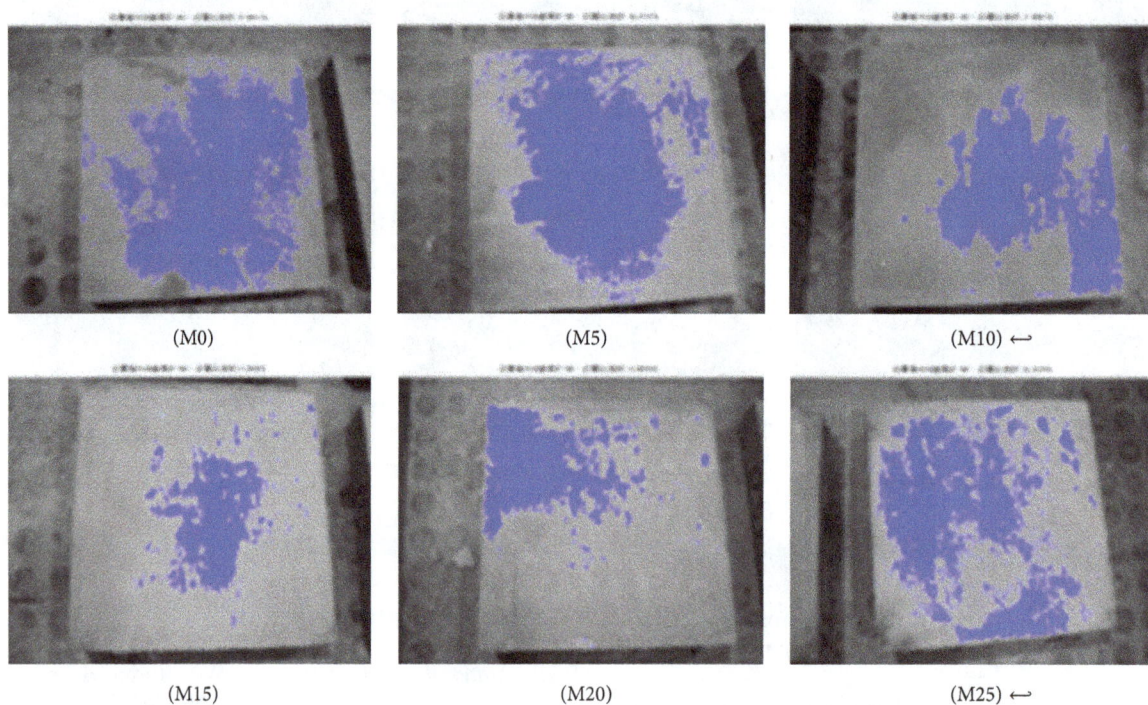

FIGURE 4: Quantity of efflorescence under carbon dioxide environmental conditions (7 days).

TABLE 1: Physical properties of metakaolin.

Item	Specific gravity	Color	Brightness (% ISO)	Surface area (m^2/g)	D10 (μm)	D50 (μm)	D90 (μm)	Bulk density (g/cm^3)
Condition	2.60	White	79–82	15	<2.0	<4.5	<25	0.03~0.04

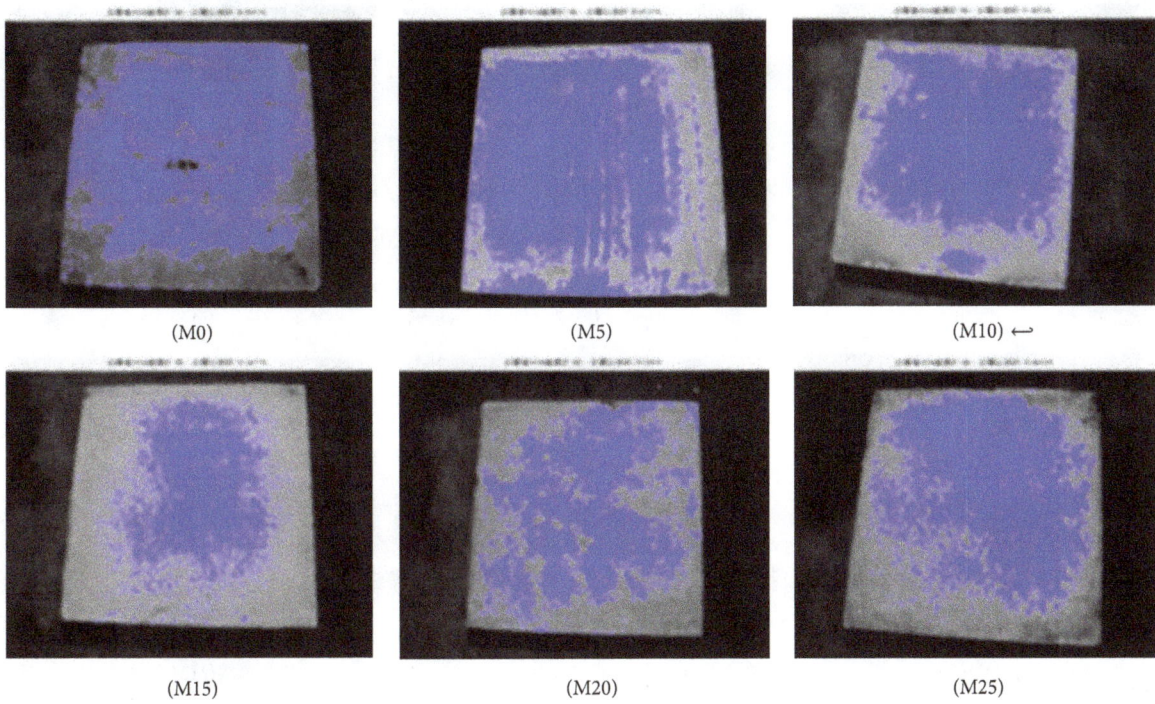

FIGURE 5: Quantity of efflorescence under carbon dioxide environmental conditions (56 days).

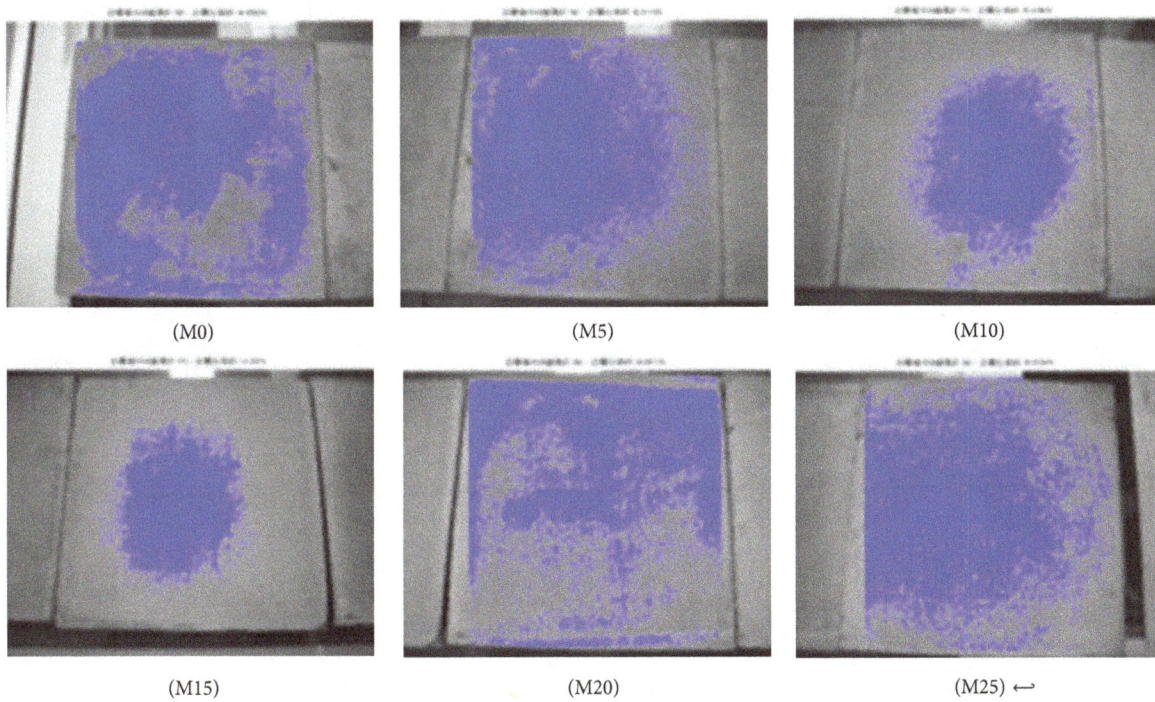

FIGURE 6: Quantity of efflorescence under low temperature environmental conditions (7 days).

TABLE 2: Chemical properties of metakaolin.

Item	SiO_2	Al_2O_3	Fe_2O_3	TiO_2	SO_4	P_2O_5	CaO	MgO	Na_2O	K_2O	L.O.I
Mass % as oxide	52–55	41–44	<1.9	<3.0	<0.05	<0.2	<0.2	<0.1	<0.05	<0.75	<0.5

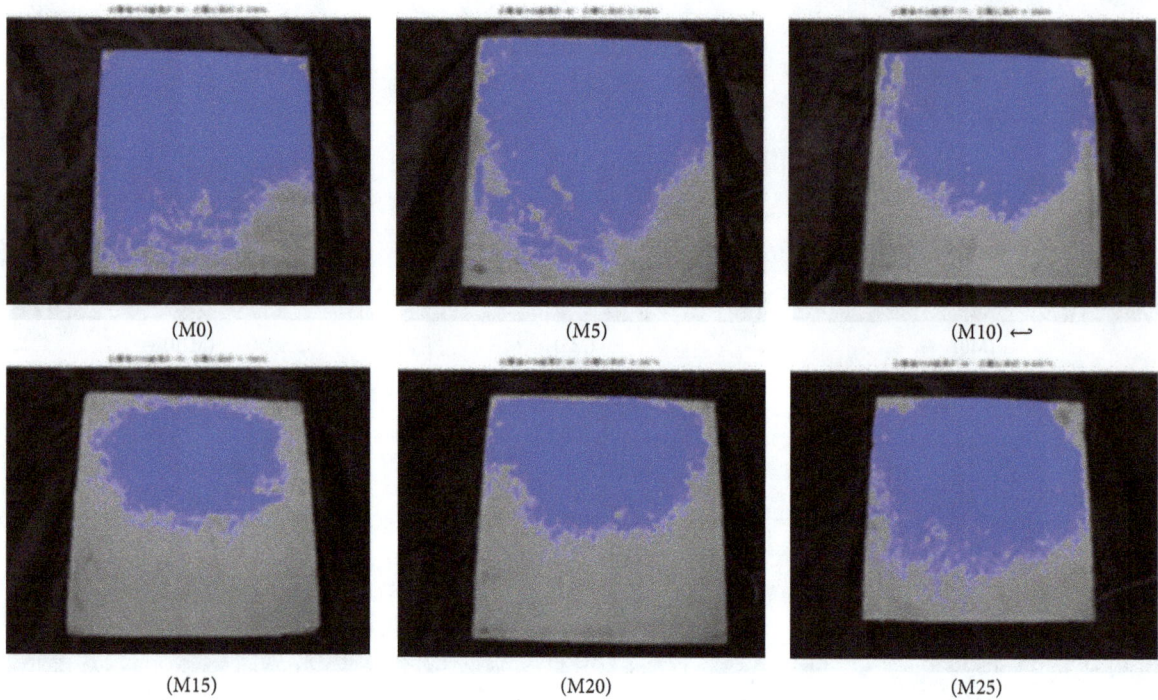

FIGURE 7: Quantity of efflorescence under low temperature environmental conditions (56 days).

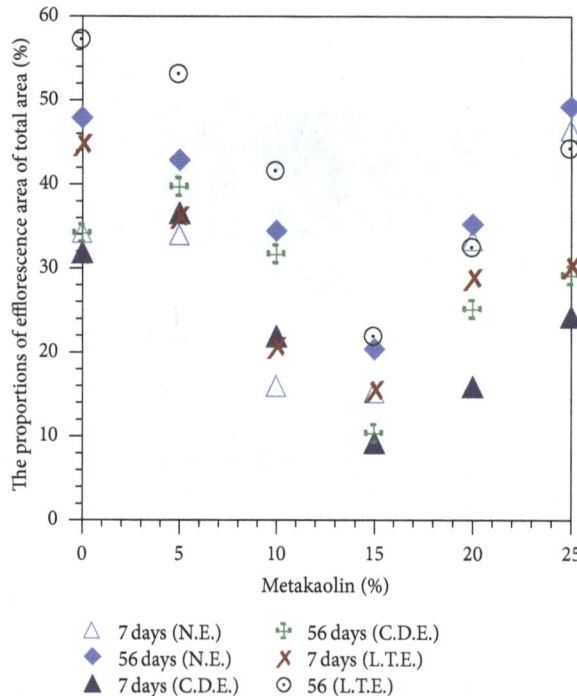

FIGURE 8: Proportional area of efflorescence versus replacement of metakaolin.

normal environment (NE), 25°C with 85% humidity; carbon dioxide environment (CDE), in a carbonization tub with 100% carbon dioxide at 15 atm pressure, 100°C, and 90% relative humidity; low temperature environment (LTE), refrigerated at −5~0°C with 2% humidity.

The mix proportions are presented in Table 3. The coding in Table 3 (M0, M5, M10, M15, M20, and M25) represents the percentage of metakaolin.

Cubic specimens (50 × 50 × 50 mm) were prepared to test the compressive strength. Additional specimens (150 × 150 ×

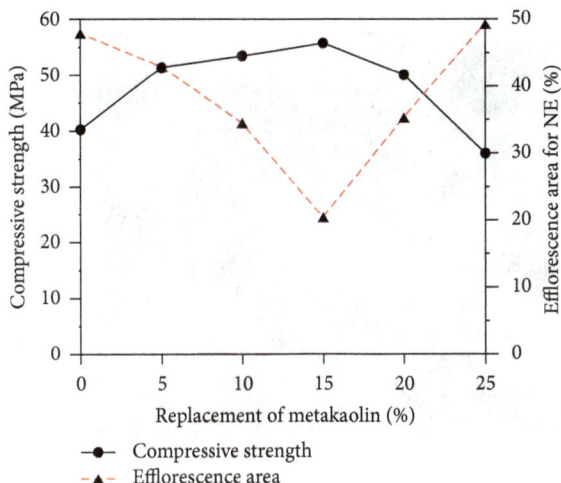

FIGURE 9: Curves comparing compressive strength versus replacement of metakaolin and area of efflorescence versus replacement of metakaolin (NE specimens).

TABLE 3: Designed mix proportions.

Mix	Water (g)	Cement (g)	Metakaolin (g)	Sand (g)
M0	417.47	834.94	0	834.94
M5	417.47	793.19	41.75	834.94
M10	417.47	751.45	83.49	834.94
M15	417.47	709.69	125.25	834.94
M20	417.47	667.96	166.98	834.94
M25	417.47	814.19	208.75	834.94

30 mm) were also prepared for the quantification of efflorescence using image analysis in MATLAB. Finally, specimens (10 × 10 × 10 mm) were sliced from the mortar specimens for observation under scanning electron microscope (SEM) and samples of mortar powder (3 g) were prepared for X-ray diffraction (XRD).

2.2. Testing Methods. Compressive strength was determined after 1, 3, 7 and 28 days of curing, according to ASTM C109-12. The extent of white efflorescence was quantified according to RGB values using MATLAB (Matrix Laboratory) image analysis of photographs (taken at 7 and 56 days) of samples exposed to the three experimental environments (NE, CDE, and LTE). MATLAB image analysis was unable to determine the thickness of efflorescence; therefore, the specimens were analyzed using the curettage method to quantify efflorescence according to weight. The curettage method indicated that we removed the efflorescence with a spatula and then weighed it.

A petrographic examination of hardened mortar was performed using SEM according to ASTM C856-11 specifications. The specimens were dried, vacuumed, and Au ion-sputtered prior to SEM investigation in order to render the surface conductive. By varying the degree of magnification (×1000 and ×3000), capillary pores of various sizes (micro- or meso-pore structure) were estimated.

We also investigated the compounds of cement-based materials following replacement with metakaolin. Because the hydration of composite materials is multiphased, all of the compositions are compounds; that is, almost no single element structures exist. As a result, we employed XRD analysis to analyze the chemical compounds within cement-based materials. These samples were ground into powder at room temperature under air-dried conditions and XRD patterns were recorded using Cu-K radiation between 20° to 80°, at a scanning speed of 0.5 s/1°. The composition of the compounds was determined by comparing XRD intensity diagrams with the peak values of corresponding compounds in the computer database.

3. Results And Discussion

3.1. Compressive Strength. Compressive strength and the percent of relative compressive strength are presented in Table 4. At the curing age of 28 days in the experiment results, the highest compressive strength was obtained from specimens with 15% metakaolin (as a replacement for cement), as illustrated in Figure 1. The compressive strength of specimens with metakaolin increased with time; however, for specimens containing 25% metakaolin, the strength characteristics failed to meet those of standard mortar specimens. The inclusion of metakaolin in cement-based composites enhances compressive strength through the filler effect in the interfacial transition zone between the cement paste and aggregate particles. In addition, CH gels are quickly removed during the hydration of cement with metakaolin and actually accelerate cementitious hydration.

3.2. Quantification of Efflorescence Using MATLAB Image Analysis. The specimens were photographed at the curing age of 7 and 56 days to quantify efflorescence using MATLAB image analysis. The areas affected by efflorescence as a ratio of the total area are summarized in Table 5. The specimens containing 15% metakaolin present a distinctly lower degree of efflorescence, compared to the other specimens, according to the presence of deposits both on and around the specimens.

As shown in Figures 2, 3, 4, 5, 6, and 7, the most obvious efflorescence was observed on specimens cured under LTE conditions. In addition, the inclusion of metakaolin decreased the extent of efflorescence in all specimens except for those with 20% and 25% metakaolin; specimens with 15% metakaolin showed the least efflorescence. These results verify that the inclusion of metakaolin can accelerate the hydration reaction with CH to produce a denser, more homogeneous material with a narrower transition zone, thereby reducing the extent of efflorescence. These results are in strong agreement with those of the relationship between efflorescence area and the replacement of metakaolin, as shown in Figure 8.

Figure 9 compares the area affected by efflorescence with compressive strength values; the data are fundamentally consistent with the visible extents of efflorescence in Figures 2, 3, 4, 5, 6, and 7, for NE specimens. As shown in Figure 9, cement-based composites produced with higher

(a) ×1000

(b) ×3000

FIGURE 10: SEM observations for M0 specimens.

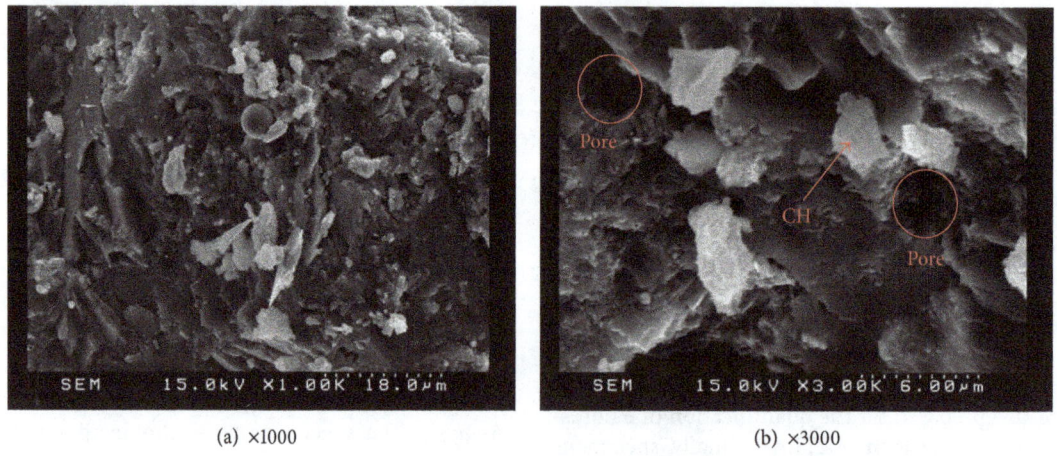

(a) ×1000

(b) ×3000

FIGURE 11: SEM images of M5 specimens.

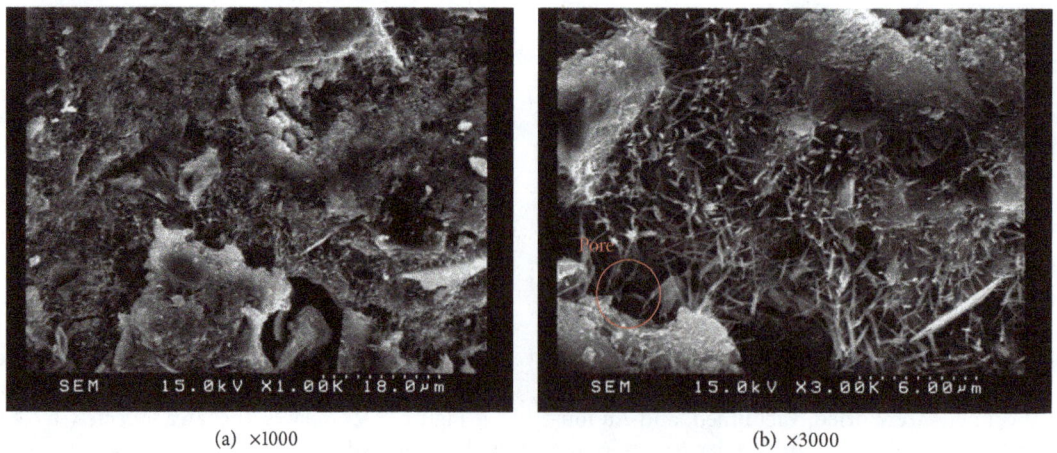

(a) ×1000

(b) ×3000

FIGURE 12: SEM images of M10 specimens.

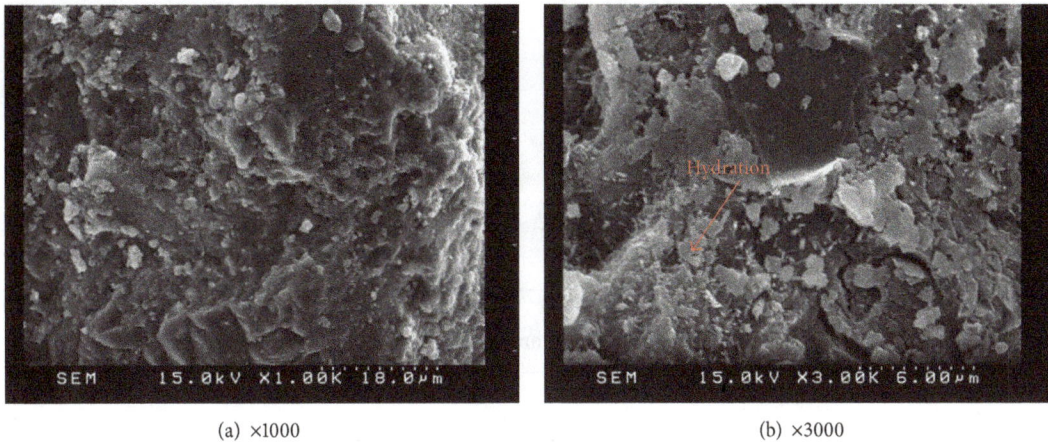

(a) ×1000

(b) ×3000

FIGURE 13: SEM images of M15 specimens.

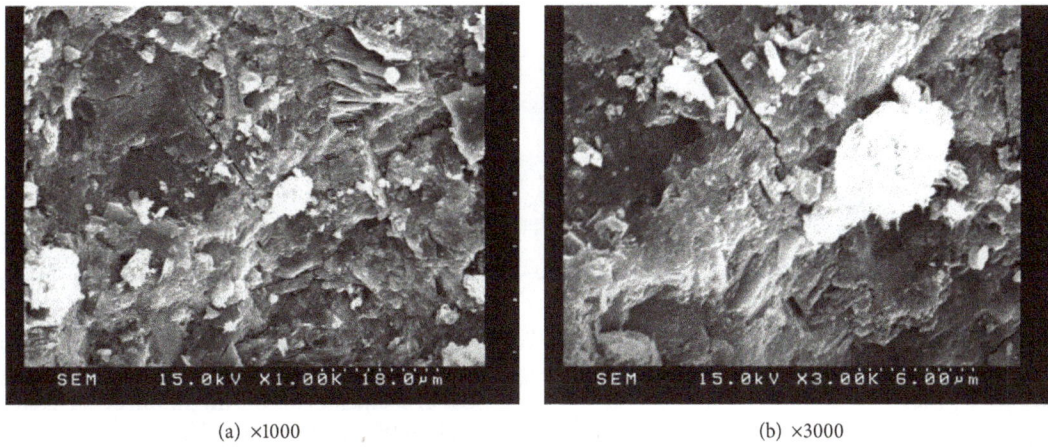

(a) ×1000

(b) ×3000

FIGURE 14: SEM images of M20 specimens.

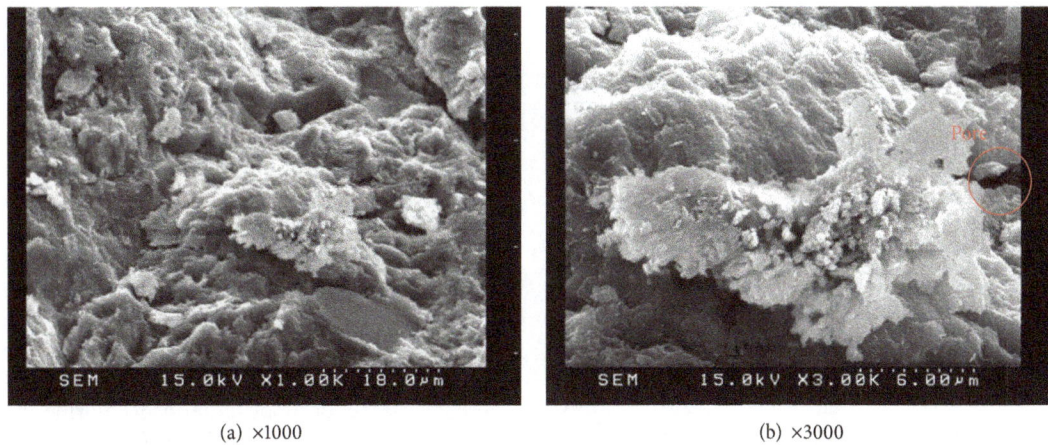

(a) ×1000

(b) ×3000

FIGURE 15: SEM images of M25 specimens.

TABLE 4: Compressive strength and percent of relative compressive strength.

Mix	Compressive strength (MPa)				Relative compressive strength (%)			
	1 day	3 day	7 day	28 day	1 day	3 day	7 day	28 day
M0	12.62	25.20	28.66	40.22	100	100	100	100
M5	14.31	28.46	32.96	51.34	113	113	115	128
M10	14.78	30.54	39.18	53.46	117	121	137	133
M15	15.03	37.16	44.36	55.72	119	147	155	139
M20	14.68	28.34	36.46	50.02	116	112	127	124
M25	11.76	20.88	26.82	35.96	93	83	94	89

TABLE 5: Proportion of white efflorescence over total area (unit: %).

Mix	Exposure environment and age in days					
	7 days (NE)	56 days (NE)	7 days (CDE)	56 days (CDE)	7 days (LTE)	56 days (LTE)
M0	34.14	47.85	31.85	34.09	44.60	57.04
M5	33.89	42.85	36.57	39.61	36.01	52.94
M10	15.87	34.42	21.90	31.59	20.53	41.38
M15	15.14	20.34	9.25	10.25	15.29	21.71
M20	33.29	35.22	15.88	25.04	28.65	32.27
M25	46.35	49.20	24.24	29.06	30.07	44.06

TABLE 6: Quantity of efflorescence using the curettage method under NE (unit: g).

	M0	M5	M10	M15	M20	M25
Weight before curettage	1352.8	1397.6	1384.8	1353.9	1410.6	1261.7
Weight after curettage	1352	1397	1384.2	1353.5	1410.1	1260.7
Weight loss	0.8	0.6	0.6	0.4	0.5	1

TABLE 7: Quantity of efflorescence using the curettage method under CDE (unit: g).

	M0	M5	M10	M15	M20	M25
Weight before curettage	1328.4	1400.2	1414.7	1378.5	1418.7	1261.8
Weight after curettage	1327.8	1400	1414.5	1378.4	1418.6	1261.4
Weight loss	0.6	0.2	0.2	0.1	0.1	0.4

TABLE 8: Quantity of efflorescence using the curettage method under LTE (unit: g).

	M0	M5	M10	M15	M20	M25
Weight before curettage	1315.3	1290.2	1346.9	1314.7	1325.2	1252.4
Weight after curettage	1314.1	1289.4	1346.3	1314.2	1324.5	1251.3
Weight loss	1.2	0.8	0.6	0.5	0.7	1.1

quantities (5% to 20%) of metakaolin provide higher strength and resistance to efflorescence, due to a denser microstructure and the controlled concentration of mobile alkalis in the pore solutions, respectively.

3.3. *Quantification of Efflorescence Using the Curettage Method.* The results for the quantification of efflorescence using the curettage method are presented in Tables 6, 7, and 8. Clearly, the addition of 15% metakaolin was most effective in inhibiting efflorescence under any exposure environment. The specimens cured under LTE conditions had a higher quantity of efflorescence than those under NE or CDE. Based on the previous study, higher humidity led to a higher quantity of efflorescence, and efflorescence increased with an increase in the size of moist particles [13]. In addition, efflorescence was proportional to alkali leaching, perhaps due to the larger volume of macropores (particularly those between 200 nm and 1000 nm) capable of increasing the diffusion coefficient for the migration of Na from geopolymer phase into the solution [14].

3.4. *Effect of Metakaolin on Microscopy Characteristics.* This study employed SEM observations to characterize the microstructural compounds produced with/without replacement metakaolin. Careful analysis of microstructures can reveal the pozzolanic reaction and the gel development. SEM magnification was set at 1,000 and 3,000 times to directly observe the development of cement hydration and pore structure. SEM observation was performed on control specimens after 56 days of aging, shown in Figure 10; large capillary pores, CH, and pore interconnectivity were observed.

SEM observation was also applied to specimens with various amounts of replacement metakaolin (5%, 10%, 15%, 20%, and 25%) at 56 days, as shown in Figures 11, 12, 13, 14, and 15. Clearly, cement-based paste specimens with replacement metakaolin developed a more compact, denser pore structure. Hydration was formed on the surface of the M5, M10, M15, and M20 specimens; the microstructure of these samples reduced the mobility of chloride and other ions, resulting in higher compressive strength and a reduction in crack stretching.

The SEM image in Figure 13 illustrates the relatively dense, homogeneous microstructure of M15 specimens with

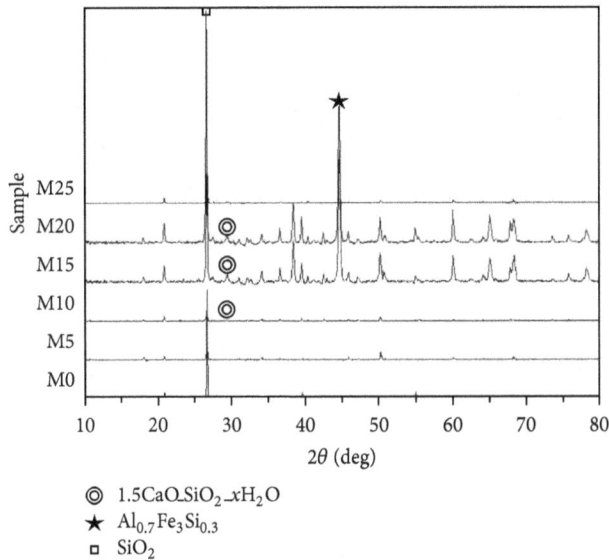

FIGURE 16: XRD results of paste specimens with metakaolin.

smooth surfaces and no distinct pores. The dense microstructure observed in SEM images is consistent with the results of compressive strength. A number of large capillary pores were observed in specimens containing 25% metakaolin, as illustrated in Figure 15, exceeding those in other samples. This may be due to the fineness of metakaolin reducing the workability of the composites, resulting in inconsistent dispersion of particles throughout the specimens. In addition, the 25% metakaolin specimens contain a lot of unreacted precursor powder due to the lack of water.

A pozzolanic reaction between metakaolin and CH occurred during the hydration of cement, resulting in the consumption of a portion of the CH. The formation of secondary calcium-silicate-hydrate (C-S-H) gel (due to pozzolanic reaction), although less dense than the primary C-S-H gel, is effective in filling and segmenting large capillary pores into small, discontinuous capillary pores through pore size refinement, thereby decreasing the total permeability of cement-based composites. The filler action of metakaolin due to its fine particle size (approximately 1 μm) compared to the particle size of cement (approximately 12 μm) further increases the density of the pore structure of metakaolin cement-based composites.

3.5. Influence of Metakaolin on Microscopy Characteristics. This study employed X-ray diffraction analysis to determine the products of hydration in cement-based materials. Figure 16 illustrates the XRD results of cement-based paste specimens with replacement metakaolin at 56 days. It was found that the amounts of the primary hydration products, C-S-H ($1.5CaO \cdot SiO_2$-xH_2O) and SiO_2, present a significant form with the addition of metakaolin (M10, M15, and M20). In contrast, $Al_{0.7}Fe_3Si_{0.3}$ content (peak value in Figure 16) increased significantly in the specimens containing 15% and 20% metakaolin. This is because metakaolin contains high quantities of $Al_{0.7}Fe_3Si_{0.3}$, which verifies that nonhydrated

particles could act as filler, changing large capillary pores into small, discontinuous capillary pores. The M15 and M20 specimens presented superior pozzolanic activity, which is consistent with the compressive strength test results, quantity of efflorescence, and SEM observations. In addition, the peak value for the M25 specimens dropped significantly, indicating an inferior hydrated reaction compared with the control specimens. This trend is similar to the results of compressive strength, in which the hydration reaction of specimens of samples with more than 20% metakaolin is lower than that of control specimens, due to the influence of reduced resistance to efflorescence.

4. Conclusion

Our results demonstrate that the replacement of cement using metakaolin at percentages of 5%, 10%, 15%, and 20% increases compressive strength with a peak increase in samples produced using 15% metakaolin. However, a distinct drop in compressive strength was observed in samples produced using 25% metakaolin. MATLAB image analysis and the curettage method both indicate that efflorescence is most pronounced when samples were cured in a low temperature environment, exceeding that produced under a normal environment or carbon dioxide environment. In the M5, M10, M15, and M20 samples, the area affected by efflorescence was lower than that of the control specimens; the M15 specimens were the least affected by efflorescence. In SEM micrographs, materials produced with metakaolin developed denser, smoother structures. XRD results indicate that the amount of main hydration products in cement-based materials with replacement metakaolin performed a significant form. Our results conclusively demonstrate that cement-based materials with 15% replacement metakaolin have superior performance with considerable potential for application in engineering.

References

[1] A. Aït-Mokhtar, R. Belarbi, F. Benboudjema, N. Burlion, B. Capra et al., "Experimental investigation of the variability of concrete durability properties," *Cement and Concrete Research*, vol. 45, pp. 21–36, 2013.

[2] E. N. Kani, A. Allahverdi, and J. L. Provis, "Efflorescence control in geopolymer binders based on natural pozzolan," *Cement and Concrete Composites*, vol. 34, no. 1, pp. 25–33, 2012.

[3] X. Y. Wang and H. S. Lee, "Modeling of chloride diffusion in concrete containing low-calcium fly ash," *Materials Chemistry and Physics*, vol. 138, pp. 917–928, 2013.

[4] V. S. Ramachandran and J. J. Beaudoin, *Handbook of Analytical Techniques in Concrete Science and Technology: Principles, Techniques and Applications*, William Andrew, 1st edition, 2001.

[5] C. L. Lee, R. Huang, W. T. Lin, and T. L. Weng, "Establishment of the durability indices for cement-based composite containing supplementary cementitious materials," *Materials and Design*, vol. 37, pp. 28–39, 2012.

[6] T. Y. Han, W. T. Lin, A. Cheng, R. Huang, and C. C. Huang, "Influence of polyolefin fibers on the engineering properties of cement-based composites containing silica fume," *Materials and Design*, vol. 37, pp. 569–576, 2012.

[7] M. I. Khan and A. M. Alhozaimy, "Properties of natural pozzolan and its potential utilization in environmental friendly concrete," *Canadian Journal of Civil Engineering*, vol. 38, no. 1, pp. 71–78, 2011.

[8] A. A. Ramezanianpour and H. B. Jovein, "Influence of metakaolin as supplementary cementing material on strength and durability of concretes," *Construction and Building Materials*, vol. 30, pp. 470–479, 2012.

[9] E. Guneyisi, M. Gesoğlu, S. Karaoğlu, and K. Mermerdas, "Strength, permeability and shrinkage cracking of silica fume and metakaolin concretes," *Construction and Building Materials*, vol. 34, pp. 120–130, 2012.

[10] H. Paiva, A. Velosa, P. Cachim, and V. M. Ferreira, "Effect of metakaolin dispersion on the fresh and hardened state properties of concrete," *Cement and Concrete Research*, vol. 42, pp. 607–612, 2012.

[11] E. Güneyisi, M. Gesoğlu, F. Karaboğa, and K. Mermerdaş, "Corrosion behavior of reinforcing steel embedded in chloride contaminated concretes with and without metakaolin," *Composites B*, vol. 45, pp. 1288–1295, 2013.

[12] A. Nadeem, S. A. Memonb, and T. Y. Lo, "Mechanical performance, durability, qualitative and quantitative analysis of microstructure of fly ash and Metakaolin mortar at elevated temperatures," *Construction and Building Materials*, vol. 38, pp. 338–347, 2013.

[13] Y. Gao, S. B. Chen, and L. E. Yu, "Efflorescence relative humidity of airborne sodium chloride particles: a theoretical investigation," *Atmospheric Environment*, vol. 41, no. 9, pp. 2019–2023, 2007.

[14] L. Zheng, W. Wang, and Y. Shi, "The effects of alkaline dosage and Si/Al ratio on the immobilization of heavy metals in municipal solid waste incineration fly ash-based geopolymer," *Chemosphere*, vol. 79, no. 6, pp. 665–671, 2010.

11

Freeze-Thaw Durability of Air-Entrained Concrete

Huai-Shuai Shang[1,2] and Ting-Hua Yi[3]

[1] *School of Civil Engineering, Qingdao Technological University, Qingdao 266033, China*
[2] *State Key Laboratory of Structural Analysis for Industrial Equipment, Dalian University of Technology, Dalian 116024, China*
[3] *School of Civil Engineering, Dalian University of Technology, Dalian 116024, China*

Correspondence should be addressed to Huai-Shuai Shang; shanghuaishuai@yahoo.com.cn

Academic Editors: S. Chen, M. Jha, Q. Q. Liang, and E. Lui

One of the most damaging actions affecting concrete is the abrupt temperature change (freeze-thaw cycles). The types of deterioration of concrete structures by cyclic freeze-thaw can be largely classified into surface scaling (characterized by the weight loss) and internal crack growth (characterized by the loss of dynamic modulus of elasticity). The present study explored the durability of concrete made with air-entraining agent subjected to 0, 100, 200, 300, and 400 cycles of freeze-thaw. The experimental study of C20, C25, C30, C40, and C50 air-entrained concrete specimens was completed according to "the test method of long-term and durability on ordinary concrete" GB/T 50082-2009. The dynamic modulus of elasticity and weight loss of specimens were measured after different cycles of freeze-thaw. The influence of freeze-thaw cycles on the relative dynamic modulus of elasticity and weight loss was analyzed. The findings showed that the dynamic modulus of elasticity and weight decreased as the freeze-thaw cycles were repeated. They revealed that the C30, C40, and C50 air-entrained concrete was still durable after 300 cycles of freeze-thaw according to the experimental results.

1. Introduction

Concrete is considered as one of the most nonhomogeneous and demanding engineering materials used by mankind. The durability [1–5] of concrete is defined as the ability to withstand damaging effects of environment without deterioration for a certain period of time. The durability of concrete involves resistance to frost, corrosion, permeation, carbonation, stress corrosion, chemical attack, and so on.

Concrete has a potential to be damaged if it is subjected to freeze-thaw cycles. The American Concrete Institute (ACI) has established specifications for protection of concrete placed during cold weather. ACI defined cold weather as the period where more than three successive days have a mean daily air temperature less than 40 F (Fahrenheit). The freeze-thaw durability of concrete is of utmost importance in countries having subzero temperature conditions, such as The Arctic Zone, Russia, Northern China, and China. Frost damage, a progressive deterioration which starts from the surface separation or scaling and ends up with complete collapse, is a major concern when concrete is used in colder regions.

The deterioration proceeds as freezing and thawing cycles are repeated, and the material gradually loses its stiffness and strength. In addition, the increasing irreversible expansion is induced. So frost damage is a very complex fatigue process. It has been a significant scientific and technical problem to improve the freeze-thaw durability and to prolong the service life of concrete.

Hong-Qiang et al. [4] and Li-kun [5] investigated the relative dynamic modulus of elasticity (RDME) and weight loss of plain concrete after different cycles of freeze-thaw. Sun et al. [6] investigated the loss of dynamic elastic modulus of high-strength concrete under the action of load and freeze-thaw cycles. Zaharieva et al. [7] investigated the influence of freeze-thaw cycles on the loss of dynamic elastic modulus of recycled aggregate concrete. The effect of sodium chloride solution, freeze-thaw cycling, and externally applied load on the relative dynamic modulus of elasticity (RDME) and weight loss of concrete was experimentally investigated by Sun et al. [8]. Cohen et al. [9] investigated the relative dynamic modulus of elasticity (RDME) and weight loss of non-air-entrained high-strength concrete after freeze-thaw cycles.

Air-entraining agent [10–13] was recommended for nearly all concretes, principally to improve resistance to freeze-thaw cycles when exposed to water and deicing chemicals in cold regions. Very little work has been documented on the freezing and thawing durability of air-entrained concrete. This paper presents experimental study on the relative dynamic modulus of elasticity and weight loss of C20, C25, C30, C40, and C50 air-entrained concretes after 0, 100, 200, 300, and 400 cycles of freeze-thaw according to "the test method of long-term and durability on ordinary concrete" GB/T50082-2009 [14]. And the influence of freeze-thaw cycles on the relative dynamic modulus of elasticity and weight loss of C20, C25, C30, C40, and C50 air-entrained concrete was analyzed according to the experimental results.

2. Experimental Procedures

2.1. Materials and Mix Proportions. In this investigation, local materials were utilized. A Chinese standard (GB175-99) [15] Portland cement 425 (which has standard compressive strength of 42.5 MPa at the age of 28 days) was used. Natural river sand with fineness modulus of 2.6 was used. Coarse aggregate was a crushed stone with diameter from 5 mm~ 20 mm. The mix proportions are listed in Table 1. The mixing started after putting all the coarse aggregate and fine aggregate into the mixer. These ingredients were mixed for about 1 min, and then the water with air-entraining agent was added in 1 minute. Finally the mixing continued for about 2 min after all water was added.

2.2. Test Specimens and Testing Programs. Concrete prisms with size of 100 mm × 100 mm × 400 mm (to determine the weight loss and the dynamic modulus of elasticity) were prepared. The specimens were cast in steel molds and compacted through external vibration and demolded 24 h later. All the specimens were cured in a condition of 20 ± 3°C and 95 percent RH for 23 days. Thereafter, the specimens were immersed in water for 4 days prior to the freeze-thaw cycles. Then when the age of the specimens was 28 days, the air-entrained concrete specimens were placed into the freeze-thaw apparatus.

In this paper, the freeze-thaw test apparatus [16] meeting the requirement of "the test method of long-term and durability on ordinary concrete" GB/T 50082-2009 was used. The freeze-thaw cycles consisted of alternately lowering the temperature of the specimens from 6°C to −15°C and raising it from −15°C to 6°C, while the temperature of the antifreeze ranged from 8 ± 2°C to −17 ± 2°C and then warms to 8 ± 2°C all within 2.5~3 hours.

The dynamic modulus of elasticity and weight loss of each specimen were measured before placing the specimens with size of 100 mm × 100 mm × 400 mm into the freeze-thaw apparatus. One specimen with size of 100 mm × 100 mm × 400 mm was placed in a rubber container. In which, standard concrete prisms were surrounded by water. Specimens were removed for testing when they were in a thawed condition

at 50-cycle or 100-cycle intervals. The dynamic modulus of elasticity and weight were recorded. Before returning the specimens into the freeze-thaw apparatus in a random order, containers were cleaned out and fresh water was added. The C20 and C25 air-entrained concrete specimens were exposed to 300 cycles of freeze-thaw, the C30, C40, and C50 air-entrained concrete specimens were exposed to 400 cycles of freeze-thaw.

3. Results and Discussions

The surface deterioration of the C30 air-entrained concrete specimens undergoing 0, 200, and 400 cycles of freeze-thaw is shown in Figures 1(a), 1(b), and 1(c). The microcracks were caused after the action of freezing and thawing cycles, and then the coarse aggregates and cement part were separated because of the action of freeze-thaw cycles. So the surface separation or scale off was caused by freeze-thaw cycles just as shown in Figures 1(b) and 1(c).

3.1. The Relative Dynamic Modulus of Elasticity. The RDME of C20, C25, C30, C40, and C50 air-entrained concrete after different cycles of freeze-thaw was given in Table 2.

The relative dynamic modulus of elasticity is defined as follows:

$$P = \frac{f_n}{f_0} \times 100, \quad (1)$$

where P is relative dynamic modulus of elasticity at n cycles of freeze-thaw, expressed as percentage, computed as the average of three specimens; f_n is dynamic modulus of elasticity at n cycles of freeze-thaw; f_0 is dynamic modulus of elasticity before freeze-thaw cycles.

After 300 cycles of freeze-thaw, the C30, C40, and C50 air-entrained concrete specimens showed a small loss of RDME, while C20 and C25 air-entrained concrete specimens showed considerable loss of RDME, as shown in Figure 2.

As seen from Table 2 and Figure 2, for C20, C25, C30, C40, and C50 air-entrained concrete, the RDME decreased slowly during the first 200 freeze-thaw cycles; the RDME of C20, C25, C30, C40, and C50 air-entrained concrete were 96.70, 90.75, 94.60, 99.05, and 97.50 percent after 200 cycles of freeze-thaw. In subsequent cycles of freeze-thaw, it is observed that the deterioration proceeds quickly. And after 300 freeze-thaw cycles, the RDME of C20 and C25 air-entrained concrete gave an obvious decrease; it decreased to about 64.95 and 62.80 percent, while the RDME of C30 and C40 air-entrained concrete only gave a decrease of 6.10, 2.65, and 9.65 percent. From 300 to 400 cycles of freeze-thaw, the RDME of C30 and C50 air-entrained concrete gave a decrease of 16.85 and 12.75 percent. The good durability of C30, C40, and C50 air-entrained concrete compared to the C20, C25 air-entrained concrete can be attributed to its higher compressive strength.

Sun et al. [8] and Zaharieva et al. [7] investigated the influence of freeze-thaw cycles on the RDME of plain concrete. The conclusion that the RDME decreased to 62 percent after 100 cycles of freeze-thaw was given by Li-kun.

TABLE 1: The mix proportion of air-entrained concrete in per cubic meter.

	Cement (MPa)	W/C	Cement (kg/m³)	Sand (kg/m³)	Coarse aggregate (kg/m³)	Water (kg/m³)	Air-entraining agent (kg/m³)	Air content (%)
C20	32.5	0.40	339.00	642.00	1185.20	133.80	0.85	5.5~6.5
C25	32.5	0.40	356.00	615.20	1188.00	141.00	0.89	5.5~6.5
C30	42.5	0.40	412.67	586.83	1186.00	164.30	1.03	5.5~6.5
C40	42.5	0.36	467.60	568.20	1148.00	166.00	1.17	5.5~6.5
C50	42.5	0.32	526.00	520.00	1154.80	168.30	1.30	5.5~6.5

(a) 0 cycles of freeze-thaw   (b) 200 cycles of freeze-thaw   (c) 400 cycles of freeze-thaw

FIGURE 1: Surface of C30 air-entrained concrete after 0, 200, and 400 cycles of freeze-thaw.

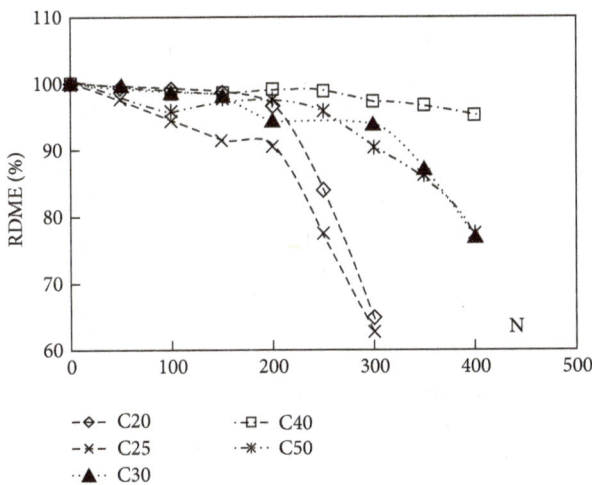

FIGURE 2: RDME of air-entrained concrete after different cycles of freeze-thaw (%).

of the freeze-thaw cycles on the RDME of plain concrete is higher than that on the RDME of the air-entrained concrete.

3.2. Weight Loss. One type of deterioration of concrete structures by cyclic freeze-thaw is surface scaling. Surface scaling is the loss of paste and mortar from the surface of concrete by the cyclic freeze-thaw or by an internal reaction of aggregate (e.g. alkali-silica reaction in concrete mixed with alkali-reactive aggregate). In extreme cases, the loss of paste can result in loosening of coarse aggregate and gradual reduction in strength of concrete structures. The weight loss will be caused by surface scaling, so the weight loss was measured.

Table 3 gives the weight of air-entrained concrete after different cycles of freeze-thaw. The weight loss for air-entrained concrete versus the number of freeze-thaw cycles is shown in Figure 3. The weight loss is defined as follows:

$$\Delta W_n = \frac{G_0 - G_n}{G_0} \times 100, \tag{2}$$

where ΔW_n is weight loss at n cycles of freeze-thaw, expressed as percentage, computed as the average of three specimens; G_n is weight at n cycles of freeze-thaw; G_0 is weight before freeze-thaw cycles.

It can be seen from Table 3 and Figure 3 that the influence of freeze-thaw cycles on the weight loss of C20 and C25 air-entrained concrete is larger than that on the weight loss of C30, C40 and C50 air-entrained concrete. After 200 cycles of freeze-thaw, the weight loss was 0.70, 0.67, and 0.25 percent for C30, C40, and C50 air-entrained concrete, while the

Hong-Qiang et al. found that the RDME decreased to 64 percent after 100 cycles of freeze-thaw. The dynamic modulus of elasticity is the proportion of stress to strain when the stress is least under dynamic loads. It can be measured by means of longitudinal vibration or flexural vibration. It reflects the elasticity performance of material, similarly to the initial tangential modulus under static loads. The loss of the dynamic modulus of elasticity with freeze-thaw cycles means the loss of the elasticity performance. Therefore, the influence

TABLE 2: RDME of air-entrained concrete after different cycles of freeze-thaw (%).

	Number of freeze-thaw cycles								
	0	50	100	150	200	250	300	350	400
C20	100	99.45	99.4	98.75	96.7	83.85	64.95	/	/
C25	100	97.60	94.35	91.55	90.75	77.35	62.8	/	/
C30	100	99.55	98.75	98.2	94.6	/	93.9	87.3	77.05
C40	100	/	98.4	98.55	99.05	98.9	97.35	96.75	95.4
C50	100	/	95.85	97.6	97.5	95.8	90.35	85.95	77.6

"/" means: "the measurements were not made."

TABLE 3: Weight of air-entrained concrete after different cycles of freeze-thaw (Kg).

	Number of freeze-thaw cycles								
	0	50	100	150	200	250	300	350	400
C20	8.930	8.920	8.860	8.770	8.720	8.660	8.540	/	/
C25	9.417	9.380	9.270	9.150	9.080	9.050	9.005	/	/
C30	9.960	9.930	9.940	9.940	9.890	/	9.900	9.840	9.685
C40	9.740	9.730	9.735	9.685	9.675	9.660	9.410	9.615	9.510
C50	9.960	/	9.925	9.940	9.935	9.890	9.870	9.800	9.585

"/" means: "the measurements were not made."

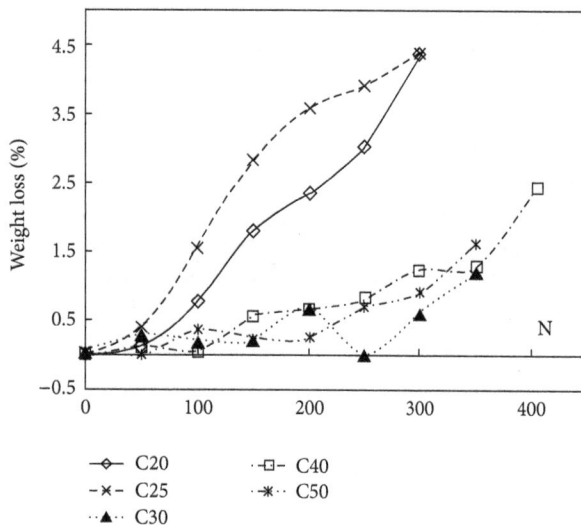

FIGURE 3: Effect of freeze-thaw cycles on weight loss of air-entrained concrete.

weight loss was 2.35 and 3.58 percent for C20 and C25 air-entrained concrete.

The test results show that the weight loss of C30 air-entrained concrete was 0.20 percent compared with 1.81 percent for plain concrete after 100 cycles of freeze-thaw [4]. For C20, C25, C30, C40, and C50 air-entrained concrete, the maximum of weight loss was only 4.38 percent after 300 cycles of freeze-thaw. While Hong-Qiang et al. [4] found that the weight loss of C30 plain concrete decreased to 3.74 percent after 125 cycles of freeze-thaw. For plain concrete, the weight increase was observed by Hong-Qiang [4] and Li-kun [5] during the first 25 freeze-thaw cycles. For air-entrained concrete, the weight increase wasnot observed.

The weight loss of concrete specimens is caused by surface separation or scale off. The weight variation during freeze-thaw cycles is due to movement in and out of water in the specimen and surface separation or scaling (surface scaling is the loss of paste and mortar from the surface of concrete by the cyclic freeze-thaw). As soon as microcracking takes place, the deteriorated zones filled with the surrounding water will cause change in the weight of the specimen. If the mass of surface separation is larger than the water absorbed by the concrete specimens, the weight of the concrete specimens will increase. The weight of the concrete specimens will decrease when the mass of surface separation is less than the water absorbed by the concrete specimens. Compared with plain concrete, the deteriorated zones filled with the surrounding water occurred in the air-entrained concrete needed much more cycles of freeze-thaw.

In actual concrete structures, concrete surface scaled markedly when exposed to deicing salt and freeze-thaw cycles caused by the change of climate. The cycling rate in the laboratory conditions was much higher than that in the natural environment because of the fast change of temperature. Thus, it is reasonable that the scaling observed during the tests was more severe, and the scaling depth of concrete specimens was over 1 mm.

3.3. The Ultrasonic Velocity. A lot of structures, like bridges, tunnels, dams, buildings, and others, were constructed with concrete material. During the life cycle of these structures, degradations can occur because of mechanical, thermal, or chemical stresses. These often lead to the development of porosity, microcracks, and cracks in the material. Knowing the concrete structure state to prevent or repair damage is needed so the nondestructive characterisation is an important way, and the ultrasonic method is often proposed.

Table 4: Decreasing percentage of the ultrasonic velocity after freeze-thaw cycles.

Number of freeze-thaw cycles	0	100	200	300	400
Loss of the ultrasonic velocity (%)	100	97.7	97.6	91.0	84.7

In this work, the ultrasonic velocity of C30 air-entrained concrete was measured with ultrasonic method according to "Testing Code of Concrete for Port and Waterwog Engineering" JTJ 270-98 [17]. Table 4 gives the decreasing percentage of the ultrasonic velocity of air-entrained concrete after different cycles of freeze-thaw. It can be seen from Table 4 that the ultrasonic velocity decreased slowly during the first 200 freeze-thaw cycles, and it gave only about a 2.4 percent decrease over the initial value. However, in subsequent freeze-thaw cycles, it is observed that the deterioration usually proceeds. And after 400 freeze-thaw cycles, it decreased to about 84.7 percent of the initial value.

3.4. Discussion. Concrete is a three-phase composite structure at microscopic scale, a cement matrix, aggregate, and the interfacial transition zone between the two. The microcracks will be caused by the action of freezing and thawing cycles; the direction and distribution of microcosmic cracks are stochastic. The microcosmic cracks manifold and become broad as freeze-thaw cycles are repeated. Air-entrained concrete contains billions of microscopic air cells when air-entraining agents were used in concrete. These relieve internal pressure on the concrete by providing tiny chambers for the expansion of water when it freezes. So, comparing the test results in this paper with the conclusion given by other authors [4, 5], the deceased percentage for the relative dynamic modulus of elasticity and weight loss of air-entrained concrete is less than that of plain concrete after the same cycles of freeze-thaw. It means that the deterioration of freeze-thaw durability for air-entrained concrete is slower than that of plain concrete. It is because the mixed air-entraining agent in concrete can make them up effectively and thus improve the freeze-thaw durability.

4. Conclusion

The effects of freeze-thaw cycles on the RDME and weight loss of C20, C25, C30, C40, and C50 air-entrained concrete were investigated. Based on the experimental work in this study and the discussion about the experimental results, the results of the investigation can be summarized as follows

    (a) The RDME decreased as the freeze-thaw cycles were repeated. After 100 cycles of freeze-thaw, the RDME decreased to 94.35 and 98.75 percent for C25 and C30 air-entrained concrete, and 64 percent for C30 plain concrete. Therefore, the freeze-thaw durability of air-entrained concrete is much higher than that of plain concrete.

    (b) After 200 cycles of freeze-thaw, the weight loss was 0.70, 0.67, and 0.25 percent for C30, C40, and C50 air-entrained concrete, and 2.35 and 3.58 percent for C20

and C25 air-entrained concrete. The weight variation during freeze-thaw cycles is due to moveming in and out of water in the specimen and surface separation or scaling.

    (c) The freeze-thaw durability of plain concrete is poor, but it can be improved greatly when air-entraining agent is mixed into concrete. It demonstrates that ordinary strength concrete can also have a high freeze-thaw durability.

Acknowledgments

This research work was jointly supported by the Science Fund for Creative Research Groups of the National Natural Science Foundation of China (Grant no. 51121005), the National Natural Science Foundation of China (Grants nos. 51208273, 51222806), a Project of Shandong Province Higher Educational Science and Technology Program (Grant no. J12LG07), and the Program for New Century Excellent Talents in University (Grant no. NCET-10-0287).

References

[1] A. E. Richardson, K. A. Coventry, and S. Wilkinson, "Freeze/thaw durability of concrete with synthetic fibre additions," *Cold Regions Science and Technology*, vol. 83-84, pp. 49–56, 2012.

[2] C. Medina, M. I. S. de Rojas, and M. Frias, "Freeze-thaw durability of recycled concrete containing ceramic aggregate," *Journal of Cleaner Production*, vol. 40, pp. 151–160, 2013.

[3] B. Mather, "Concrete durability," *Cement and Concrete Composites*, vol. 26, no. 1, pp. 3–4, 2004.

[4] C. Hong-Qiang, Z. Lei-shun, and L. Ping-xian, "The influence of freeze-thaw to concrete strength," *Henan Science*, vol. 21, no. 2, pp. 214–216, 2003 (Chinese).

[5] Q. Li-kun, *Study on the strength and deformation of concrete under multiaxial stress after high-temperature of freeze-thaw cycling [Ph.D. thesis]*, Dalian University of Technology, Liaoning, China, 2003.

[6] Sun, Zhang, Yan, and Mu, "Damage and damage resistance of high strength concrete under the action of load and freeze-thaw cycles," *Cement and Concrete Research*, vol. 29, no. 9, pp. 1519–1523, 1999.

[7] R. Zaharieva, F. Buyle-Bodin, and E. Wirquin, "Frost resistance of recycled aggregate concrete," *Cement and Concrete Research*, vol. 34, no. 10, pp. 1927–1932, 2004.

[8] W. Sun, R. Mu, X. Luo, and C. Miao, "Effect of chloride salt, freeze-thaw cycling and externally applied load on the performance of the concrete," *Cement and Concrete Research*, vol. 32, no. 12, pp. 1859–1864, 2002.

[9] Cohen, Yixia, and Dolch, "Non-air-entrained high-strength concrete-is it frost resistant?" *ACI Materials Journal*, vol. 89, no. 4, pp. 406–415, 1992.

[10] M. Molero, S. Aparicio, G. Al-Assadi, M. J. Casati, M. G. Hernández, and J. J. Anaya, "Evaluation of freeze-thaw damage in concrete by ultrasonic imaging," *NDT & E International*, vol. 52, pp. 86–94, 2012.

[11] R. Sahin, M. A. Tasdemir, R. Guel, and C. Celik, "Optimization study and damage evaluation in concrete mixtures exposed to

slow freeze-thaw cycles," *Journal of Materials in Civil Engineering*, vol. 19, no. 7, pp. 609–615, 2007.

[12] C. Atkins, "Physical deterioration mechanisms," in *Concrete Durability: A Practical Guide to the Design of Durable Concrete Structures*, M. Soutsos, Ed., ThomasTelford, London, UK, 2010.

[13] G.-F. Peng, Q. Ma, H.-M. Hu, R. Gao, Q.-F. Yao, and Y.-F. Liu, "The effects of airentrainment and pozzolans on frost resistance of 50-60 MPa grade concrete," *Construction and Building Materials*, vol. 21, no. 5, pp. 1034–1039, 2007.

[14] National Standard of the People's Republic of China, "The test method of long-term and durability on ordinary concrete," Tech. Rep. GB/T50082-2009, National Standard of the People's Republic of China, Beijing, China, 2009.

[15] National Standard of the People's Republic of China, "Portland cement and ordinary portland cement," Tech. Rep. GB175-99, National Standard of the People's Republic of China, Beijing, China, 1999.

[16] H. Shang, Y. Song, and L. Qin, "Experimental study on strength and deformation of plain concrete under triaxial compression after freeze-thaw cycles," *Building and Environment*, vol. 43, no. 7, pp. 1197–1204, 2008.

[17] National Standard of the People's Republic of China, "Testing code of concrete for port and Waterwog engineering," Tech. Rep. JTJ 270-98, National Standard of the People's Republic of China, Beijing, China, 1999.

12

Methodology for Assessing the Probability of Corrosion in Concrete Structures on the Basis of Half-Cell Potential and Concrete Resistivity Measurements

Lukasz Sadowski

Institute of Building Engineering, Wroclaw University of Technology, Plac Grunwaldzki 11, 50-377 Wroclaw, Poland

Correspondence should be addressed to Lukasz Sadowski; lukasz.sadowski@pwr.wroc.pl

Academic Editors: S. Chen and Q. Q. Liang

In recent years, the corrosion of steel reinforcement has become a major problem in the construction industry. Therefore, much attention has been given to developing methods of predicting the service life of reinforced concrete structures. The progress of corrosion cannot be visually assessed until a crack or a delamination appears. The corrosion process can be tracked using several electrochemical techniques. Most commonly the half-cell potential measurement technique is used for this purpose. However, it is generally accepted that it should be supplemented with other techniques. Hence, a methodology for assessing the probability of corrosion in concrete slabs by means of a combination of two methods, that is, the half-cell potential method and the concrete resistivity method, is proposed. An assessment of the probability of corrosion in reinforced concrete structures carried out using the proposed methodology is presented. 200 mm thick 750 mm × 750 mm reinforced concrete slab specimens were investigated. Potential E_{corr} and concrete resistivity ρ in each point of the applied grid were measured. The experimental results indicate that the proposed methodology can be successfully used to assess the probability of corrosion in concrete structures.

1. Introduction

Corrosion of the steel reinforcement in concrete is a crucial problem for the construction industry since it poses the most serious risk to the structural integrity of reinforced concrete structures. Inspection and monitoring techniques are needed to assess the corrosion of the reinforcement in order to maintain, protect, and repair buildings and bridge decks so that they remain safe [1, 2]. In the last few years much attention has been given to developing techniques for predicting the remaining service life of concrete structures [3]. Most of the reported research in this area focuses on the corrosion of concrete reinforcement [4].

Several electrochemical techniques for monitoring and assessing the corrosion of steel in concrete structures were presented in [5–7]. The most popular method of *in situ* corrosion testing is the half-cell potential measurement, the idea of which is illustrated in Figure 1(a). The use of this method and the interpretation of its results are described in ASTM C876 [8]. Such corrosion potential measurements, however, should be supplemented with other nondestructive testing methods [9–12]. The use of half-cell potential measurements for determining the probability of corrosion in concrete was extensively described by Flis et al. [13], Grantham et al. [14], and Žvica [15]. The latter also presented dependences between the rate of reinforcement corrosion and temperature. The effectiveness of the test was studied in [16]. It should be noted that half-cell potential values merely provide information about the probability of corrosion and not about the rate of corrosion.

It is well known that the probability of corrosion in concrete structures depends on the ionic conductivity of the concrete electrolyte, the humidity, the temperature, and the quality of the concrete cover. The ionic conductivity is measured quantitatively as the resistivity of the concrete [17]. Concrete resistivity ρ ranges widely from 10^1 to 10^6 Ω m, depending on mainly the moisture content [18] and the material of the concrete [19, 20]. One of the promising techniques of measuring concrete resistivity is shown in Figure 1(b). As shown by Feliu et al. [21], concrete resistivity

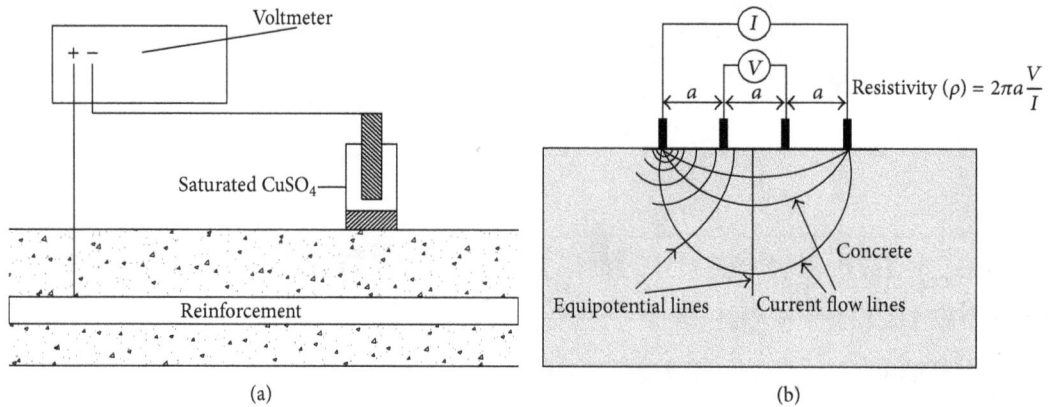

FIGURE 1: Existing corrosion methods: (a) half-cell potential measurement [27] and (b) concrete resistivity measurement [28].

TABLE 1: Dependence between potential and corrosion probability [8].

Potential E_{corr}	Probability of corrosion
$E_{corr} < -350\,mV$	Greater than 90% probability that reinforcing steel corrosion is occurring in that area at the time of measurement
$-350\,mV \leq E_{corr} \leq -200\,mV$	Corrosion activity of the reinforcing steel in that area is uncertain
$E_{corr} > -200\,mV$	90% probability that no reinforcing steel corrosion is occurring in that area at the time of measurement (10% risk of corrosion)

TABLE 2: Dependence between concrete resistivity and corrosion probability [34].

Concrete resistivity ρ, kΩcm	Probability of corrosion
$\rho < 5$	Very high
$5 < \rho < 10$	High
$10 < \rho < 20$	Low to moderate
$\rho > 20$	Low

ρ is inversely proportional to the corrosion rate. This was confirmed by Glass et al. [22] who showed that the effect of mortar resistivity is strongly dependent on the relative humidity of the environment, while López et al. [23] showed that the amount of pores in concrete determines its resistivity ρ and corrosion rate. The four-point resistivity method enables one to determine the severity of corrosion in a quick and nondestructive manner. Morris et al. [24] found that rebars undergo active corrosion when concrete resistivity ρ is below 10 kΩcm, whereas at concrete resistivity ρ above 30 kΩcm the probability of their corrosion is low. Extensive research on the resistivity technique, covering experimental analyses [25] and an analysis of the effects of geometry and material properties [26], was done by Zhang et al.

It should be noted that in the literature there are only a few papers in which the probability of corrosion is determined using both concrete resistivity measurements and half-cell potential mapping. One of them is the paper by Millard and Sadowski [29] in which in order to determine the degree of corrosion of the reinforcement, the resistivity of the concrete is measured by the electrodes used in the half-cell potential method. The combined use of half-cell potential and resistivity measurements was presented in [27]. In [28] it was shown that the combination of the method described in [29] and the conventional method of measuring concrete resistivity could be a reliable tool for directly determining the corrosion rate of the reinforcement in concrete. In paper [30] for this purpose Vedalakshmi et al. used the galvanostatic pulse technique with electrochemical impedance spectroscopy and the weight-loss method. Rhazi [31] measured concrete resistivity ρ in the same locations where half-cell potentials were measured, but it should be noted that the measurements were carried out on a concrete bridge deck covered with asphalt.

Considering that in the literature it is hard to find a systematic methodology for assessing the probability of corrosion in concrete slabs through combined half-cell potential and concrete resistivity measurements, this paper proposes such a methodology based on the combined use of the four-point Wenner concrete resistivity method and the half-cell potential method.

2. Methodology for Assessing Probability of Corrosion in Concrete Slabs

The proposed methodology for assessing the probability of corrosion in concrete slabs through a combination of two methods, that is, the half-cell potential method and the concrete resistivity method, is illustrated graphically in Figure 2.

Before measurements, the surface of the concrete slab should be prepared by brushing and polishing with abrasive paper, and a grid of n measuring points spaced at every 75 mm should be marked on the slab surface. Then the

Methodology for Assessing the Probability of Corrosion in Concrete Structures on the Basis of Half-Cell Potential and
Concrete Resistivity Measurements

103

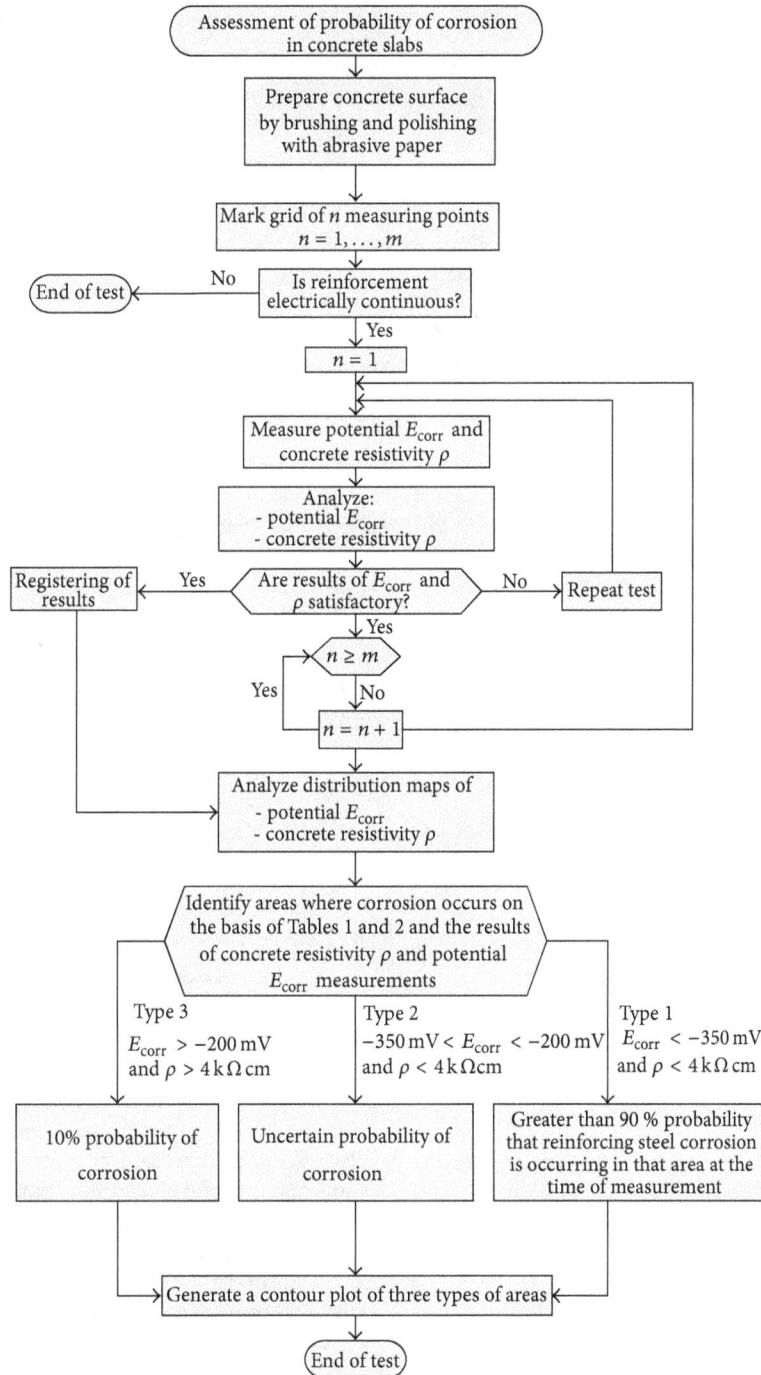

FIGURE 2: Assessment of probability of corrosion in concrete slabs through half-cell potential and concrete resistivity measurements.

electrical continuity of the reinforcement is checked in three randomly selected grid points. Subsequently potential E_{corr} and concrete resistivity ρ are measured. If the results of the E_{corr} and ρ measurements are satisfactory, they are processed using the specialized software and maps of the distribution of the parameter values on the slab surface are produced. The distribution maps of potential E_{corr} and concrete resistivity ρ should be examined. Table 2 summarizes the typical interpretation of half-cell potential readings [8]. The dependences between reinforcement corrosion probability and concrete resistivity measured by the four-point Wenner method are shown in accordance with [32] in Table 3. If there is a probability of corrosion, one should identify the areas in which corrosion may occur. On the basis of Tables 1 and 2 and the results of the concrete resistivity ρ and potential E_{corr} measurements, three areas of different types can be generated:

(i) area type 1—low values of both parameters, greater than 90% probability that reinforcing steel corrosion is occurring in that area at the time of measurement;

TABLE 3: Potential E_{corr}.

Nr	Potential E_{corr}, mV								
	A	B	C	D	E	F	G	H	I
1	−192	−168	−154	−151	−144	−156	−178	−192	−288
2	−240	−167	−169	−167	−168	−187	−201	−216	−276
3	−234	−168	−198	−189	−178	−189	−200	−215	−281
4	−201	−178	−201	−190	−189	−191	−200	−199	−291
5	−168	−165	−181	−191	−192	−189	−194	−192	−360
6	−231	−174	−182	−204	−201	−199	−194	−198	−365
7	−203	−164	−198	−219	−232	−231	−199	−196	−369
8	−216	−167	−197	−207	−216	−245	−234	−192	−384
9	−192	−168	−201	−231	−240	−245	−219	−216	−432

TABLE 4: Concrete resistivity ρ.

Nr	Concrete resistivity ρ, kΩcm								
	A	B	C	D	E	F	G	H	I
1	6.91	5.62	4.56	4.56	4.34	4.12	4.08	4.39	3.77
2	9.11	5.96	4.63	4.57	4.56	4.34	4.23	4.38	3.76
3	9.89	5.98	4.85	4.78	4.67	4.22	5.03	5.02	3.75
4	9.88	5.99	5.99	4.23	4.78	4.89	5.65	5.33	3.71
5	9.99	6.02	6.98	4.8	4.76	5.02	5.43	5.35	3.74
6	9.45	6.25	6.71	5.43	4.23	5.03	5.68	5.67	3.67
7	9.77	6.66	6.77	5.55	7.51	5.42	5.43	5.21	3.62
8	9.73	6.71	6.88	6.01	7.01	5.98	5.67	4.88	3.61
9	8.79	6.91	6.93	6.96	6.98	6.91	5.96	4.71	3.14

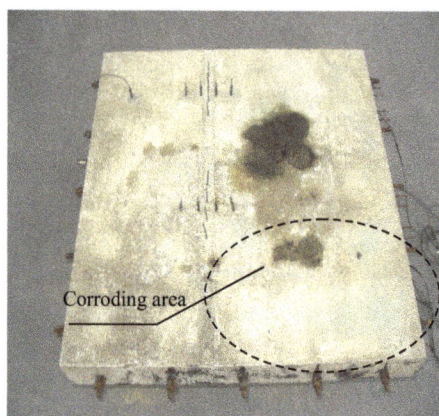

FIGURE 3: General view of concrete slab specimen.

(ii) area type 2—low values of concrete resistivity ρ and high values of measured potential E_{corr}, an uncertain probability of corrosion;

(iii) area type 3—high values of both parameters, a 10% probability of corrosion.

Finally, a contour plot of three types of areas (type 1, type 2, and type 3) should be generated.

The corrosion probability assessment can be practically verified through test pits made in selected places and visual inspection [33].

3. Exemplary Application of the Proposed Methodology

3.1. Materials and Methods. An exemplary application of the proposed methodology is presented below. 200 mm thick 750 mm × 750 mm concrete slab specimen was investigated (Figure 3). The concrete had been designed to strengthen class C 20/25. Portland cement CEM I 42.5R, well-graded sand, and crushed blue granite with a maximum total grading of 5 mm, consistency S3, and w/c = 0.5 had been used to cast the slab specimens. A reinforcement mesh made of A-III 34GS steel rebars 10 mm in diameter spaced at every 150 mm with a 50 mm cover had been embedded inside each slab. The concrete slab specimen was placed in a normal atmosphere and was subjected to a two-hour spray wetting with a sodium chloride solution cycle followed by twenty hours drying cycle to generate corroding area presented in Figure 3.

Measurements were carried out after 90 days of concrete curing, except for compressive strength tests which were done after 28 days. The concrete slabs cured at an air temperature of +18°C (±3°C) and a relative air humidity of 60%. It is important to measure concrete resistivity in constant temperature-humidity conditions since, as shown in [34], with each degree Celsius relative humidity increases by 3% [35]. After the concrete slabs were labelled, a 750 mm × 750 mm test area was demarcated on each of them and a grid of points spaced at every 75 mm was marked on each of the slabs. The columns were denoted with letters from A to I, and

Methodology for Assessing the Probability of Corrosion in Concrete Structures on the Basis of Half-Cell Potential and Concrete Resistivity Measurements

105

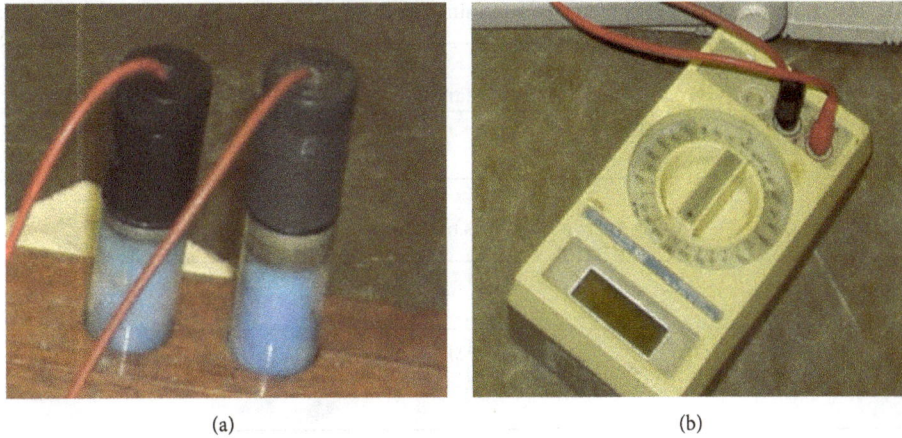

FIGURE 4: Half-cell potential measurements: (a) copper/copper sulphate electrode and (b) digital voltmeter.

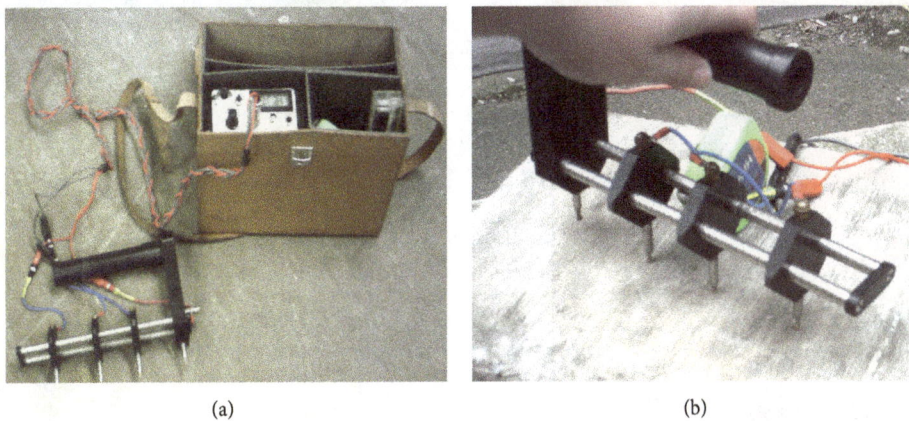

FIGURE 5: Concrete resistivity measurement: (a) test setup and (b) measurement.

the rows were numbered from 1 to 9. A total of 81 measuring points were marked on the surface.

Prior to half-cell potential measurements, the electrical continuity of the reinforcement was checked at three randomly selected grid points. The potential differences in the three points were measured with a digital voltmeter, and all the measured values were found to be below 1 mV. The copper/copper sulphate electrode used in this test is shown in Figure 4(a), and the digital voltmeter with high input impedance is shown in Figure 4(b). Before the measurements, the concrete surface was prepared by brushing and polishing with abrasive paper. The area where each measurement was to be taken was wetted with tap water to ensure better electrical contact. Then the reference electrode and the reinforcement mesh were connected to the high-impedance voltmeter, and the reference electrode was placed on the surface of the concrete.

The concrete resistivity measurements were carried out at frequencies in a range of 50–1000 Hz. Before the measurements, the equipment was calibrated using a 1 kΩ calibration bar. A single deviation was not larger than 2% and the average deviation amounted to less than 1%. Concrete resistivity ρ was measured in each point of the grid (Figure 5).

3.2. Results

3.2.1. Half-Cell Potential Measurements.
Exemplary results of the half-cell potential measurements are shown in Table 3. The results of the half-cell potential measurements were plotted on a contour map for visual interpretation (Figure 6). It is evident that potential E_{corr} is low (< -350 mV) in the area around measuring points 7 to 9 from E to I, which indicates a 95% probability of corrosion. In the other measuring points, potential E_{corr} is high (-350 mV $\leq E_{corr} \leq -200$ mV), which indicates a 10% or uncertain probability of corrosion.

3.2.2. Concrete Resistivity Measurements.
The results of the concrete resistivity measurements are shown in Table 4. The results of the concrete resistivity measurements were plotted on an equipotential contour map for visual interpretation (Figure 7). It is evident that concrete resistivity ρ is low (<5 kΩcm) in the area around measuring points 5 to 9 from A to I, which indicates a very high probability of corrosion. In the other measuring points, concrete resistivity ρ is high (>5 kΩcm), indicating a high or moderate probability of corrosion.

3.3. Statistical Analyses of Test Results.
Selected statistical characteristics of the parameters determined by half-cell

TABLE 5: Selected statistical characteristics of parameters determined by half-cell potential and concrete resistivity measurements.

	Statistical characteristics			
	Average	Standard deviation	Minimum	Maximum
Potential E_{corr}, mV	−211.09	53.46	−432	−144
Concrete resistivity ρ, kΩcm	5.68	1.66	3.14	9.99

TABLE 6: Proposed types of corrosion probability.

		Potential E_{corr}, mV		
		$E_{corr} < -350$	$-350 \leq E_{corr} \leq -200$	$-200 \leq E_{corr}$
Concrete resistivity ρ, kΩcm	$\rho < 4$	Type 1	Type 2	Type 2
	$4 < \rho < 5$			Type 3
	$\rho > 5$			Type 3

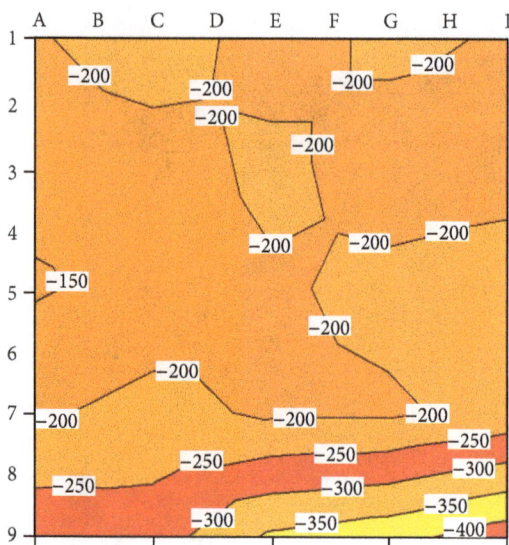

FIGURE 6: Contour plot of potential E_{corr}.

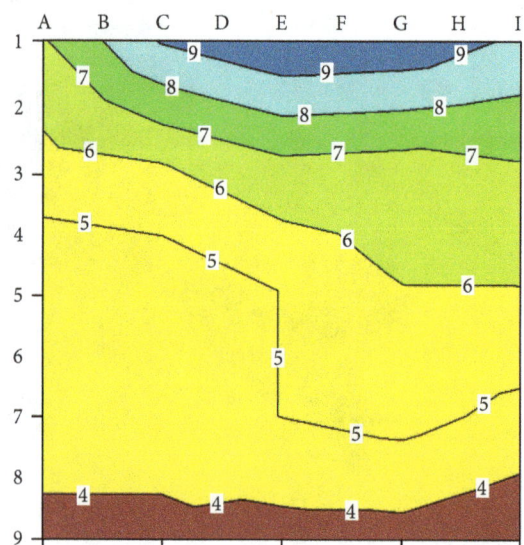

FIGURE 7: Contour plot of concrete resistivity ρ.

potential and concrete resistivity measurements are shown in Table 5. As it appears from the histograms presented in Figure 8 and statistical characteristics shown in Table 5, the half-cell potential method yielded potential E_{corr} ranging from −432 to −144 mV. The average value of potential E_{corr} was −211.09 mV with standard deviation of 53.46 mV.

Concrete resistivity ρ determined by concrete resistivity measurements ranged from 3.14 to 9.99 kΩcm. The average value of concrete resistivity ρ was 5.68 kΩcm with standard deviation of 1.66 kΩcm.

3.4. Discussion. The dependence between concrete resistivity ρ and potential E_{corr}, measured on the concrete slab surface, is shown in Figure 9. One should note that potential E_{corr} sharply increases for resistivity ρ below 4 kΩcm while above 4 kΩcm, it oscillates between −150 and −250 mV.

On the basis of Tables 1 and 2 and the results of the concrete resistivity ρ and potential E_{corr} measurements, three areas of different types were generated (Table 6) as it has been presented in Section 2. A contourplot of the three areas is

shown in Figure 10. The area around measuring points 6 to 9 from C to I is of type 1, which means that there is a 90% probability of corrosion. In the measuring points, potential E_{corr} is below −250 mV and concrete resistivity ρ is above 4 kΩcm. The area around measuring points 3 to 9 from A to E is of type 2, which means that there is a uncertain probability of corrosion. In the measuring points, potential E_{corr} is between −150 mV and −350 mV and concrete resistivity ρ is above 4 kΩcm. The area around measuring points 1 to 7 from A to I is of type 3, which means that there is an 10% probability of corrosion. In the measuring points, potential E_{corr} is above −250 mV and concrete resistivity ρ is above 5 kΩcm.

4. Conclusion

A methodology for assessing the probability of corrosion in concrete slabs based on a combination of two nondestructive methods, that is, the half-cell potential method and the concrete resistivity method, was briefly described. Comparative tests were carried out using the two methods to determine the probability of corrosion in model test concrete

Methodology for Assessing the Probability of Corrosion in Concrete Structures on the Basis of Half-Cell Potential and Concrete Resistivity Measurements

107

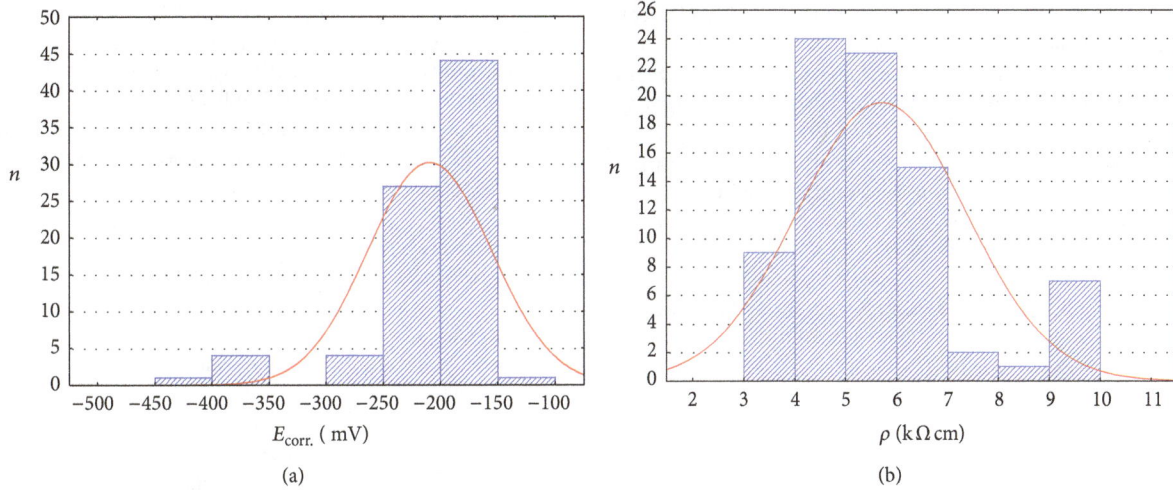

FIGURE 8: Histogram of (a) potential E_{corr} and (b) concrete resistivity ρ.

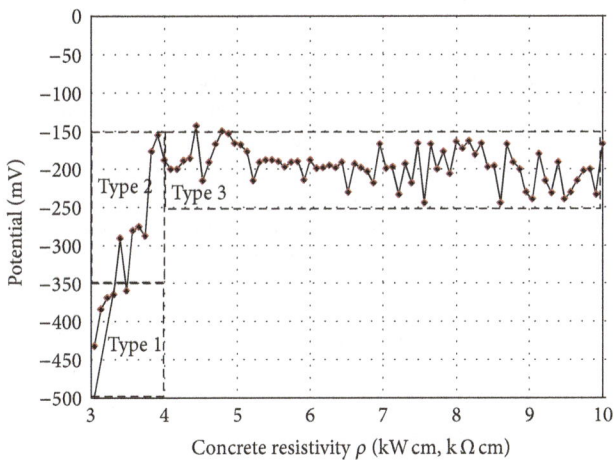

FIGURE 9: Concrete resistivity ρ versus potential E_{corr}.

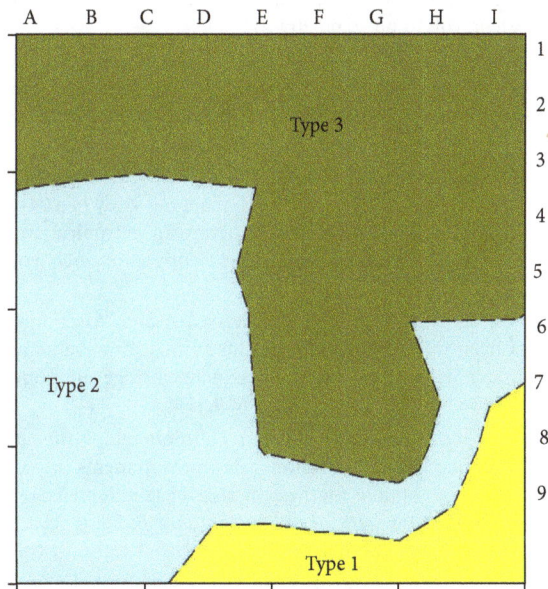

FIGURE 10: A contour plot of three areas of different types.

slab specimens. The experimental results showed that the two nondestructive techniques can be used together in order to obtain maximum information about the probability of corrosion in a tested structure.

This study was motivated by the engineer's need for a combination of the half-cell potential mapping technique and concrete resistivity measurements to more accurately assess the probability of corrosion. The combined techniques can be used in both the field and the laboratory environment. Moreover, they can be automated and integrated into monitoring systems for new or existing reinforced concrete structures. However, it is still recommended to perform additional tests for other rebar diameters, different aggregate grading, and a wider range of covers.

Acknowledgments

This research was carried out in collaboration with the Construction and Infrastructure Group in the University of Liverpool under the project "Modern Nondestructive Testing Methods to Determine the Corrosion Rate in Concrete Structures" funded by the Erasmus LLP Programme.

References

[1] J. Berger, S. Bruschetini-Ambro, and J. Kollegger, "An innovative design concept for improving the durability of concrete bridges," *Structural Concrete*, vol. 12, no. 3, pp. 155–163, 2011.

[2] M. Beck, A. Burkert, J. Harnisch et al., "Deterioration model and input parameters for reinforcement corrosion," *Structural Concrete*, vol. 13, no. 3, pp. 145–155, 2012.

[3] A. Lowinska-Kluge and T. Blaszczynski, "The influence of internal corrosion on the durability of concrete," *Archives of Civil and Mechanical Engineering*, vol. 12, no. 2, pp. 219–227, 2012.

[4] K. Osterminski and P. Schießl, "Design model for reinforcement corrosion," *Structural Concrete*, vol. 13, no. 3, pp. 156–165, 2012.

[5] H. Song and V. Saraswalhy, "Corrosion monitoring of reinforced concrete structures—a review," *International Journal of Electrochemical Science*, vol. 2, pp. 1–28, 2007.

[6] C. Andrade, C. Alonso, J. Gulikers et al., "Test methods for on-site corrosion rate measurement of steel reinforcement in concrete by means of the polarization resistance method," *Materials and Structures*, vol. 37, no. 273, pp. 623–643, 2004.

[7] G. Sposito, P. Cawley, and P. B. Nagy, "Potential drop mapping for the monitoring of corrosion or erosion," *NDT and E International*, vol. 43, no. 5, pp. 394–402, 2010.

[8] "*ASTM C876-91*: standard test method for half-cell potentials of uncoated reinforcing steel in concrete," 1999.

[9] J. Hoła and M. Ksiazek, "Research on usability of sulphur polymer composite for corrosion protection of reinforcing steel in concrete," *Archives of Civil and Mechanical Engineering*, vol. 9, no. 1, pp. 47–59, 2009.

[10] M. Kosior-Kazberuk and W. Jezierski, "Evaluation of concrete resistance to chloride ions penetration by means of electric resistivity monitoring," *Journal of Civil Engineering and Management*, vol. 11, no. 2, pp. 109–114, 2005.

[11] J. Hoła and K. Schabowicz, "State-of-the-art non-destructive methods for diagnostic testing of building structures—anticipated development trends," *Archives of Civil and Mechanical Engineering*, vol. 10, no. 3, pp. 5–18, 2010.

[12] K. Liam, S. Roy, and D. Northwood, "Chloride ingress measurements and corrosion potential mapping study of a 24-year-old reinforced concrete jetty structure in a tropical marine environment research," *Magazine of Concrete Research*, vol. 44, pp. 205–215, 1992.

[13] J. Flis, H. W. Pickering, and K. Osseo-Asare, "Assessment of data from three electrochemical instruments for evaluation of reinforcement corrosion rates in concrete bridge components," *Corrosion*, vol. 51, no. 8, pp. 602–609, 1995.

[14] M. G. Grantham, B. Herts, and J. Broomfield, "The use of linear polarisation corrosion rate measurements in aiding rehabilitation options for the deck slabs of a reinforced concrete underground car park," *Construction and Building Materials*, vol. 11, no. 4, pp. 215–224, 1997.

[15] V. Žvica, "Possibility of improvement of potentiodynamic method for monitoring corrosion rate of steel reinforcement in concrete," *Bulletin of Materials Science*, vol. 24, no. 5, pp. 555–558, 2001.

[16] S. Erdogdu, I. L. Kondratova, and T. W. Bremner, "Determination of chloride diffusion coefficient of concrete using open-circuit potential measurements," *Cement and Concrete Research*, vol. 34, no. 4, pp. 603–609, 2004.

[17] N. Bowler and Y. Huang, "Electrical conductivity measurement of metal plates using broadband eddy-current and four-point methods," *Measurement Science and Technology*, vol. 16, no. 11, pp. 2193–2200, 2005.

[18] T. Gorzelańczyk, "Moisture influence on the failure of self-compacting concrete under compression," *Archives of Civil and Mechanical Engineering*, vol. 11, no. 1, pp. 45–60, 2011.

[19] R. Polder, C. Andrade, B. Elsener et al., "Test methods for on site measurement of resistivity of concrete," *Materials and Structures*, vol. 33, pp. 603–611, 2000.

[20] B. Elsener, L. Bertolini, P. Pedeferri, and R. Polder, *Corrosion of Steel in Concrete—Prevention, Diagnosis, Repair*, Wiley-Vch, Berlin, Germany, 2004.

[21] S. Feliu, J. A. Gonzalez, S. Feliu, and M. C. Andrade, "Confinement of the electrical signal for in situ measurement of polarization resistance in reinforced concrete," *ACI Materials Journal*, vol. 87, no. 5, pp. 457–460, 1990.

[22] G. K. Glass, C. L. Page, and N. R. Short, "Factors affecting the corrosion rate of steel in carbonated mortars," *Corrosion Science*, vol. 32, no. 12, pp. 1283–1294, 1991.

[23] W. López, J. A. González, and C. Andrade, "Influence of temperature on the service life of rebars," *Cement and Concrete Research*, vol. 23, no. 5, pp. 1130–1140, 1993.

[24] W. Morris, A. Vico, M. Vazquez, and S. R. de Sanchez, "Corrosion of reinforcing steel evaluated by means of concrete resistivity measurements," *Corrosion Science*, vol. 44, no. 1, pp. 81–99, 2002.

[25] J. Zhang, P. J. M. Monteiro, and H. F. Morrison, "Noninvasive surface measurement of corrosion impedance of reinforcing bar in concrete—part 1: experimental results," *ACI Materials Journal*, vol. 98, no. 2, pp. 116–125, 2001.

[26] J. Zhang, P. J. M. Monteiro, H. F. Morrison, and M. Mancio, "Noninvasive surface measurement of corrosion impedance of reinforcing bar in concrete—part 3: effect of geometry and material properties," *ACI Materials Journal*, vol. 101, no. 4, pp. 273–280, 2004.

[27] B. Elsener, C. Andrade, J. Gulikers, R. Polder, and M. Raupach, "Half-cell potential measurements—potential mapping on reinforced concrete structures," *Materials and Structures*, vol. 36, no. 261, pp. 461–471, 2003.

[28] L. Sadowski, "New non-destructive method for linear polarisation resistance corrosion rate measurement," *Archives of Civil and Mechanical Engineering*, vol. 10, no. 2, pp. 109–116, 2010.

[29] S. Millard and L. Sadowski, "Novel method for linear polarisation resistance corrosion measurement," *e-Journal of Nondestructive Testing & Ultrasonics*, vol. 14, 2009.

[30] R. Vedalakshmi, L. Balamurugan, V. Saraswathy, S. H. Kim, and K. Y. Ann, "Reliability of Galvanostatic Pulse Technique in assessing the corrosion rate of rebar in concrete structures: laboratory vs field studies," *KSCE Journal of Civil Engineering*, vol. 14, no. 6, pp. 867–877, 2010.

[31] J. Rhazi, "Test method for evaluating asphalt-covered concrete bridge decks using ground penetrating radar," in *PIERS Proceedings*, Marrakesh, Morocco, 2011.

[32] J. Bungey and S. Millard :, *Testing of Concrete in Structures*, Chapman & Hall, Glasgow, UK, 1996.

[33] E. Bardal and J. Drugli :, "Corrosion detection and diagnosis," in *Materials Science and Engineering, Vol. 3, Encyclopedia of Life Support Systems*, R. D. Rawlings, Ed., 2004.

[34] W. Elkey and E. Sellevold, *Electrical Resistivity of Concrete*, Publication no. 80, Norwegian Road Research Laboratory, Oslo, Norway, 1995.

[35] L. Sadowski, "Non-destructive investigation of corrosion current density in steel reinforced concrete by artificial neural networks," *Archives of Civil and Mechanical Engineering*, vol. 13, no. 1, pp. 104–111, 2013.

13

Mechanical Properties of Recycled Concrete in Marine Environment

Jianxiu Wang,[1] **Tianrong Huang,**[1] **Xiaotian Liu,**[1] **Pengcheng Wu,**[2,3] **and Zhiying Guo**[2]

[1] *College of Civil Engineering, Tongji University, Shanghai 200092, China*
[2] *College of Marine Environment and Engineering, Shanghai Maritime University, Shanghai 200135, China*
[3] *Shanghai International Shipping Service Center Development Co., Shanghai 200120, China*

Correspondence should be addressed to Jianxiu Wang; 260070860@qq.com

Academic Editors: P. Melin and S. Torii

Experimental work was carried out to develop information about mechanical properties of recycled concrete (RC) in marine environment. By using the seawater and dry-wet circulation to simulate the marine environment, specimens of RC were tested with different replacement percentages of 0%, 30%, and 60% after immersing in seawater for 4, 8, 12, and 16 months, respectively. Based on the analysis of the stress-strain curves (SSCs) and compressive strength, it is revealed that RC' peak value and elastic modulus decreased with the increase of replacement percentage and corroding time in marine environment. And the failure of recycled concrete was speeded up with more obvious cracks and larger angles of 65° to 85° in the surface when compared with normal concrete. Finally, the grey model (GM) with equal time intervals was constructed to investigate the law of compressive strength of recycled concrete in marine environment, and it is found that the GM is accurate and feasible for the prediction of RC compressive strength in marine environment.

1. Introduction

The use of recycled materials as an aggregate in concrete has become popular recently in terms of reducing the consumption of natural aggregate and for the environmental advantage of the disposal of waste materials. To make this technology feasible, a significant amount of experimental works has been conducted. And it has proved that some properties of recycled concrete may be generally lower than those of normal concrete, but they are still sufficient for practical application in some constructions and buildings [1, 2].

The most important mechanical properties of recycled concrete are the compressive strength, the tensile and the flexural strengths, the bond strength, and elastic modulus of such concrete. For the peak value of e stress-strain curve that yields the compressive strength, several investigations have been performed for the stress-strain relation of recycled concrete in recent years. Bairagi et al. [3] discovered that similar trends existed in the stress-strain curves of recycled concrete, and the curvature of each curve gradually improved

with the increase of replacement percentage. Topcu [4] found in his investigation that the values of compressive strength, toughness, plastic energy capacity and elastic energy, and the elastic modulus decrease with the increase of recycled coarse aggregated amount. Rqhl and Atkinson [5] reached a conclusion that the peak strain increases as the recycled aggregates increase after they investigated the complete stress-strain curve of recycled concrete with different replacement content. Xiao et al. [6] found that the failure mode of recycled aggregate concrete is a shear mode, and the replacement percentage has a considerable influence on the stress-strain curves of recycled aggregate concrete. After a series of tests, Adom-Asamoah and Afrifa [7] discovered that the trends in the development of compressive and bending strengths of plain phyllite concrete were similar to those of conventional concrete, but the compressive and bending strengths of phyllite concrete mixes were on the average 15–20% lower than those of the corresponding granite concrete mixes. Somna et al. [8] learned that the modulus of elasticity of recycled aggregate concrete with and without ground bagasse ash was lower than that of conventional concrete

TABLE 1: Physical properties of NCA and RCA.

Physical index	NCA	RCA
Grading (mm)	5–32.5	5–32.5
Bulk density (kg/m^3)	1466	1305
Apparent density (kg/m^3)	2812	2498
Water absorption (%)	0.45	9.15
Crush index (%)	4.12	14.9

FIGURE 1: Test setup.

by approximately 19% after their investigation. However, there are few reports which were concerned about the mechanical properties of recycled concrete in marine environment until now, which prohibits a wider application of recycled concrete in the practical design of civil engineering structures.

In this study, experiments are conducted to provide a comprehensive analytical evaluation of the mechanical properties of RC in marine environment. The results are compared with normal concrete in normal environment. The stress-strain relations of RC with different replacement content and corroding time are explicitly analyzed, and the elastic modulus and failure behaviors are also included. And a grey model (GM) with equal time intervals was constructed to investigate the decaying law of RC compressive strength in marine environment, by which the durability of RC in marine environment was revealed. The results presented in this paper are significant to efficiently use recycled concrete in practical engineering in marine environment.

2. Experimental Descriptions

2.1. Materials. Ordinary Portland cement with 28d compressive strength of 42.5 MPa was used in this research. And the fine aggregate used was river sand with a fineness modulus of 2.8. The coarse aggregate includes natural coarse aggregate (NCA) and recycled coarse aggregate (RCA) obtained from the building demolition in Pudong Avenue, Shanghai, China. Their physical properties are shown in Table 1.

2.2. Mix Proportions. Ye et al. [9] suggest that common mixing methods are only fit for the recycled concrete C25 and the lower strength grades in China (C25 represent that the compressive strengths is not less than 25 MPa after 28 days of standard curing). Therefore, this research was using the C25 mix proportion, and the water/cement ratio was kept constant as 0.55. The mixture was divided into three groups. The main difference between these three groups is the replacement percentage, which is 0%, 30%, and 60%, respectively. The case of replacement percentage 0% is normal concrete, which serve as the reference concrete. The mix proportions of concretes are shown in Table 2.

2.3. Preparation of Specimens. The preparations of specimens were performed in the Laboratory for Concrete Material Research at Shanghai Maritime University in Shanghai,

China. All mixings were conducted under laboratory conditions. The sand, cement, and coarse aggregates were placed and dry mixed for about 2 minutes before water was added. After 3 minutes of mixing followed water was added, a slump test was run to determine its workability. The mixture in each group was cast in 100 × 100 × 100 mm cubes in three steel moulds and then compacted on vibration table. They were demolded a day after casting and were cured in a fog room (20 ± 2°C, 95% relative humidity) for 28 days. The cube specimens were used to obtain the cube compressive strength of the RAC.

After 28-day curing, all specimens are submerged in the seawater for 8 hours and out for 16 hours every day, which is to simulate the tidal zone in marine environment. The duration of this dry-wet circulation lasted for 4, 8, 12, and 16 months, respectively. The specimens were cleaned up and dried naturally when the circulation ended. And each group to be tested includes three specimens in the same curing condition. The seawater used was picked up from the Huanghai Sea, which is one of the four key seas of China. Its content is listed in Table 3 [10].

2.4. Test Setup and Test Method. The loading setup as shown in Figure 1 was microcomputer-controlled electrohydraulic servotester. During the experiment, the axial compression and the vertical deformation of the test specimens were automatically collected by the computer installed. The measured deformation is the deformation of the top of the specimen. Each specimen was preloaded before the actual loading in order to lessen the impacts on the test results due to the loose of the specimen end. While preloading, 30–40% of the estimated peak loading (based on the test results for the cube compressive strength) is applied, and the loading is repeated three times.

3. Test Results and Discussion

3.1. Stress-Strain Curves. The typical stress-strain curves (SSCs) of RC in marine environment with different RCA contents and corroding time are shown in Figure 2.

TABLE 2: Mix proportions of concretes (kg/m^3).

No.	Replacement percentage	Water/cement	Cement	Sand	NCA	RCA	Mixing water
NC	0	0.55	425	520	1305	—	234
RC-30	30	0.55	425	500	874	375	234
RC-60	60	0.55	425	480	496	745	234

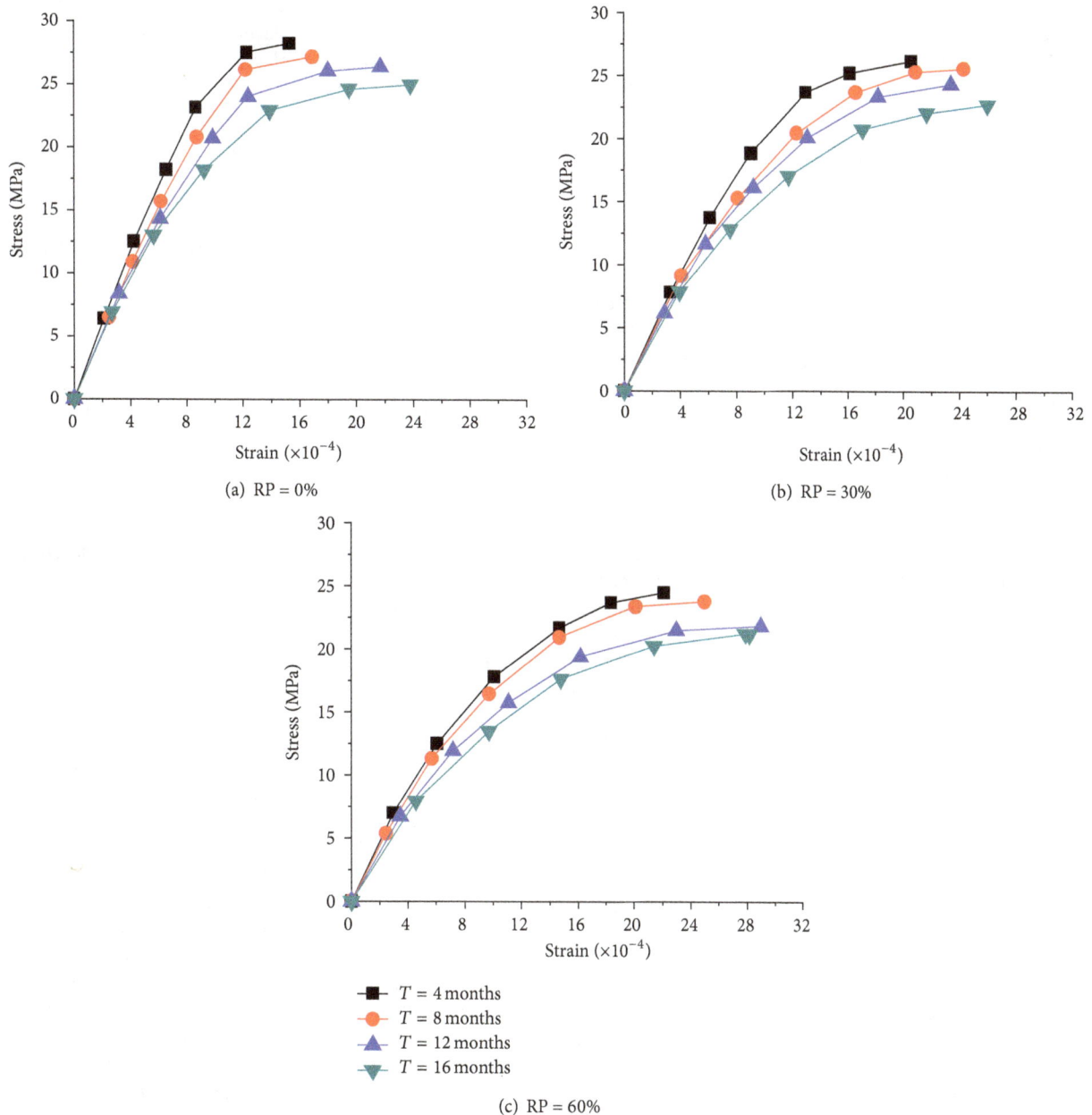

(a) RP = 0%

(b) RP = 30%

(c) RP = 60%

FIGURE 2: Typical stress-strain curves of RC in marine environment.

Figure 2 illustrates that the shape of the stress-strain curve for all the RCs in marine environment was similar to that of the natural aggregate concrete, irrespective of the RCA replacement percentage and corroding time, which leads to the conclusion that the theory of plasticity would be still suitable for structural design process. Roughly speaking, the SSC includes three characteristic parts. The first part is the linear portion; the second part is the nonlinear portion of the ascending branch, and the third part represents the descending branch.

Another notable fact of the SSC is that the RCA replacement percentage and corroding time have remarkable influences on the SSC of RC. The ultimate stress of the SSC decreases with the increasing amount of replacement

TABLE 3: Content of seawater (g/L).

Content	NaCl	$MgCl_2$	$MgSO_4$	$CaSO_4$	K_2SO_4	$CaCO_3$
Amount	27.2	3.8	1.7	1.2	0.9	0.1

TABLE 4: Decreasing of compressive strength with corroding time.

Samples	Compressive strength (MPa) (%)		
	NC	RC-30	RC-60
4 months	28.92 (100)	27.34 (100)	25.79 (100)
8 months	28.39 (98)	26.66 (98)	24.99 (98)
12 months	26.86 (93)	25.70 (94)	23.43 (91)
16 months	25.80 (89)	23.49 (86)	22.11 (86)

percentage. And the ultimate stress of the SSC decreases with the corroding time for the same series, which means that the compressive strength decreases with the corroding time. This also demonstrates that the plastic deformation and residual strength of RC decreases, and the destruction process accelerates with the increase of replacement percentage and corroding time.

3.2. Compressive Strength.
Based on the test procedure in 2.3, every group has three peak values, and its mean value yields the compressive strength. The results of all tests are shown in Figure 3.

Figure 3 indicates that the compressive strength of RC decreases with the replacement percentage. The decreasing range of compressive strength is 1.16 to 2.31 MPa at different corroding time when the replacement percentage increases from 0% to 30%. And this decreasing range goes up from 3.13 to 3.69 MPa when the replacement percentage increases from 0% to 60%. Because the NC is normal concrete and RC-30 is recycled concrete with its replacement percentage as 30%, it can be found that the suitable replacement percentage should not exceed 30% in marine environment, which agreed with those obtained by Kasai [11].

Meanwhile, the compressive strength of RC decreases gradually as the corroding time increases. It presents the process of the decreasing of RC in marine environment quantitatively in Table 4.

From Table 4, it can be seen that the compressive strength decreases at about 2% and not less than 25 MPa when the corroding time is within 8 months. And when the time increases to 12 months and 16 months, the compressive strength becomes smaller than 25 MPa with a decreasing range of 4% to 8%. It can be interpreted that the destruction process accelerates after internal bonding is gradually destroyed. Therefore, the corroding of seawater cost, time, and it has promising future for the application of recycled concrete.

3.3. Modulus of Elasticity.
The elastic modulus E_c of the recycled concrete was determined from compressive strength by the following empirical equation [12]:

$$E_C = 5.639\sqrt{f_c} - 4.952, \tag{1}$$

where E_C represents the modulus of elasticity of concrete (GPa); f_c is the compressive strength of concrete (MPa).

The elastic modulus of RC is shown in Figure 4, versus different replacement percentage, and Figure 5, versus different corroding time (T and RP represent corroding time and replacement percentage, resp.).

Figure 4 shows that the elastic modulus of the RC in marine environment is lower than that of normal concrete (i.e., RP = 0%), and it decreases with increasing replacement percentage. When the replacement percentage is 60%, the elastic modulus is reduced by about 7.5%. This is caused by the application of the RAC with a lower elastic modulus than that of the natural coarse aggregate.

And Figure 5 indicates that the corroding of seawater has considerable influence on the elastic modulus of the RC. The elastic modulus of the RC in marine environment is decreasing with the corroding time, and its decreasing accelerates when the corroding time increases. When the corroding time is 8 months, the elastic modulus drops by 2%, but the elastic modulus is reduced by 9% when the corroding time increases to 16 months. Thus, it can be seen that the elastic modulus decreases gradually in marine environment as the seawater penetrates into the inner of recycled concrete.

3.4. Failure Behavior

3.4.1. Normal Concrete.
In the early stage of loading, the test specimens did not show any cracks. As the compression loading increases, small vertical microcracks were gradually forming in the surface of test specimens. When reaching the peak stress, several discontinuous short vertical cracks appeared, and they combined into inclined macrocracks in the end, which means the destruction of the concrete. The inclination angle of the macrocracks with respect to the vertical loading plumb is about 59–62°, which fits well with the research of normal concrete in common environment by Xiao et al. [6].

3.4.2. Recycled Concrete.
As in the normal concretes the specimens of RC did not show any cracks in the early loading stage. However, very short ad thin vertical microcracks appeared when the compression loading exceeded the peak value. By continuing the test, an inclined macrocrack was formed quickly through the specimen, and the load went down in an instance. Sound of cracking was heard for some samples. And some vertical or slightly inclined branch cracks could be seen on some specimens when the loading keeps constant. All the test specimens displayed an inclined failure plane, with an inclination angle of about 65–85° with respect to the loading plumb. The inclination angle of the failure plane of RC in marine environment is not only significantly larger than that of the normal concrete; but also larger than recycled concrete in normal condition. Therefore, the plastic deformation of RC in marine environment is less than that of normal concrete and concrete in normal condition. The typical failure process of RC in marine environment is shown in Figure 6.

(a) $T = 4$ months

(b) $T = 8$ months

(c) $T = 12$ months

(d) $T = 16$ months

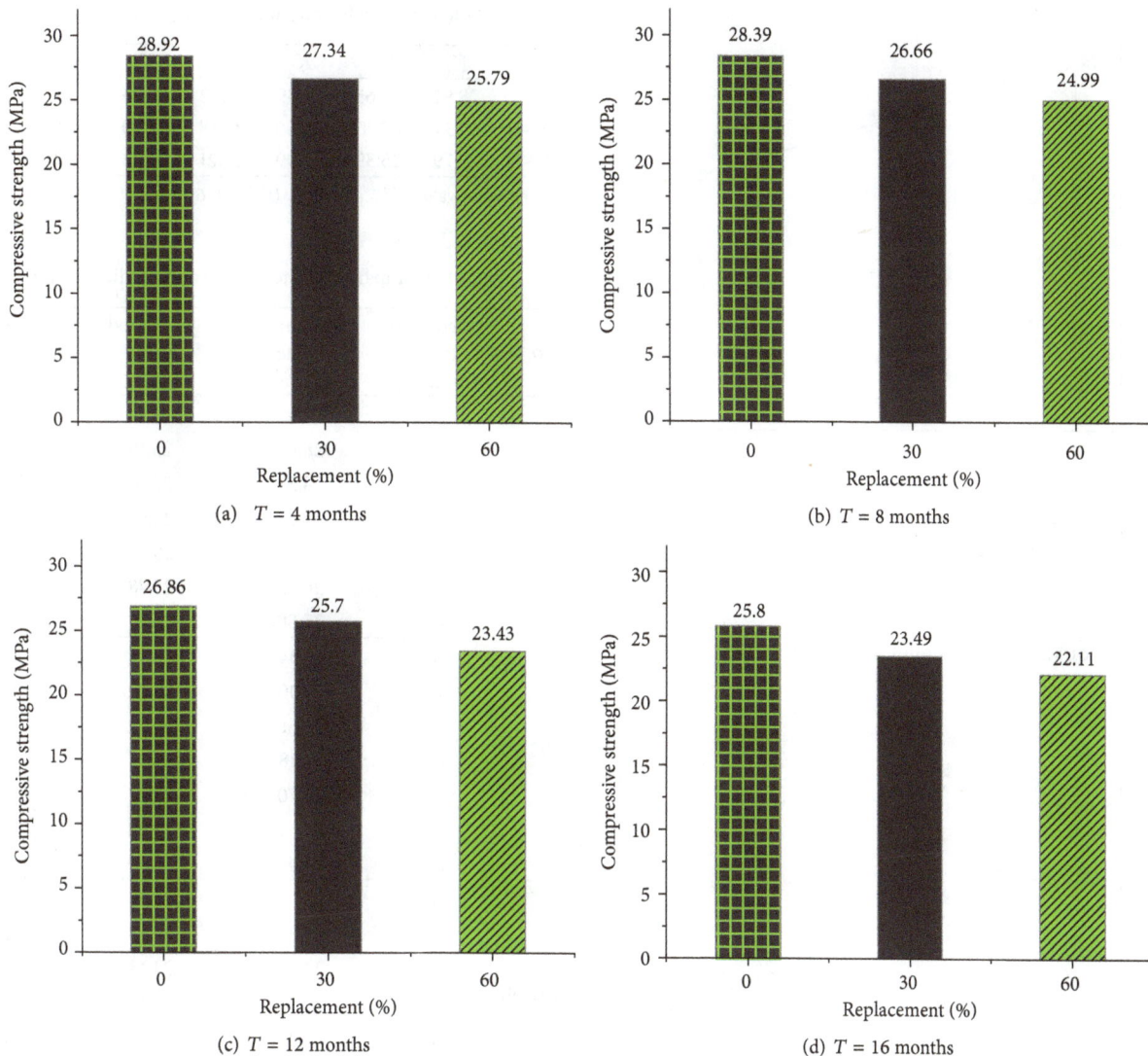

FIGURE 3: Compressive strength of RC in marine environment.

4. Application of Grey Model (GM) in Prediction

4.1. Information about GM. Grey model (GM) was developed by Professor Deng in the 1980s. The basic idea of the grey theory is like this. (1) Measured data are processed by accumulation so that the influence of the stochastic factor in the sequence of the data is decreased, and the internal law in the sequence of the data becomes obvious. (2) The sequence of the data becomes a variable in the grey model, which has a combination of the differential, difference, and approximate index [12]. The steps for building the grey model are given in the appendix.

The grey theory does not require a large amount of data and the typical distributing order, but it has high prediction precision; thus, it is widely used for prediction in agriculture, geology, and civil engineering, and so forth [13–17]. All the results indicate that the grey theory can predict reasonably

well compared to the measured value, and the difference is in an acceptable range.

The decaying of compressive strength of RC is a complex process in marine environment, which is affected by many factors. This process cannot be explicitly expressed by a mathematical formula. But it is not totally unknown like a "black box" whose internal structure, parameters, and characteristics are unknown. The grey theory model is applicable under these conditions and had been successfully used for prediction of compressive strength of normal concrete in seawater environment in 2004 by Lin and others [18]. However, few researchers have used GM to predict compressive strength of recycled concrete in marine environment up to now.

4.2. Constitution of GM and Prediction. Based on test results of compressive strength, the data used to build the GM is listed in Table 5.

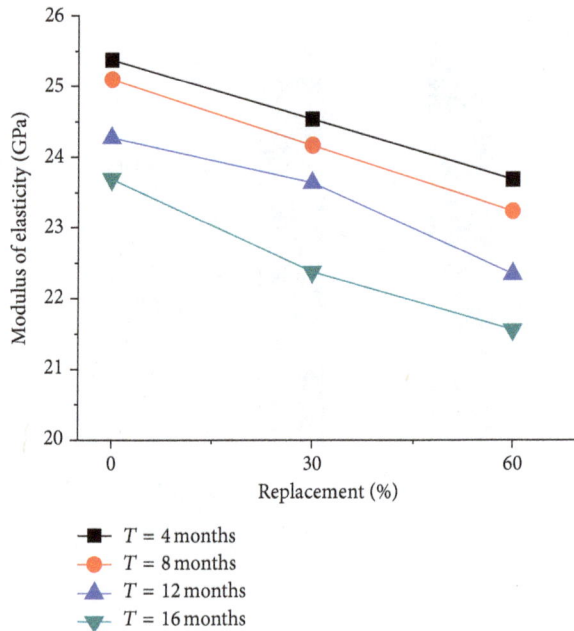

FIGURE 4: Elastic modulus with different RP.

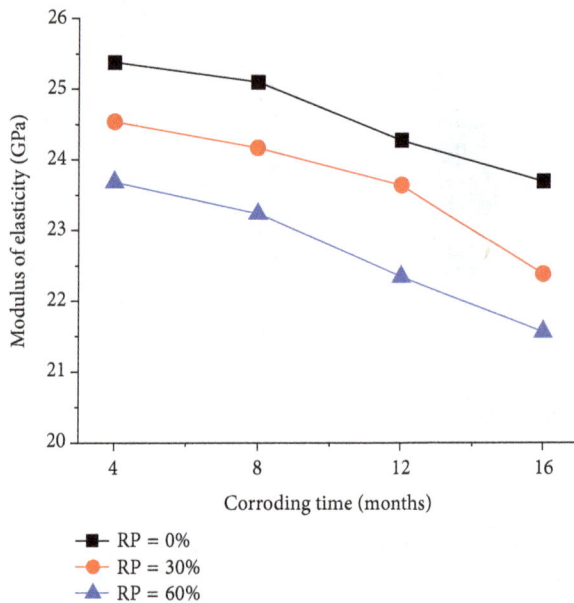

FIGURE 5: Elastic modulus with different CT.

According to Table 5 and the steps of the appendix, the grey model for NC, RC-30, and RC-60 is constructed as follows, respectively:

$$\hat{x}^{(1)}_{(k+1)_A} = -1312.84113 \exp\left(-0.02219k\right) + 1341.7611,$$

$$\hat{x}^{(1)}_{(k+1)_B} = -1036.1320 \exp\left(-0.02675k\right) + 1063.4687, \quad (2)$$

$$\hat{x}^{(1)}_{(k+1)_C} = -906.8300 \exp\left(-0.02858k\right) + 932.6167.$$

The prediction value is obtained by the following equation:

$$\hat{x}^{(0)}_{(k)} = \hat{x}^{(1)}_{(k)} - \hat{x}^{(1)}_{(k-1)} \quad (k = 2, 3, \ldots, N). \quad (3)$$

TABLE 5: Data for building the grey model (MPa).

No.	K_1	K_2*	K_3	K_4*	K_5	K_6*	K_7
NC	28.92	28.65	28.39	27.62	26.86	26.33	25.80
RC-30	27.34	27.00	26.66	26.18	25.70	24.60	23.49
RC-60	25.79	25.39	24.99	24.21	23.43	22.77	22.11

*Is the mean value of k_{i-1} and k_{i+1} ($i = 2, 4, 6$).

TABLE 6: Comparison of test value and predicted values.

No.	Corroding time (months)	Measured value (MPa)	Predicted value (MPa)	Relative error (%)
NC	4	28.92	28.81	0.38
	6	28.65	28.18	1.67
	8	28.39	27.56	2.92
	10	27.62	26.95	2.43
	12	26.86	26.36	1.86
	14	26.33	25.78	2.08
	16	25.80	25.22	2.25
RC-30	4	27.34	27.34	0
	6	27.00	26.62	1.41
	8	26.66	25.92	2.78
	10	26.18	25.24	3.59
	12	25.70	24.57	4.39
	14	24.60	23.92	2.72
	16	23.49	23.29	0.85
RC-60	4	25.79	25.55	0.93
	6	25.39	24.83	2.2
	8	24.99	24.13	3.44
	10	24.21	23.45	3.14
	12	23.43	22.79	2.73
	14	22.77	22.14	2.72
	16	22.11	21.52	2.67

By using the above equations, predicted value of compressive strength in marine environment was acquired, and the comparison of test value and predicted value was listed in Table 6.

Table 6 shows that predicted value is well confirming to the test value, which demonstrates that grey model is feasible for the prediction of compressive strength of recycled concrete in marine environment. The predicted value is very similar to the test value with a low relative error below 5%. In order to know long-term effect of seawater, the predicted GM value was also applied to draw regressive equation; thus, the law of decaying of compressive strength of RC in marine environment was illustrated in Figure 7 (R^2 represents the correlation coefficient).

Figure 7 shows that the compressive strength of RC decreases with corroding time in marine environment linearly. Its slope is about -0.29 to -0.33. The compressive strength of RC is decreasing coherently, which is also testifying to the validity of the test results. According to

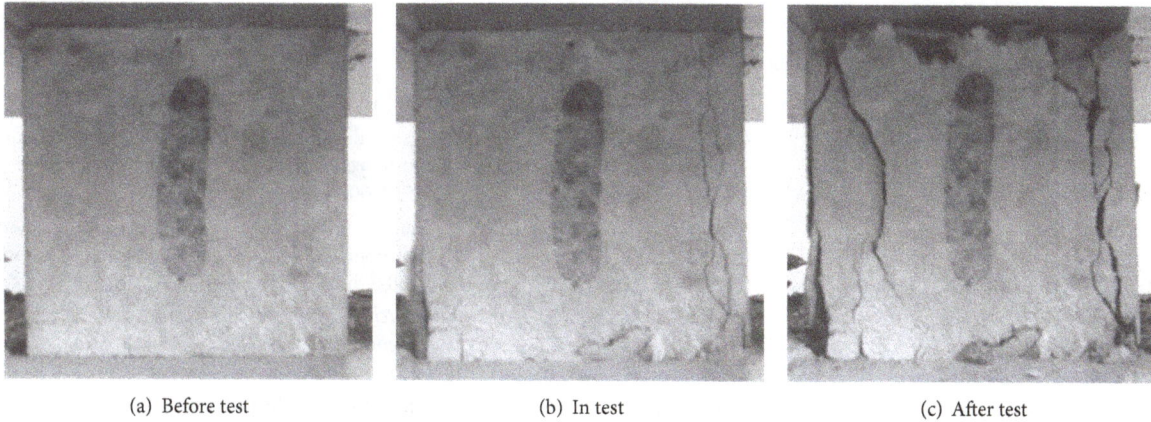

(a) Before test

(b) In test

(c) After test

FIGURE 6: Failure process of recycled concrete in marine environment.

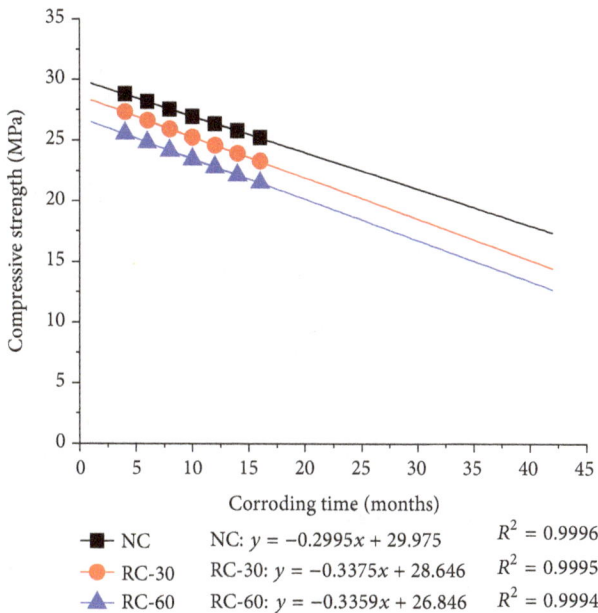

FIGURE 7: Regressive fitting of RC compressive strength in marine environment.

the regressive equation, it was found that the compressive strength of NC, RC-30, and RC-60 drops by 33.2%, 38.2%, 40.8%, respectively, after 40-month corroding in marine of environment. This has reminded us that prevention measures should be taken to improve the durability of recycled concrete in marine environment.

5. Conclusions

In this paper, test results for the mechanical properties of recycled concrete in marine environment are presented and discussed. From this investigation, the conclusions can be drawn as follows.

(1) The shape of the stress-strain curve for all the recycled concrete in marine environment was similar to that of the natural aggregate concrete, irrespective of the RCA replacement percentage and corroding time. However, the RCA replacement percentage and corroding time have remarkable influences on the SSC of RC.

(2) The compressive strength of RC decreases with the replacement percentage, and the suitable replacement percentage should not exceed 30% in marine environment.

(3) The compressive strength of RC also decreases with the corroding time. It decreases at 2% when the corroding time is within 8 months, and it decreases from 4% to 8% when corroding time exceeds 8 months.

(4) The elastic modulus of the RC in marine environment is lower than that of normal concrete, and it decreases with increasing replacement percentage and corroding time. The elastic modulus is reduced by about 7.5% when the replacement percentage is 60%, and it dropped by 2% and 9% when the corroding time is within or over 8 months, respectively.

(5) The failure behavior of normal concrete in marine environment is similar to that in common environment. But the failure process of RC in marine environment takes less time with a larger inclination angle of 65–85° between the failure plane and the loading plumb.

(6) The grey model is feasible for the prediction of RC compressive strength in marine environment. Its accuracy is very high with a relative error less than 5%. The regressive equation shows that the compressive strength of NC, RC-30, and RC-60 drop by 33.2%, 38.2%, 40.8%, respectively, after 40-month-corroding in marine of environment.

Appendix

Constitution of Grey Model

Let the sequence of initial time-interval data be:

$$x^{[0]} = \left\{ x^{(0)}(t_1), x^{(0)}(t_2), \ldots, x^{(0)}(t_n) \right\} \qquad \text{(A.1)}$$

with time step $t_{i+1} - t_i = k = $ constant and $i = 1, \ldots, n-1$.

By accumulative addition, a smooth sequence is obtained:

$$x^{[1]} = \left\{ x^{(1)}(1), x^{(1)}(2), \ldots, x^{(1)}(n) \right\}, \qquad \text{(A.2)}$$

where $x^{[1]}(k) = \sum_{i=1}^{k} x^{(1)}(t_i)$.

Based on the above expression, we can obtain the differential equation

$$\frac{dx^{(1)}}{dt} + ax^{(1)} = c, \qquad \text{(A.3)}$$

where $c = \sum_{i=1}^{n-1} b_i x^{(1)}(i+1)$ and the denoting coefficient vector $\hat{a} = [a, c]^T$.

The solution is

$$\hat{x}^{(1)}(k+1) = \left[x^{(0)}(t_1) - \frac{c}{a} \right] \exp(-ak) + \frac{c}{a}, \qquad \text{(A.4)}$$

where coefficients a and c are the parameters to be identified and can be obtained by the least square method.

Acknowledgments

This work was supported by the Shanghai Science and Technology Committee (no. 07ZZ99) and National Natural Science Funding of China (no. 41072205), which is gratefully acknowledged. The authors would like to thank Z. Guo for her contribution to the realization of the experimental works, which were carried out at Laboratory for Concrete Material Research in Shanghai Maritime University. The reviewers were also gratefully acknowledged for their contribution in perfecting this paper.

References

[1] T. C. Hansen, *Recycling of Demolished Concrete and Masonry*, E&FNSPON, London, UK, 1992.

[2] ACI Committee 555, "Removal and reuse of hardened concrete," *ACI Materials Journal*, vol. 99, no. 3, pp. 300–325, 2002.

[3] N. K. Bairagi, K. Ravande, and V. K. Pareek, "Behaviour of concrete with different proportions of natural and recycled aggregates," *Resources, Conservation and Recycling*, vol. 9, no. 1-2, pp. 109–126, 1993.

[4] I. B. Topcu, "Using waste concrete as aggregate," *Cement and Concrete Research*, vol. 25, no. 7, pp. 1385–1390, 1995.

[5] M. Rqhl and G. Atkinson, "The influence of recycled aggregate concrete on the stress-strain relation of concrete," *Darmstadt Concrete*, vol. 26, no. 14, pp. 36–52, 1999 (German).

[6] J. Xiao, J. Li, and C. Zhang, "Mechanical properties of recycled aggregate concrete under uniaxial loading," *Cement and Concrete Research*, vol. 35, no. 6, pp. 1187–1194, 2005.

[7] M. Adom-Asamoah and R. O. Afrifa, "A study of concrete properties using phyllite as coarse aggregates," *Materials and Design*, vol. 31, no. 9, pp. 4561–4566, 2010.

[8] R. Somna, C. Jaturapitakkul, P. Rattanachu, and W. Chalee, "Effect of ground bagasse ash on mechanical and durability properties of recycled aggregate concrete," *Materials & Design*, vol. 36, pp. 597–603, 2012.

[9] J. Ye, L. Zhang, and X. Tang, "The development of recycled concrete techniques," *Sino-Foreign Highway*, vol. 28, no. 5, pp. 232–234, 2008 (Chinese).

[10] A. Ge, *Research on the corroding characteristic of reinforced concrete in seawater [M.S. thesis]*, China Ocean University, Shandong, China, 2004.

[11] Y. Kasai, "Guidelines and the present state of the reuse of demolished concrete in Japan," in *Proceedings of the 3rd International RILEM Symposium on Demolition and Reuse of Concrete and Masonry*, vol. 3, pp. 93–104, 1993.

[12] J. Deng, *Basic Methodology of Grey System*, Publishing House of Huazhong University of Science & Technology, Wuhan, China, 1987.

[13] Z. C. Lin and W. S. Lin, "The application of grey theory to the prediction of measurement points for circularity geometric tolerance," *International Journal of Advanced Manufacturing Technology*, vol. 17, no. 5, pp. 348–360, 2001.

[14] H. H. Wu, A. Y. H. Liao, and P. C. Wang, "Using grey theory in quality function deployment to analyse dynamic customer requirements," *International Journal of Advanced Manufacturing Technology*, vol. 25, no. 11-12, pp. 1241–1247, 2005.

[15] M. Wu, S. Qiu, J. Liu, and L. Zhao, "Prediction model based on the grey theory for tackling wax deposition in oil pipelines," *Journal of Natural Gas Chemistry*, vol. 14, no. 4, pp. 243–247, 2005.

[16] Q. Wu, W. Zhou, S. Li, and X. Wu, "Application of grey numerical model to groundwater resource evaluation," *Environmental Geology*, vol. 47, no. 7, pp. 991–999, 2005.

[17] Y. Q. Tang, Z. D. Cui, J. X. Wang, L. P. Yan, and X. X. Yan, "Application of grey theory-based model to prediction of land subsidence due to engineering environment in Shanghai," *Environmental Geology*, vol. 55, no. 3, pp. 583–593, 2008.

[18] Y. Lin, T. Wang, and M. Dong, "Study on forecast model for strength of concrete in seawater," *China Harbor Engineering*, vol. 16, no. 4, pp. 35–41, 2004 (Chinese).

14

Predicting Subcontractor Performance Using Web-Based Evolutionary Fuzzy Neural Networks

Chien-Ho Ko

Department of Civil Engineering, National Pingtung University of Science and Technology, 1 Shuefu Road, Neipu, Pingtung 912, Taiwan

Correspondence should be addressed to Chien-Ho Ko; fpecount@yahoo.com.tw

Academic Editors: S. Chen and E. Lui

Subcontractor performance directly affects project success. The use of inappropriate subcontractors may result in individual work delays, cost overruns, and quality defects throughout the project. This study develops web-based Evolutionary Fuzzy Neural Networks (EFNNs) to predict subcontractor performance. EFNNs are a fusion of Genetic Algorithms (GAs), Fuzzy Logic (FL), and Neural Networks (NNs). FL is primarily used to mimic high level of decision-making processes and deal with uncertainty in the construction industry. NNs are used to identify the association between previous performance and future status when predicting subcontractor performance. GAs are optimizing parameters required in FL and NNs. EFNNs encode FL and NNs using floating numbers to shorten the length of a string. A multi-cut-point crossover operator is used to explore the parameter and retain solution legality. Finally, the applicability of the proposed EFNNs is validated using real subcontractors. The EFNNs are evolved using 22 historical patterns and tested using 12 unseen cases. Application results show that the proposed EFNNs surpass FL and NNs in predicting subcontractor performance. The proposed approach improves prediction accuracy and reduces the effort required to predict subcontractor performance, providing field operators with web-based remote access to a reliable, scientific prediction mechanism.

1. Introduction

A construction project involves various work items that need to be accomplished by subcontractors, including earthwork, formwork, concrete pouring, plastering, rebar, and mechanical and electrical tasks. Subcontractor performance directly influences project cost, duration, quality, and safety [1–3]. Project success cannot be achieved without appropriate performance on the part of the subcontractors [4]. The general contractor's key responsibility is selecting subcontractors with the capacity to perform the required work [5–7]. When selecting a subcontractor, general contractors frequently use the subcontractor's previous performance as a reference for their future outcome [8]. However, this approach leaves much to be desired and general contractors could benefit significantly from techniques which would allow greater accuracy in predicting subcontractor future performance [9].

Many studies have been devoted to enhancing the performance assessment of construction subcontractors. Ekström et al. [10] used source credibility theory to assess subcontractor performance in architecture/engineering/construction

(AEC) using a weighted rating tool. Mbachu [11] investigated the key criteria for assessing subcontractor performance at the construction stage. Their research found that a subcontractor's previous performance is the most critical criterion for selecting high-performing subcontractors at the prequalification stage and for assessing their performance at the construction stage. Lean construction, a relatively new research area in the construction industry, has also been used to enhance subcontractor performance assessment [12]. Maturana et al. [13] conducted weekly assessments of subcontractor performance, rating quality, schedule fulfillment, safety, and cleanliness in terms of "good," "regular," or "bad." Evaluation results were fed back to the general contractor for continuous improvement based on lean principles.

While most studies have focused on enhancing subcontractor performance assessment, a few investigations have investigated methods of predicting subcontractor performance. Le-Hoai et al. [14] applied multiple regression analysis techniques to integrate significant variables including subcontractor selection to predict project length. Park [15] investigated critical success factors for whole life performance assessment, placing the identified factors into a criteria matrix to aid decision making for selecting subcontractors at the bid stage. Another investigation conducted by Elazouni and Metwally [16] developed a decision support system that assigns work items to subcontractors under constraints and predicts the project's final profit. Previous studies have considered subcontracting and subcontractor performance as factors for project success, but construction projects involve a variety of subcontractors. A successful outcome relies directly on the aggregated success of these subcontractors. One way to achieve project success is to predict subcontractor performance and base subcontractor selection on their predicted ability to adequately perform the work.

The process of predicting subcontractor performance is complex, full of uncertainty, and highly contextualized, and it thus relies on decisions by experts [17]. Artificial intelligence (AI) is concerned with building computer systems that solve problems intelligently by emulating human behavior [18], making AI suitable for predicting subcontractor performance. The most popular AI paradigms are genetic algorithms (GAs), fuzzy logic (FL), and neural networks (NNs) [19]. These three techniques simulate different aspects of biological behaviors. The GA is a stochastic searching process based on natural selection and natural genetics [20]; FL simulates high level human decision-making processes [21]; NNs model brain functions [22]. Each method offers certain benefits for problem solving, and combining GAs, FL, and NNs provides potentially combines these benefits to provide a promising direction for predicting subcontractor performance.

The objective of this research is to develop Evolutionary Fuzzy Neural Networks (EFNNs) to predict subcontractor performance. In EFNNs, a floating point codification is used to encode parameters required in Fuzzy Neural Networks (FNNs). A multi-cut-point crossover is adopted to explore the parameters required in NNs and FL. To improve implementation convenience, a web-based system is developed to facilitate decision-making processes.

This research first introduces practices used to predict subcontractor performance in the construction industry. Section 3 explains the development of the EFNNs, with a detailed discussion of evolutionary processes, floating number codifications, and the multi-cut-point crossover operator. Applicability of the EFNNs is validated in Section 4, comparing the performance of EFNNs, FL, and NNs in real cases. Finally, the paper concludes with suggestions for future research directions.

2. Performance Prediction Practice

Historical performance serves as an important indicator for general contractors use in selecting subcontractors [23]. Predicting subcontractor performance can be treated as a process in which previous patterns are applied to the present condition. In this situation, the mapping between the previous behavior and later performance is unknown. In addition, subcontractor performance is affected by various knowable and unknowable factors, such as management ability, site working condition, and subjective assessment [24, 25]. Thus, predicting subcontractor performance is a complex process based on uncertain information, thus requiring the knowledge and experience of experts. Current practice predicts subcontractor performance through the subjective perception of the manager. Diverse backgrounds and work experience may result in significant prediction discrepancies. AI techniques, which involve machine learning and optimization to mimic human decision-making process, may provide a scientific approach to overcome these drawbacks.

3. Evolutionary Fuzzy Neural Networks

3.1. Architecture. Figure 1 shows the EFNNs architecture as a synergism of GAs, FL, and NNs. In EFNNs, NNs are used to learn the complex association between a subcontractor's previous performance and future status from historical data; FL is used to simulate high-level managerial decision making processes; GAs are used to achieve the optimal parameters required in NNs and FL, including distributions of the membership function, NN topology, and defuzzification parameters. Prediction results are stored in the database.

3.2. Adaptation Process. EFNNs optimize the required parameters using GAs. Figure 2 displays the evolution process, which is explained next.

3.2.1. Initializing Population. The EFNNs adaptation process first randomly generates a set of initial solutions. Each solution encodes variables into a floating Fuzzy Neural Network (FNN) string to simulate a natural chromosome. Every FNN string comprises of two segments: an MF substring and an NN substring.

MF Substring. A Summit and Width Representation Method (SWRM) method [26] is used to encode membership functions (MFs) using floating numbers. The SWRM defines the distributions of uneven MFs by its summits and widths as shown in Figure 3. In Figure 3(a), the summits of the MF are su_1 and su_2 while the left and right widths are wi_1 and wi_2. A triangular MF can be regarded as a special case of a trapezoidal MF when $su_1 = su_2$ (see Figure 3(b)). For modeling problems using either trapezoidal MFs or triangular MFs, a complete MF set includes two shoulder MFs (see Figure 3(c)). The complete MF set shown in Figure 3(c) can thus be encoded using the SWRM, as demonstrated in Figure 4.

Using the SWRM, the required length of the floating numbers of MF substring RL^{MF} for encoding MFs is carried out as follows:

$$RL^{MF} = rn^{cMF} \times \left(n^{su} \times rl^{su} + n^{wi} \times rl^{wi}\right), \qquad (1)$$

where rn^{cMF} is the required number of the complete MF sets, n^{su} is the number of summits in a complete MF set, rl^{su} is the required length for a summit depending on the demand, n^{wi} is the number of widths in one complete MF set, and rl^{wi} is the required length for a width depending on the demand.

The mapping from a domain $[lb^x, ub^x]$ to a required length rl^x for variable x can be written as

$$10^{rl^x-1} < \left(ub^x - lb^x\right) \times 10^{rp} \leq 10^{rl^x} - 1, \qquad (2)$$

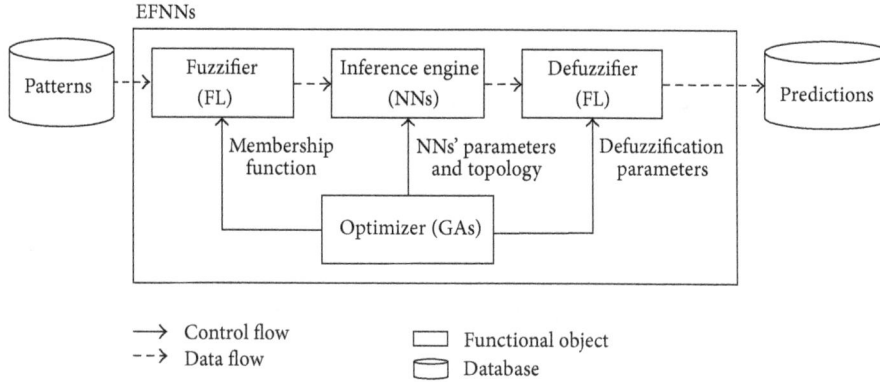

FIGURE 1: Architecture of Evolutionary Fuzzy Neural Networks.

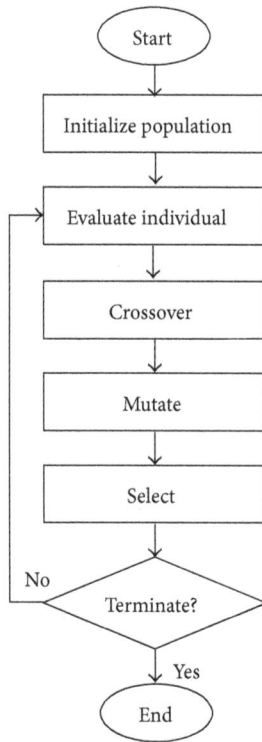

FIGURE 2: Evolutionary process of Evolutionary Fuzzy Neural Networks.

where rp is the required number of places after the decimal point and lb^x and ub^x are the lower and upper bound values of the variable x. Taking log functions on both sides of previous right-hand parts yields

$$rl^x = \left\lceil \frac{\log\left((ub^x - lb^x) \times 10^{rp} + 1\right)}{\log(10)} \right\rceil. \qquad (3)$$

The length of the floating numbers required for variables displayed in Figure 4 can be calculated using (3).

NN Substring. A Block Representation Method (BRM) [26] is used to represent the NN floating numbers. The BRM describes the NN by its topology and network parameters

(see Figure 5). The NN topology consists of the input layer and its neurons, the fuzzification layer and its neurons, many hidden layers and their hidden neurons, and the defuzzification layer and its neurons. The NN parameters include interconnections, weight values, bias values, and the slopes of the activation functions.

The number of hidden layers and their hidden neurons of the NN are randomly generated using the BRM. A random number of hidden layers rn^{hl} is generated in $[lb^{hl}, ub^{hl}]$ where lb^{hl} and ub^{hl} are the lower and upper bounds of the hidden layers. The method then generates rn^{hl} random numbers (rn^{hn}) to determine the hidden neurons of each hidden layer. Each random number rn^{hn} is generated between lb^{hn} and ub^{hn} where lb^{hn} and ub^{hn} denote the lower and upper bounds of the hidden neurons. According to the generated topologies, the BRM calculates the required spaces to represent the NN. The method for encoding NNs is shown in Figure 6.

In Figure 6, the "Subblock A" represents the relationship between the fuzzification layer (front layer) and the first hidden layer (back layer). The height of the subblock (h_A^{sb}) directly indicates the number of hidden neurons in the back layer. Each row of the "Subblock A" represents one neuron of the back layer. The length of the NN substring, L^{NN}, is expressed as

$$L^{NN} = \sum_{i=\text{Sub-block A}}^{\text{Sub-block}(rn^{hl}+1)} \left(h_i^{sb} \times rcn_i^{sb}\right), \qquad (4)$$

where h_i^{sb} is the height of subblock i and rcn_i^{sb} is the required width of subblock i. The required column number (width) of subblock i (noted with rcn_i^{sb}) can be calculated using (5). The length of the floating numbers for the variables can be calculated using (3). Consider

$$rnc_i^{sb} = nn_i^{fl} \times \left(rcn_i^{in} + rcn_i^{we}\right) + rcn_i^{bi} + rcn_i^{as}. \qquad (5)$$

The MF substring encodes the distribution of MFs, and the NN substring encodes the NN parameters. To find the optimum combination of MFs and NNs, the MF substring and NN substring are combined. A complete chromosome, an FNN string L^{FNN}, is defined by (6). Via evolutionary

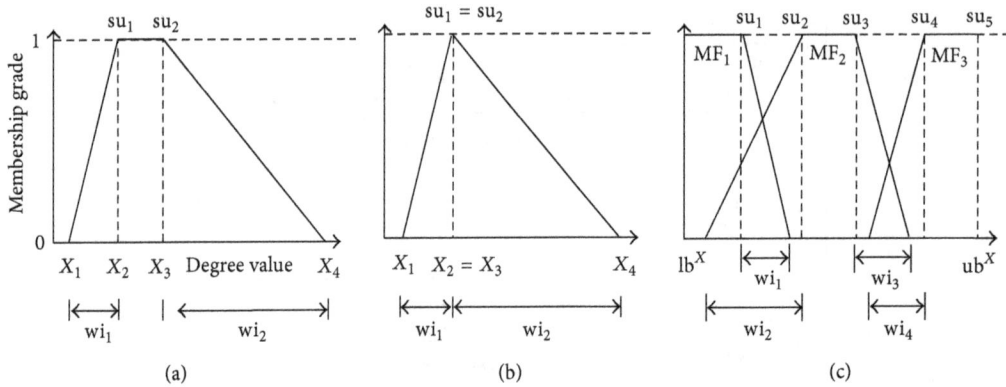

FIGURE 3: Membership functions: (a) trapezoidal MF; (b) special case of trapezoidal MF; (c) complete MF Set.

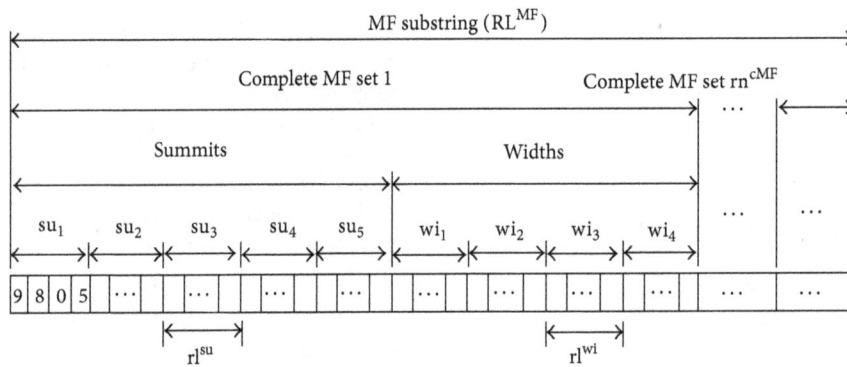

FIGURE 4: Summit and Width Representation Method (SWRM).

processes, the combined chromosome concurrently identifies the optimum decision variables, where

$$L^{\mathrm{FNN}} = \mathrm{RL}^{\mathrm{MF}} + L^{\mathrm{NN}}. \qquad (6)$$

3.2.2. Evaluating Individual Chromosomes. The adaptation process is designed to obtain EFNNs with high accuracy and good generalization properties. The EFNNs accuracy on input patterns can be improved by increasing network complexity. However, an accurate fit of the network to the input patterns does not mean that the overall problem behaviors are captured well [27]. A large network size also entails a higher computational cost and generally suffers from overfitting of data in input patterns and deterioration of generalization properties [28]. Thus, the objective of the adaptation process is to preserve acceptable levels of prediction accuracy using the fittest shapes of MFs with the minimum NN topology and optimum NN parameters. This is posed as an optimization problem. The objective function of the EFNNs is a combination of prediction accuracy and network complexity as follows:

$$v^{\mathrm{ob}} = c^{\mathrm{aw}} \times s^{\mathrm{er}} + c^{\mathrm{cw}} \times \mathrm{mc}, \qquad (7)$$

where v^{ob} is the objective value, c^{aw} is the accuracy weighting coefficient, s^{er} is the error signal, c^{cw} is the complexity weighting coefficient, and mc is the network complexity.

3.2.3. Crossover. The crossover repeatedly exchanges high performance notations in attempting to improve performance. It operates on a pair of parent chromosomes and produces two children by exchanging the parent features. EFNNs use a three-cut-point crossover to exchange the distribution of MFs and NN information, as shown in Figure 7. A complete chromosome consists of two substrings: an MF substring and an NN substring. Two points, noted as a and b, are randomly generated for the MF substrings. Child 1 inherits alleles between the a and b segments of parent 2. Child 2 inherits alleles between the a and b segments of parent 1. Complementary portions of the a and b segments are retained for the other child. To explore the topology and parameters of the NNs, the third cut-point c is randomly generated for the NN substring. The produced children exchange the right-hand features after the cut-point from their parents.

3.2.4. Mutation. The mutation produces spontaneous random changes in various chromosomes, thus protecting against premature loss of important notations. The purpose of mutation is to improve performance by adjusting the value of the summits and widths of MFs, along with interconnections, weights, biases, and activation slopes. It alters one or more genes with a probability (p^{ge}), which is smaller than or equal to the mutation rate (p^{mu}). Mutation operation compares each gene's p^{ge} with p^{mu}. If $p^{\mathrm{ge}} \le p^{\mathrm{mu}}$, then value of the gene is changed to another unrepeated number, as shown in Figure 8.

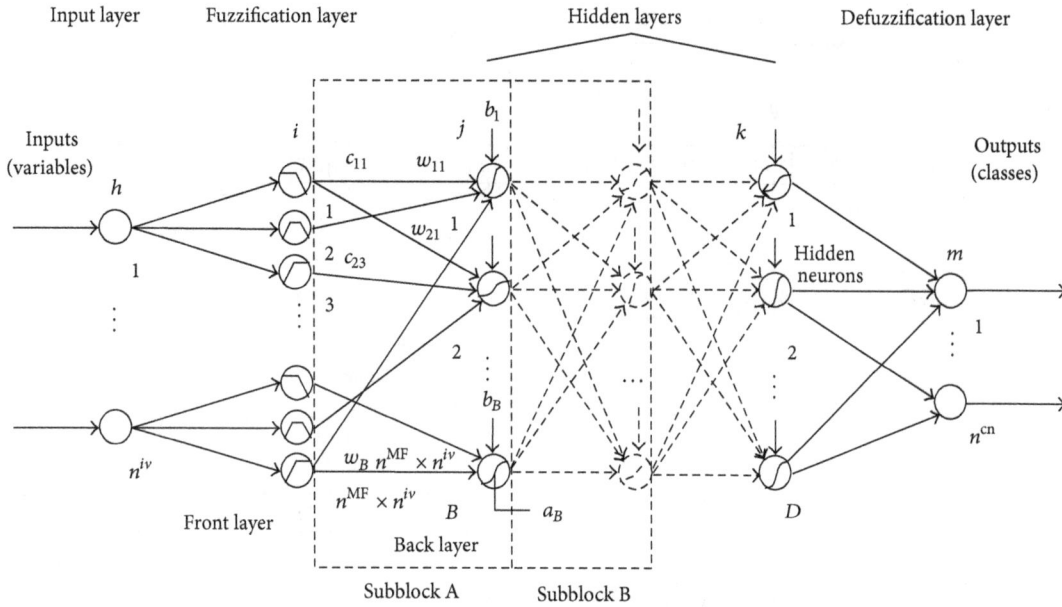

FIGURE 5: Neural networks structure.

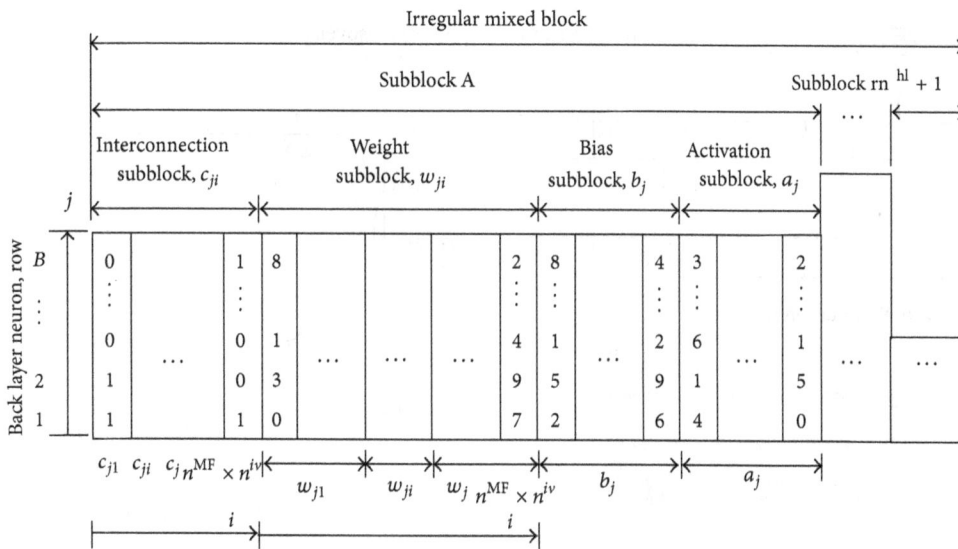

FIGURE 6: Block Representation Method (BRM).

3.2.5. Selection. The selection process emulates the survival-of-the-fittest mechanism found in nature. It selects a new population with respect to the probability distribution based on fitness for survival. The probability distribution is established using the roulette wheel method [29], constructed as follows:

(1) calculate the total fitness for the enlarged sampling space;

(2) calculate the selection probability for each chromosome;

(3) calculate the cumulative probability for each chromosome.

4. Application

4.1. Case Study. To validate feasibility of the proposed EFNNs, a real construction company in Taiwan is studied. Establishing in 1956, the company is ISO 9002 certified, with a capitalization of about 11 million USD. Based on Wu's [23] findings, a subcontractor's previous three performances are used to predict its next performance. Historical subcontractor performance records are extracted from Wu [23] and are shown in Table 1. The 34 subcontractor performances shown in the table are real cases based on 14 subcontractors. Of the 34 input patterns, 22 are used to evolve EFNNs, while the unseen 12 test sets are used to validate the generalization of EFNNs.

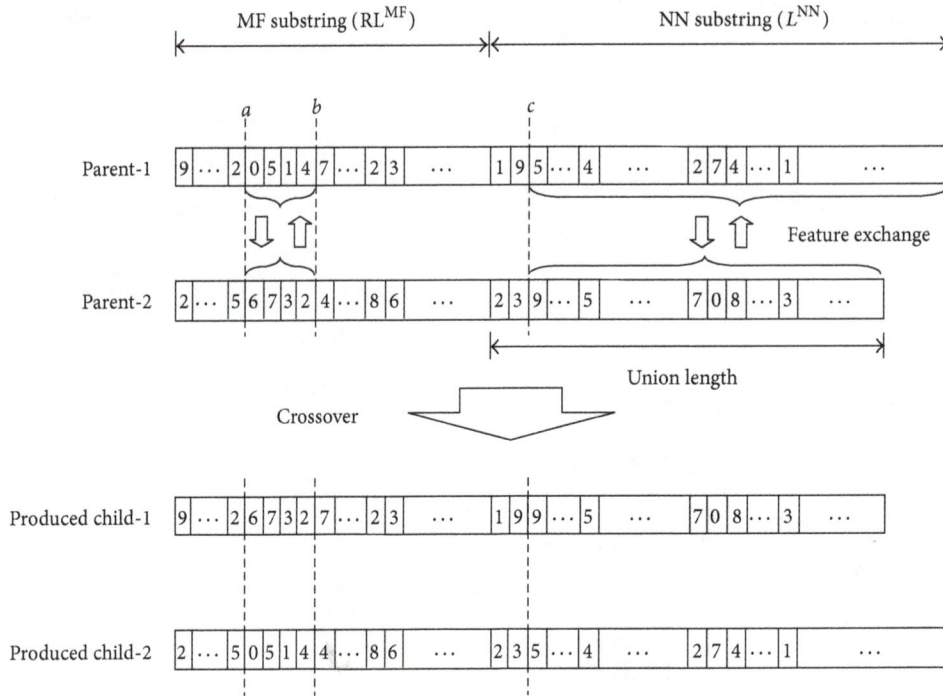

FIGURE 7: Three-cut-points crossover.

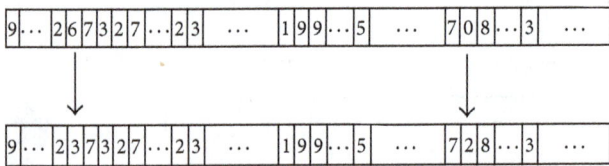

FIGURE 8: Mutation operation.

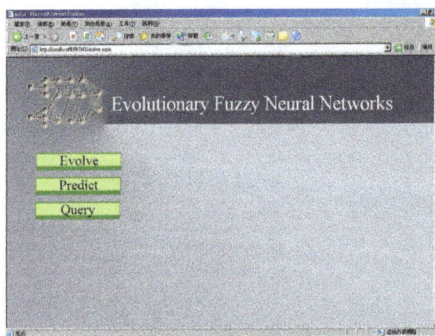

FIGURE 9: Application main interface.

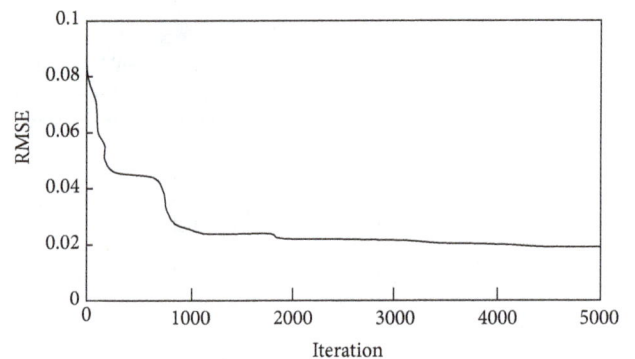

FIGURE 10: Evolutionary process.

FL, and NNs into an integrated network and thus contain parameters of these three AI techniques, which are summarized in Table 2. The system parameters and database can be manipulated using the query module. Figure 10 displays the evolutionary process for the case study. The optimum solution is derived at iteration 4947.

4.3. Subcontractor Performance Prediction. Table 3 compares the generalization ability of the proposed method with that of FL and NNs. The test data are not included in the training process. Prediction accuracy is visualized in Figure 11, which shows that subcontractor performance is predicted more accurately using EFNNs than by using FL and NNs. The performance of each method is calculated using the root mean square error (RMSE). In the table, the generalization ability of the EFNNs outperforms that of the NNs, and EFNNs

4.2. Web-Based Evolutionary Fuzzy Neural Networks. Web-based EFNN software was developed to automate the evolutionary process. The main interface of the web-based system is shown in Figure 9. Three modules are provided in the system. The evolutionary module is used to implement the EFNN evolutionary process. The prediction module can be used to predict subcontractor performance using the network obtained by the evolutionary module. EFNNs fuse GAs,

TABLE 1: Subcontractor performance historical records.

| Pattern no. | Performance | Input | | | Output |
		Last 3	Last 2	Last 1	Normalized performance
Input patterns					
1	80	72	76	76	0.8333
2	86	76	76	80	0.8958
3	74	80	76	76	0.7708
4	70	76	76	74	0.7292
5	68	56	62	66	0.7083
6	66	60	66	68	0.6875
7	66	70	72	68	0.6875
8	58	62	66	60	0.6042
9	56	66	60	58	0.5833
10	80	76	74	76	0.8333
11	86	74	76	80	0.8958
12	88	76	80	86	0.9167
13	76	86	80	80	0.7917
14	70	66	68	66	0.7292
15	70	68	66	70	0.7292
16	76	66	70	70	0.7917
17	74	66	70	76	0.7708
18	76	70	76	74	0.7917
19	80	74	76	76	0.8333
20	66	62	58	62	0.6875
21	68	58	62	66	0.7083
22	76	76	74	76	0.7919
Test patterns					
23	66	62	56	60	0.6875
24	68	56	60	66	0.7083
25	66	60	66	68	0.6875
26	66	66	68	66	0.6875
27	70	68	66	66	0.7292
28	76	66	66	70	0.7917
29	74	66	70	76	0.7708
30	76	70	76	74	0.7917
31	76	76	74	76	0.7917
32	80	74	76	76	0.8333
33	86	76	76	80	0.8958
34	88	76	80	86	0.9167

Note: Last 1 denotes the subcontractor's latest performance, and so forth. Normalized performance is divided by 96.

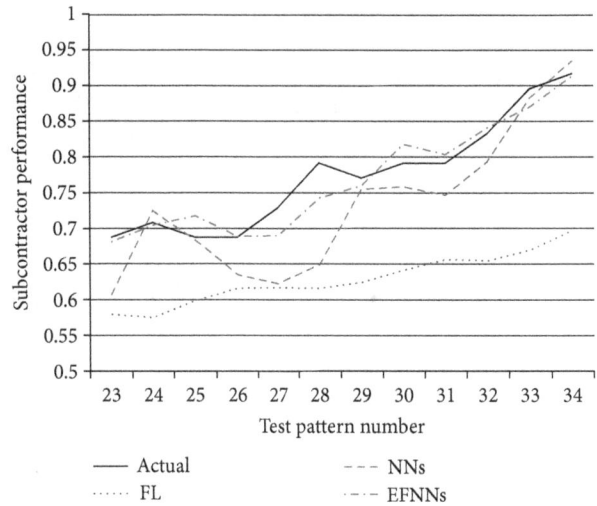

FIGURE 11: Comparison of prediction results.

TABLE 2: EFNN parameters.

Technique	Parameter	Value
GAs	Population size	50
	Crossover rate	0.5
	Mutation rate	0.01
	Terminal condition	5000 iterations
FL	Number of fuzzy sets	5
	Defuzzification function	Average output method
	MF shape	Trapezoidal shape
NNs	Connection weight	0.0–1.0
	Bias	−1–0
	Activation function	Heaviside
	Hidden layers	1–6
	Hidden neurons	1–6

terms of MF identification, fuzzy rule acquisition, composition operator determination, NN topology configuration, or NN parameter recognition. Thus the proposed approach both improves prediction accuracy and reduces the time required to develop a tool for performance prediction.

5. Conclusions

Subcontractor performance is considered an important indicator for general contractors to select subcontractors. To facilitate such decision-making, this study hybridizes NNs, FL, and GAs to develop the EFNNs. Parameters required in NNs and FL are encoded using floating numbers. A multi-cut-point crossover is used to explore the optimum combination of parameters and maintain solution legitimacy. Twelve test cases not used in the evolutionary process are applied to validate the performance of the proposed method. Application results show that the proposed EFNNs outperform NNs and FL in predicting subcontractor performance. Furthermore, users do not need to configure parameters such as membership function distributions, NN parameters and

significantly outperform FL. In addition, applying EFNNs for predicting subcontractor performance requires no effort in

TABLE 3: Comparison of prediction results.

Pattern no.	Subcontractor performance	FL predicted performance	NNs predicted performance	EFNNs predicted performance
23	0.6875	0.5793	0.6064	0.6810
24	0.7083	0.5748	0.7247	0.7046
25	0.6875	0.5986	0.6837	0.7175
26	0.6875	0.6158	0.6350	0.6892
27	0.7292	0.6169	0.6225	0.6898
28	0.7917	0.6158	0.6491	0.7428
29	0.7708	0.6245	0.7550	0.7601
30	0.7917	0.6413	0.7588	0.8180
31	0.7917	0.6567	0.7469	0.8038
32	0.8333	0.6546	0.7943	0.8409
33	0.8958	0.6694	0.8826	0.8702
34	0.9167	0.6968	0.9353	0.9132
RMSE		0.1527	0.0624	0.0234

Real performance score is multiplied by 96.

topology, and defuzzification parameters, thus reducing the effort required to develop prediction tools. A web-based application is developed to automate the evolutionary process, thus increasing user convenience. Subcontractor performance is associated with previous outcomes, and predicting future performance depends on identifying this association. The proposed web-based EFNNs system can be used to automatically establish this association, thus enhancing the efficiency of managerial decision-making. The proposed method is one of the first attempts to apply AI methods to predicting subcontractor performance. Future studies may explore different approaches to further enhance prediction accuracy and application convenience.

Acknowledgments

This research was funded by Grants NSC 98-2221-E-020-035-MY2 from the National Science Council (Taiwan), whose support is gratefully acknowledged. Any opinions, findings, conclusions, or recommendations expressed in the paper are those of the author and do not reflect the views of the National Science Council. The author would also like to thank the graduate student Yu-Xian Lin who helped in programming the web-based software.

References

[1] V. W. Y. Tam, L. Y. Shen, and J. S. Y. Kong, "Impacts of multi-layer chain subcontracting on project management performance," *International Journal of Project Management*, vol. 29, no. 1, pp. 108–116, 2011.

[2] E. H. Sparer and J. T. Dennerlein, "Determining safety inspection thresholds for employee incentives programs on construction sites," *Safety Science*, vol. 51, no. 1, pp. 77–84, 2013.

[3] C. H. Ko and J. D. Kuo, "Making formwork construction lean," *Journal of Civil Engineering and Management*, 2013.

[4] K. Manley, "Implementation of innovation by manufacturers subcontracting to construction projects," *Engineering, Construction and Architectural Management*, vol. 15, no. 3, pp. 230–245, 2008.

[5] R. M. Choudhry, J. W. Hinze, M. Arshad, and H. F. Gabriel, "Subcontracting practices in the construction industry of Pakistan," *Journal of Construction Engineering and Management*, vol. 138, no. 12, pp. 1353–1359, 2012.

[6] R. M. Choudhry, D. Fang, and S. M. Ahmed, "Safety management in construction: best practices in Hong Kong," *Journal of Professional Issues in Engineering Education and Practice*, vol. 134, no. 1, pp. 20–32, 2008.

[7] A. Hartmann, F. Y. Y. Ling, and J. S. H. Tan, "Relative importance of subcontractor selection criteria: evidence from Singapore," *Journal of Construction Engineering and Management*, vol. 135, no. 9, pp. 826–832, 2009.

[8] V. Albino and A. C. Garavelli, "A neural network application to subcontractor rating in construction firms," *International Journal of Project Management*, vol. 16, no. 1, pp. 9–14, 1998.

[9] M. M. Kumaraswamy and J. D. Matthews, "Improved subcontractor selection employing partnering principles," *Journal of Management in Engineering*, vol. 16, no. 3, pp. 47–57, 2000.

[10] M. A. Ekström, H. C. Björnsson, and C. I. Nass, "Accounting for rater credibility when evaluating AEC subcontractors," *Construction Management and Economics*, vol. 21, no. 2, pp. 197–208, 2003.

[11] J. Mbachu, "Conceptual framework for the assessment of subcontractors' eligibility and performance in the construction industry," *Construction Management and Economics*, vol. 26, no. 5, pp. 471–484, 2008.

[12] C. H. Ko, "Application of lean production system in the construction industry: an empirical study," *Journal of Engineering and Applied Sciences*, vol. 5, no. 2, pp. 71–77, 2010.

[13] S. Maturana, L. F. Alarcón, P. Gazmuri, and M. Vrsalovic, "On-site subcontractor evaluation method based on lean principles and partnering practices," *Journal of Management in Engineering*, vol. 23, no. 2, pp. 67–74, 2007.

[14] L. Le-Hoai, Y. D. Lee, and A. T. Nguyen, "Estimating time performance for building construction projects in Vietnam," *KSCE Journal of Civil Engineering*, vol. 17, no. 1, pp. 1–8, 2013.

[15] S. H. Park, "Whole life performance assessment: critical success factors," *Journal of Construction Engineering and Management*, vol. 135, no. 11, pp. 1146–1161, 2009.

[16] A. M. Elazouni and F. G. Metwally, "D-sub: decision support system for subcontracting construction works," *Journal of Construction Engineering and Management*, vol. 126, no. 3, pp. 191–200, 2000.

[17] K. Warwick, *Artificial Intelligence: The Basics*, Routledge, New York, NY, USA, 2012.

[18] G. Phillips-Wren, "AI tools in decision making support systems: a review," *International Journal on Artificial Intelligence Tools*, vol. 21, no. 2, Article ID 1240005, 2012.

[19] C. H. Ko, *Evolutionary fuzzy neural inference model (EFNIM) for decision-making in construction management [Ph.D. thesis]*, National Taiwan University of Science and Technology, Taipei, Taiwan, 2002.

[20] J. H. Holland, *Adaptation in Natural and Artificial Systems*, The University of Michigan Press, Ann Arbor, Mich, USA, 1975.

[21] L. A. Zadeh, "Fuzzy sets," *Information and Control*, vol. 8, no. 3, pp. 338–353, 1965.

[22] S. O. Haykin, *Neural Networks and Learning Machines*, Prentice Hall, New York, NY, USA, 2008.

[23] T. K. Wu, *Performance evaluation and prediction model for construction subcontractor [M.S. thesis]*, National Taiwan University of Science and Technology, Taipei, Taiwan, 2001.

[24] C. H. Ko, M. Y. Cheng, and T. K. Wu, "Evaluating sub-contractors performance using EFNIM," *Automation in Construction*, vol. 16, no. 4, pp. 525–530, 2007.

[25] S. Ulubeyli, E. Manisali, and A. Kazaz, "Subcontractor selection practices in international construction projects," *Journal of Civil Engineering and Management*, vol. 16, no. 1, pp. 47–56, 2010.

[26] M. Y. Cheng and C. H. Ko, "A genetic-fuzzy-neuro model encodes FNNs using SWRM and BRM," *Engineering Applications of Artificial Intelligence*, vol. 19, no. 8, pp. 891–903, 2006.

[27] A. P. Piotrowski and J. J. Napiorkowski, "A comparison of methods to avoid overfitting in neural networks training in the case of catchment runoff modeling," *Journal of Hydrology*, vol. 476, pp. 97–111, 2013.

[28] K. El Hindi and M. Al-Akhras, "Smoothing decision boundaries to avoid overfitting in neural network training," *Neural Network World*, vol. 21, no. 4, pp. 311–325, 2011.

[29] M. Gen and R. Cheng, *Genetic Algorithms and Engineering Design*, Wiley, New York, NY, USA, 1999.

15

SGC Tests for Influence of Material Composition on Compaction Characteristic of Asphalt Mixtures

Qun Chen[1] and Yuzhi Li[2]

[1] School of Traffic and Transportation Engineering, Central South University, Railway Campus, Changsha 410075, China
[2] Changsha University of Science and Technology, Changsha 410076, China

Correspondence should be addressed to Qun Chen; chenqun631@csu.edu.cn

Academic Editors: J. Assaad and F. Pacheco Torgal

Compaction characteristic of the surface layer asphalt mixture (13-type gradation mixture) was studied using Superpave gyratory compactor (SGC) simulative compaction tests. Based on analysis of densification curve of gyratory compaction, influence rules of the contents of mineral aggregates of all sizes and asphalt on compaction characteristic of asphalt mixtures were obtained. SGC Tests show that, for the mixture with a bigger content of asphalt, its density increases faster, that there is an optimal amount of fine aggregates for optimal compaction and that an appropriate amount of mineral powder will improve workability of mixtures, but overmuch mineral powder will make mixtures dry and hard. Conclusions based on SGC tests can provide basis for how to adjust material composition for improving compaction performance of asphalt mixtures, and for the designed asphalt mixture, its compaction performance can be predicted through these conclusions, which also contributes to the choice of compaction schemes.

1. Introduction

Compaction characteristic of asphalt mixtures is often used to describe how easy or difficult it is to compact a mixture on a roadway. For the hot-mix asphalt mixtures, compaction has a great influence on its strength and durability. Good compaction can make asphalt mixtures acquire enough carrying capacity to meet the need of heavy traffic. However, compaction mechanism of asphalt concrete is very complicated and there are many influence factors. Any error in the process of compaction may do harm to the quality and field performance of the whole pavement. Influence factors of compaction performance of asphalt mixtures include many aspects; a lot of researches aim at temperature, rolling machines, and so on [1, 2]. Xiao and Wang [3] analyzed compaction performance of the multilevel interlocked dense type asphalt mixture through the lab rutting test. Stakston et al. [4] and Aho et al. [5] analyzed influence of aggregate shape on compaction characteristic of mixture through tests. Hussain and Timothy [6] proposed the concept of densification energy index through analysis of densification curve characteristic of gyratory compaction. Li et al. [7]

researched influence of material composition (gradation, the content of asphalt) and compaction pressure on compaction characteristic through gyratory compaction tests, but it only compared the compaction characteristic of several gradations such as AC-13I, AK-13A, and Superpave gradations and did not analyze the influence rule of gradation variety on compaction characteristic systematically. Leiva and West [8] analyzed the basic compaction parameters (such as the compaction energy index, the slope of densification curve, and the number of gyrations required to reach 92% of the theoretical maximum density) and their relations through the lab gyratory compaction tests, providing a basis for analysis of influence of material composition (gradation, aggregate shape, binder grade, and so on) on compaction characteristic.

The internal material composition of asphalt mixture is the base of its all exterior characteristics; so compaction characteristic of asphalt mixture is mainly decided by its material composition. However, there are few deep and careful researches about influence rules of material composition on compaction characteristic. This paper will systematically analyze influence rules of material composition on compaction characteristic of asphalt mixtures through the SGC

TABLE 1: Superpave gyratory compaction parameters (AASHTO PP28-00) [10].

Designed ESAL/10^6	Compaction parameters			Typical road applications
	N_{ini}	N_{des}	N_{max}	
<0.3	6	50	75	Light-traffic roads, such as local roads and county roads
0.3~<3	7	75	115	Distributed roads, entrance roads into blocks, city roads of medium traffic, and parts of county roads
3~<30	8	100	160	Two-lane roads, multilane roads, entrance roads into cities, city roads of medium or heavy traffic, state roads, and country roads
≥30	9	125	205	Cross-state roads, roads of heavy traffic, and truck-exclusive roads

simulative compaction tests. This research aims at the surface layer asphalt mixture (13-type gradation); other types of mixtures will be discussed in the later other researches. In addition, this research only analyzes the influence rule and trend of compaction characteristic and does not judge if the compaction performance of some mixture is good or bad. The arrangement of this paper is as follows: Section 2 introduces the simulative compaction principle of Superpave gyratory compactor, Section 3 is the gyratory compaction tests and data analysis, and Section 4 is the conclusions.

2. Simulative Compaction Principle of Superpave Gyratory Compactor

Among many test instruments for researching compaction characteristic of asphalt mixtures, Superpave gyratory compactor can better simulate the compaction process of mixtures under rolling and vehicles. Through observation of height variety of test specimens in the lab gyratory compaction test, the densification characteristic of asphalt mixtures during construction and after traffic is open can be evaluated [9]. In this paper, Superpave gyratory compactor is used to analyze the influence of all kinds of material compositions on compaction characteristic through the lab tests, which cannot be done in the field tests. Strategic Highway Research Program (SHRP) researchers have several purposes in the development of lab compaction methods. The most important one is to compact the test specimen to the field density simulatively. The larger compaction equipment is needed to adapt to the mineral aggregates of big size. Compaction performance tests are needed to recognize the unstable mixtures and other compaction problems. SHRP researchers also considered the weight of equipment. Because the current compaction equipments did not meet these needs, so they developed the Superpave gyratory compactor (SGC).

SGC can satisfy the need of simulative compaction and its weight is also very light. The diameter of test specimens is 6 inches (150 mm) and can fit the mixtures composed of mineral aggregates of maximum size 50 mm (nominal maximum size 37.5 mm). Gyratory compaction molding can simulate the action on mixtures of construction machines and vehicles in the process of paving, rolling and traffic loading. Similar to the other mix design methods, the mixture is designed under

a certain compaction level. In the Superpave design method, the compaction level is a function of the designed gyratory compaction number N_{des} which is used to distinguish the difference of compaction effort. N_{des} is a function of traffic level; the traffic level can be denoted by the designed ESAL (equivalent single axle load). The values of N_{des} are listed in Table 1.

The initial gyratory number N_{ini} is equivalent to compaction effort of paving; the degree of compaction (ratio of compaction density to theoretical maximum density) needs to be under 89% to avoid soft asphalt mixtures. The designed gyratory number N_{des} is equivalent to compaction effort of rolling and traffic loading until the mixture is stable; the degree of compaction is demanded to be equal to 96% to attain the designed air voids 4%. The maximum gyratory number N_{max} is equivalent to compaction effort of lasting action of traffic loads. If the minimum air void is less than 2%, the mixture can be considered to have been destroyed; so the degree of compaction is demanded to be under 98% under the maximum gyratory number N_{max} [11]. Figure 1 illustrated how the density of mixtures increases with gyratory number. In general, if the semilogarithmic coordinates are applied, the densification curve between N_{ini} and N_{des} takes on the form of line, but the curve between N_{des} and N_{max} is not a perfect line [12]. So the whole densification curve can be divided into two sections; each section reflects the characteristic of different compaction phases.

This paper mainly analyzes the densification curve between gyratory number N_{ini} and N_{des}, namely, analyzing the compaction characteristic of mixtures from paving and rolling to attaining stability under traffic loads, also, namely, analyzing influence factors and rules of how easy or difficult the mixture is compacted to the designed air voids. For the densification curve between N_{des} and N_{max} (viz., compaction process from the stability phase to the minimum air voids), the main problem is to judge if mixtures will be destroyed (viz., air voids less than 2%), which may be discussed in the later other researches and is not studied in this paper. This paper only researches compaction characteristic of the first phase. Because the densification curve between N_{ini} and N_{des} on the semilogarithmic coordinates takes on the form of line; so this research will mainly use the semilogarithm coordinates for simplifying analysis.

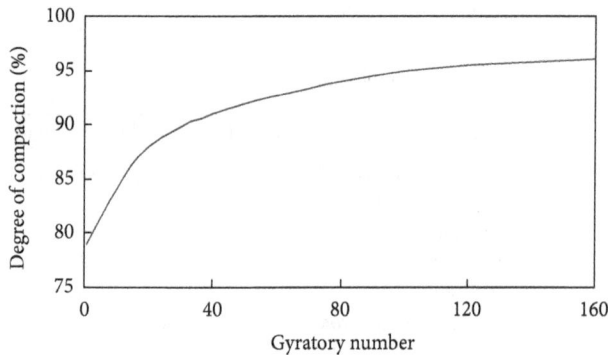

FIGURE 1: Relations of degree of compaction and gyratory number.

3. Gyratory Compaction Tests and Data Analysis

Gyratory number 125 under a pressure of 600 kPa mainly analyzes the densification curve between gyratory number N_{ini} and N_{des}, namely, studying compaction characteristic from paving and rolling to attaining stability under traffic loads, also, namely, analyzing how easy or difficult the mixture is compacted to the designed air voids. Here, according to Table 1, N_{ini} takes 8; N_{des} takes 100.

3.1. Gyratory Compaction Tests. This research aims at asphalt mixtures of 13-type gradation. According to analysis needs, the content of mineral aggregates and asphalt of all test schemes is made in Table 2. 3.5% and 4.5% for the content of the bitumen are because they stood for two typical level of bitumen content of general mixtures. Mineral aggregates are basalts; mineral powder is limestone (size less than 0.15 mm). Heavy-traffic asphalt AH-90 is used in the tests. The compaction temperature for asphalt mixtures is about 160°C.

Each test scheme molds 3 test specimens; the mix volumetric properties (average values of 3 specimens) are measured after gyratory number 125 (Table 3).

Record the specimen height of each gyratory number in the compaction process (SGC automatically records), calculate the specimen bulk volume density of each gyratory number according to (1), and calculate the degree of compaction according to (2).

The bulk volume density ρ of the specimen height h is approximately

$$\rho = \frac{\rho_b}{m_0/(\pi r^2 h_t)} \left(\frac{m_0}{\pi r^2 h}\right). \tag{1}$$

In (1), ρ_b is the specimen bulk volume density under gyratory number 125, m_0 is the dry weight of specimens, r is the specimen radius (7.5 cm), and h_t is the specimen height of gyratory number 125.

Degree of compaction γ is

$$\gamma = \frac{\rho}{G_{mm}} \times 100\%. \tag{2}$$

In (2), G_{mm} is the theoretical maximum density of asphalt mixtures.

Then for each test scheme, average the degree of compaction of 3 specimens (Table 4). Densification curve of each test scheme can also be drawn.

3.2. Data Analysis. Data analysis indexes include the slope of densification curve between N_{ini} and N_{des}, the degree of compaction of gyratory number 8, the degree of compaction of gyratory number 100 and the number of gyrations required to reach 92% of the theoretical maximum density (G_{mm}).

On the semilogarithmic coordinates, curve slope reflects the increasing speed of mix density in the compaction process. The degree of compaction of gyratory number 8 can show if the mixture is a soft asphalt mixture. The degree of compaction of gyratory number 100 can show the air void of the mixture when it attains stability under traffic loads. Because many specifications [10] all demand pavements to be compacted to a density of 92% G_{mm}, the number of gyrations required to reach 92% G_{mm} can be found in the densification curve; if this number is big, construction compaction is difficult.

The following comparison data are from SGC tests (see Table 4).

3.2.1. Compare Test Schemes (3) and (8) and Analyze Influence of Variety of Binder Content on Compaction Characteristic. Densification curves of schemes (3) and (8) are drawn in Figure 2.

It is known from Table 2 that the gradations of test schemes (3) and (8) are the same, but binder content of test scheme (3) is 1% more than that of test scheme (8).

Slope Comparison. The slope of densification curve of scheme (3) is a little higher than that of scheme (8); so a higher binder content will make density-increasing speed become faster.

Density of Gyratory Number 8. Mixture (3) with a higher content of asphalt has a higher density (86.7% G_{mm}, Table 4); mixture (8) with a low content of asphalt has a small density (84.9% G_{mm}). There is a difference of 1.8% G_{mm} between mixtures (3) and (8), which shows more of asphalt improved initial workability of mixtures.

Density of Gyratory Number 100. Mixture (3) with a higher content of asphalt has a higher density (97.5% G_{mm}); mixture (8) with a low content of asphalt has a small density (94.2% G_{mm}). There is a difference of 3.3% G_{mm} between mixtures (3) and (8).

The Number of Gyrations Required to Reach 92% G_{mm}. It is known from Figure 2 that gyratory number of scheme (3) (higher binder content) is smaller, which shows increasing binder content will facilitate construction compaction.

3.2.2. Analyze the Results of Test Schemes (1), (2), (3), and (4) and Study the Influence of Ratio of Coarse Aggregates to Fine Aggregates on Compaction Characteristic. Densification curves of mixtures (1), (2), (3) and (4) are drawn in Figure 3.

TABLE 2: Content of mineral aggregates and asphalt of all test schemes.

Test schemes	Percentage content of mineral aggregates of each size section (%)									Asphalt-aggregate ratio (%)
	13.2~16 mm	9.5~13.2 mm	4.75~9.5 mm	2.36~4.75 mm	1.18~2.36 mm	0.6~1.18 mm	0.3~0.6 mm	0.15~0.3 mm	Mineral powder	
(1)	5	10	15	12	12	12	12	12	10	4.5
(2)	8	16	24	8	8	8	9	9	10	4.5
(3)	11	22	33	4	5	5	5	5	10	4.5
(4)	14	28	42	1	1	1	1	2	10	4.5
(5)	8	16	24	9	9	10	10	10	4	4.5
(6)	12	23	34	3	3	3	3	3	16	4.5
(7)	11	22	33	5	6	6	6	6	5	3.5
(8)	11	22	33	4	5	5	5	5	10	3.5
(9)	22	33	11	4	5	5	5	5	10	4.5
(10)	8	16	24	10	16	16	0	0	10	4.5
(11)	8	16	24	4	0	4	14	20	10	4.5

TABLE 3: Mix volumetric properties of all test schemes.

Test schemes	Bulk volume density (g/cm^3)	Theoretical maximum density (g/cm^3)	Air voids (%)	Percent voids in coarse mineral aggregate (%)
(1)	2.401	2.553	5.93	74.8
(2)	2.486	2.555	2.69	58.3
(3)	2.51	2.558	1.89	42.1
(4)	2.401	2.560	6.24	29.5
(5)	2.43	2.554	4.86	59.2
(6)	2.522	2.559	1.45	39.1
(7)	2.432	2.595	6.25	43.3
(8)	2.461	2.595	5.17	42.7
(9)	2.527	2.555	1.12	41.6
(10)	2.432	2.565	5.17	59.2
(11)	2.412	2.542	5.14	59.5

TABLE 4: Degree of compaction of each gyratory number (%) (parts of data).

Gyratory number	Test schemes										
	(1)	(2)	(3)	(4)	(5)	(6)	(7)	(8)	(9)	(10)	(11)
8	85.3	88.5	86.7	80.0	86.3	86.3	83.6	84.9	88.2	84.0	87.1
12	86.7	89.9	88.7	82.0	87.8	88.5	85.3	86.5	90.1	85.7	88.4
20	88.6	91.9	91.2	84.7	89.6	91.2	87.3	88.5	92.3	87.9	89.9
30	89.9	93.3	93.0	86.8	91.0	93.2	88.9	90.1	94.0	89.6	91.1
40	90.9	94.3	94.2	88.3	91.9	94.5	90.0	91.2	95.1	90.8	92.0
50	91.5	95.0	95.1	89.4	92.6	95.5	90.8	92.0	96.0	91.6	92.6
60	92.1	95.5	95.8	90.4	93.2	96.3	91.5	92.6	96.7	92.3	93.1
70	92.6	96.0	96.3	91.1	93.6	96.8	91.9	93.1	97.2	92.9	93.5
80	93.0	96.3	96.8	91.8	94.0	97.3	92.4	93.5	97.7	93.4	93.8
90	93.3	96.6	97.2	92.3	94.3	97.6	92.8	93.9	98.0	93.8	94.1
100	93.5	96.8	97.5	92.7	94.6	97.9	93.1	94.2	98.4	94.1	94.3
110	93.8	97.1	97.8	93.2	94.8	98.2	93.3	94.5	98.5	94.4	94.6
125	94.1	97.3	98.1	93.8	95.1	98.6	93.7	94.8	98.9	94.8	94.9

It is known from Table 2 that schemes (1), (2), (3), and (4) have an equal binder content and an equal mineral powder content, but the content of coarse aggregates increases gradually and that of fine aggregates decreases gradually.

Compare mixtures (1) with (2); the content of fine aggregates less than 4.75 mm decreases from 60% to 42% while coarse aggregates increases from 30% to 48%.

Slope Comparison. Densification curves of schemes (1) and (2) are nearly parallel; so their density-increasing speeds are nearly the same.

Density of Gyratory Number 8. The density of mixture (1) is 85.3% G_{mm}; the density of mixture (2) is 88.5% G_{mm}. There is a difference of 3.2% G_{mm} between mixtures (1) and (2); overmany fine aggregates will make the initial air voids become large.

Density of Gyratory Number 100. The density of mixture (1) is 93.5% G_{mm}; the density of mixture (2) is 96.8% G_{mm}. There is a difference of 3.3% G_{mm} between mixtures (1) and (2);

overmany fine aggregates will also make the final air voids become large.

The Number of Gyrations Required to Reach 92% G_{mm}. It is known from Figure 3 that gyratory number of scheme (2) is much smaller than that of scheme (1), which shows that overmany fine aggregates will make specific surface area become large and make oil film become thin, thus making compaction become more difficult.

Compare mixtures (2) with (3); the content of fine aggregates less than 4.75 mm decreases from 42% to 24% while coarse aggregates increases from 48% to 66%.

Slope Comparison. Slope of densification curve of scheme (3) is higher than that of test scheme (2), which shows that when the content of coarse aggregates adds to a certain amount, the density-increasing speed becomes faster.

Density of Gyratory Number 8. The density of mixture (3) is 86.7% G_{mm}; the density of mixture (2) is 88.5% G_{mm}. There is a difference of 3.2% G_{mm} between mixtures (2) and (3). When

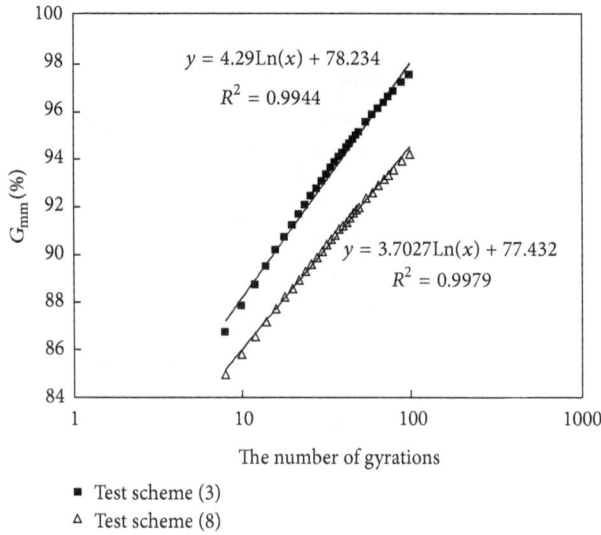

FIGURE 2: Densification curves of schemes (3) and (8).

- Test scheme (3)
- Test scheme (8)

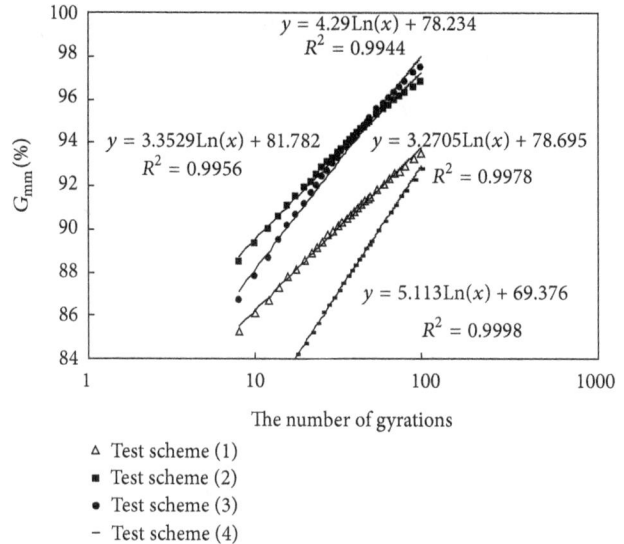

△ Test scheme (1)
- Test scheme (2)
- Test scheme (3)
- Test scheme (4)

FIGURE 3: Densification curves of schemes (1), (2), (3), and (4).

the content of coarse aggregates adds to a certain amount, the initial air voids start to become large.

Density of Gyratory Number 100. The density of mixture (3) is 97.5% G_{mm}; the density of mixture (2) is 96.8% G_{mm}. There is a difference of 0.7% G_{mm} between mixtures (2) and (3). The final air voids continue to decrease.

The Number of Gyrations Required to Reach 92% G_{mm}. It is known from Figure 3 that gyratory number of scheme (3) is a little more than that of scheme (2), which shows that compaction characteristic of mixtures will change and construction compaction become difficult when the content of fine aggregates decreases to a certain amount.

Compare mixtures (3) with (4); the content of fine aggregates less than 4.75 mm decreases from 24% to 6% while coarse aggregates increases from 66% to 84%.

Slope Comparison. Slope of densification curve of scheme (4) is higher. The content of coarse aggregates continues to add; the density-increasing speed continues to become greater.

Density of Gyratory Number 8. The density of mixture (4) is 80.0% G_{mm}; the density of mixture (3) is 86.7% G_{mm}. There is a difference of 6.7% G_{mm} between mixtures (3) and (4). When coarse aggregates continue to increase, the initial density of mixtures continues to fall (air voids become larger).

Density of Gyratory Number 100. The density of mixture (4) is 92.7% G_{mm}; the density of mixture (3) is 97.5% G_{mm}. There is a difference of 4.8% G_{mm} between mixtures (3) and (4). When coarse aggregates continue to increase, the final density of mixtures starts to fall (air voids become large) owing to lack of enough fine aggregates filled in coarse ones.

The Number of Gyrations Required to Reach 92% G_{mm}. It is known from Figure 3 that gyratory number of scheme (4) is much more than that of scheme (3), which shows that

overmany coarse aggregates and oversmall fine aggregates will make construction compaction become very difficult because a strong interlocking strength is formed between coarse aggregates and so the strong inner frictional resistance needs to be conquered in the compaction process.

From the above analysis, variety rules of compaction performance with the content of fine aggregates are depicted in Figure 4.

3.2.3. Compare the Results of Test Schemes of (5), (10), and (11). Densification curves of schemes (5), (10), and (11) are drawn in Figure 5.

Compare mixtures (10) with (11); they have the same contents of coarse aggregates above 4.75 mm, the same contents of mineral powder, and the same contents of asphalt, but mixture (10) has more of thicker fine aggregates while mixture (11) has more of thinner fine aggregates.

Slope Comparison. Slope of densification curve of test scheme (10) is higher; the mixture having more of thicker fine aggregates has a higher density-increasing speed.

Density of Gyratory Number 8. The density of mixture (10) is 84.0% G_{mm}; the density of mixture (11) is 87.1% G_{mm}. The density of mixture (10) is 3.1% G_{mm} lower than that of mixture (11); the mixture having more of thicker fine aggregates has a low initial density and a high air void.

Density of Gyratory Number 100. Being nearly the same, the density of mixture (10) is 94.1% G_{mm}; and that of mixture (11) 94.3% G_{mm}.

The Number of Gyrations Required to Reach 92% G_{mm}. It is known from Figure 5 that gyratory number of scheme (10) is more, which is because the thicker fine aggregates have a stronger inner frictional resistance.

Compare mixtures (5) with (11); they have the same contents of coarse aggregates above 4.75 mm and the same

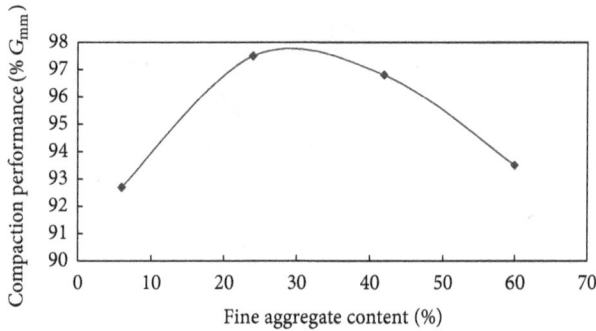

FIGURE 4: Variety rule of compaction performance with fine aggregate content.

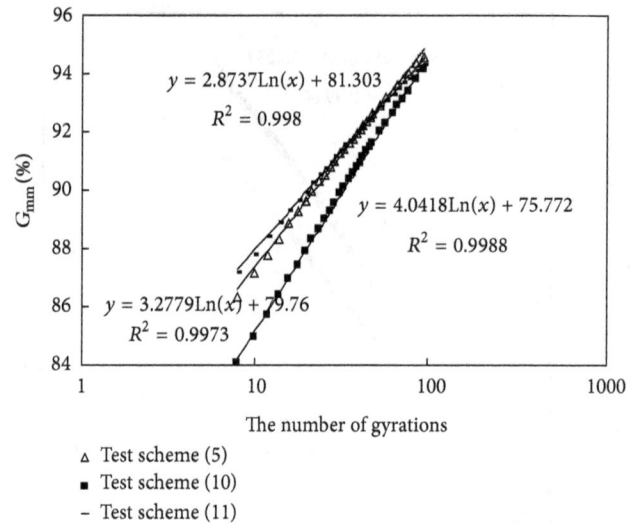

△ Test scheme (5)
■ Test scheme (10)
– Test scheme (11)

FIGURE 5: Densification curves of schemes (5), (10), and (11).

contents of asphalt, but mixture (5) has more of thicker fine aggregates.

Slope Comparison. Slope of densification curve of test scheme (5) is a little higher.

Density of Gyratory Number 8. The initial density of mixture (5) is a little lower than that of mixture (11).

Density of Gyratory Number 100. There is no big difference between the two mixtures.

The Number of Gyrations Required to Reach 92% G_{mm}. There is no big difference between the two mixtures.

Compare mixtures (5) with (10); they have the same contents of coarse aggregates and the same contents of asphalt; mixture (10) has more of thicker fine aggregates and mineral powder but less of thinner fine aggregates.

Slope Comparison. Slope of densification curve of test scheme (10) is a little higher.

Density of Gyratory Number 8. The density of mixture (5) is 86.3% G_{mm}; and that of mixture (10) 84.0% G_{mm}. The mixture having more of thicker fine aggregates has a higher density-increasing speed.

Density of Gyratory Number 100. There is no big difference between the two mixtures.

The Number of Gyrations Required to Reach 92% G_{mm}. Gyratory number of scheme (10) is more; this is because scheme (10) has a big amount of thicker fine aggregates which produce a strong inner frictional resistance; so construction compaction performance of mixture (10) is worse than that of mixture (5); this also shows that the effect of thicker fine aggregates on compaction performance is prominent.

3.2.4. Compare the Results of Schemes (3) and (9). Densification curves of schemes (3) and (9) are drawn in Figure 6.

Mixtures (3) and (9) have the same amount of fine aggregates, mineral powder, and asphalt, but mixture (9) has more of thicker coarse aggregates. Their final ail voids have no big difference, but the initial air void of mixture (9) is a little lower, which shows the proportion of each size section among coarse aggregates above 4.75 mm has no obvious influence

on compaction characteristic, may be more of thicker coarse aggregates will contribute to compaction; this is probably because there are less contact points between coarse aggregate particles.

3.2.5. Compare the Results of Schemes (7) and (8). Densification curves of schemes (7) and (8) are drawn in Figure 7.

Mixtures (7) and (8) have the same amount of coarse aggregates and asphalt, but mixture (7) has a less amount of mineral powder. It is known from Figure 7 that the slopes of two curves are nearly equal; namely, their density-increasing speeds are nearly equal. But mixture (8) is more easy to compact (needing less of gyrations for attaining a density of 92% G_{mm}) than mixture (7) because the more mineral powder caused a smaller air void, which also shows that an appropriate amount of mineral powder will contribute to compaction.

3.2.6. Compare the Results of Schemes (2) and (5). Densification curves of schemes (2) and (5) are drawn in Figure 8.

Similarly to analysis of (5), compare mixture (2) with (5), they have the same amount of coarse aggregates and asphalt, but mixture (2) has a higher amount of mineral powder (10%) and mixture (5) has a lower amount of mineral powder (4%). It is known from Figure 8 that the slopes of two curves are nearly equal, but mixture (2) has a smaller air void and mixture (2) is more easy to compact than mixture (5) because it needs less of gyrations for attaining a density of 92% G_{mm}, which shows that an appropriate amount of mineral powder will improve workability of mixtures and contribute to field construction compaction.

3.2.7. Compare the Results of Schemes (3) and (6). Densification curves of schemes (3) and (6) are drawn in Figure 9.

Mixtures (3) and (6) have the same amount of asphalt and the contents of coarse aggregates are also nearly equal, but mixture (6) has a higher amount of mineral powder (16%). It is known from Figure 9 that two curves almost overlap and

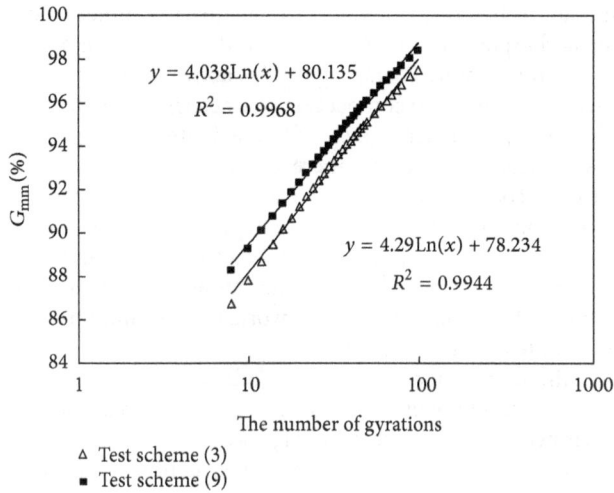

FIGURE 6: Densification curves of schemes (3) and (9).

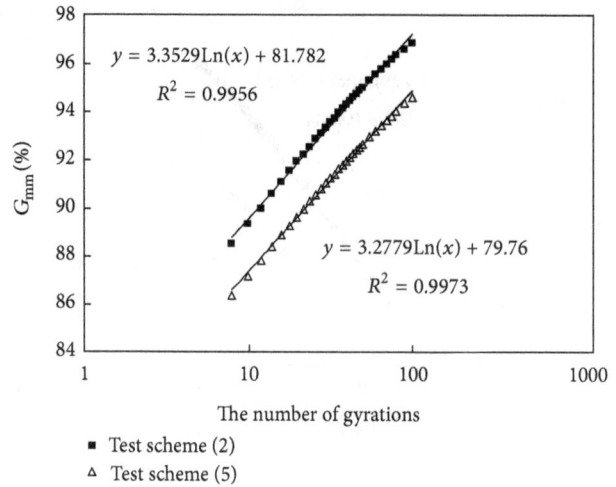

FIGURE 8: Densification curves of schemes (2) and (5).

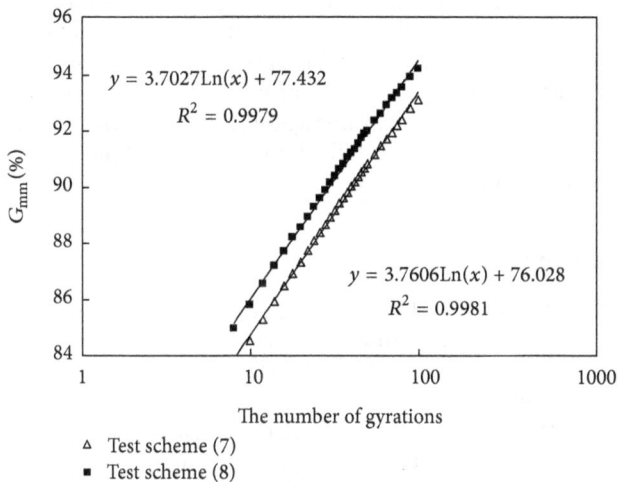

FIGURE 7: Densification curves of schemes (7) and (8).

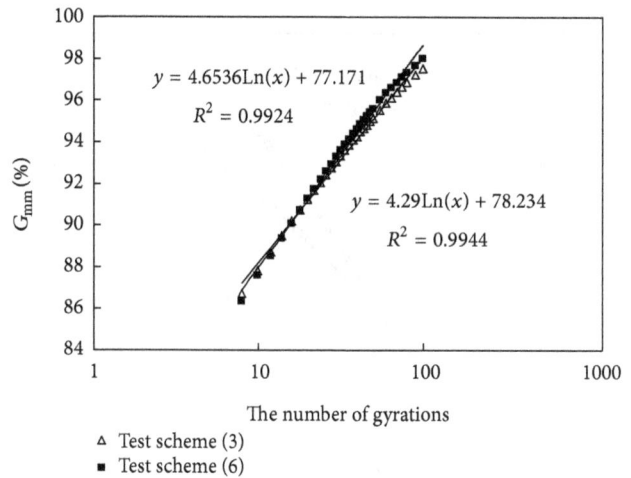

FIGURE 9: Densification curves of schemes (3) and (6).

scheme (6) has a little higher slope and a little lower final air void. Mineral powder continuing to add after it has added to a certain amount will not do good to compaction any longer; this is because overmuch mineral powder will make mixtures dry and do harm to field compaction.

3.2.8. Compare the Results of Schemes (3) and (7). Densification curves of schemes (3) and (7) are drawn in Figure 10.

Mixtures (3) and (7) have the same amount of coarse aggregates, but mixture (3) has a higher amount of mineral powder and a higher amount of asphalt. It is known from Figure 10 that mixture (3) has a smaller air void and the slope of densification curve of scheme (3) is a little higher. The density-increasing speed of mixture (3) is faster and the number of gyrations required to reach 92% G_{mm} is also much smaller than that of scheme (7).

3.2.9. Compare the Results of Schemes (5) and (6). Densification curves of schemes (5) and (6) are drawn in Figure 11.

Mixtures (5) and (6) have the same amount of asphalt; mixture (6) has a higher amount of coarse aggregates above 4.75 mm (69%), a higher amount of mineral powder (16%), and a lower amount of fine aggregates (15%); comparatively, mixture (5) has a higher amount of fine aggregates (48%) and a lower amount of mineral powder (only 4%). It is known from Figure 11 that the final air void of mixture (6) is very small; its curve slope is big, so the density-increasing speed is fast; the number of gyrations required to reach 92% G_{mm} is small, it is more easy to compact than mixture (5) but has an oversmall final air void; mixture (5) has overmany fine aggregates but lacks enough mineral powder; so it is more difficult to compact and the final air void is big.

4. Conclusions

The slope of densification curve cannot be used as the only criterion for judging whether compaction is difficult or easy; otherwise, the inaccurate conclusion that more coarse aggregates make compaction easier may be drawn. The

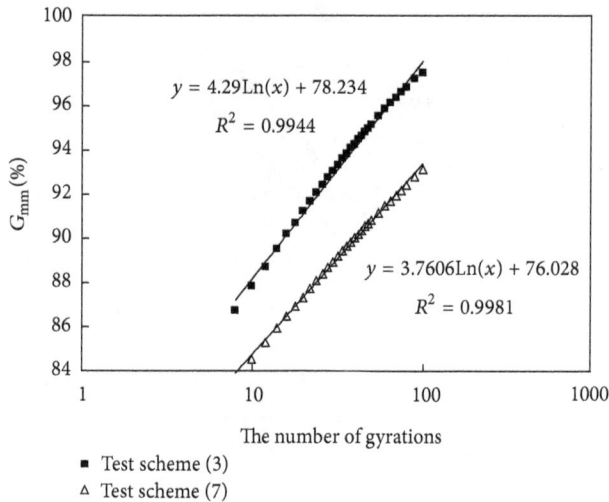

FIGURE 10: Densification curves of schemes (3) and (7).

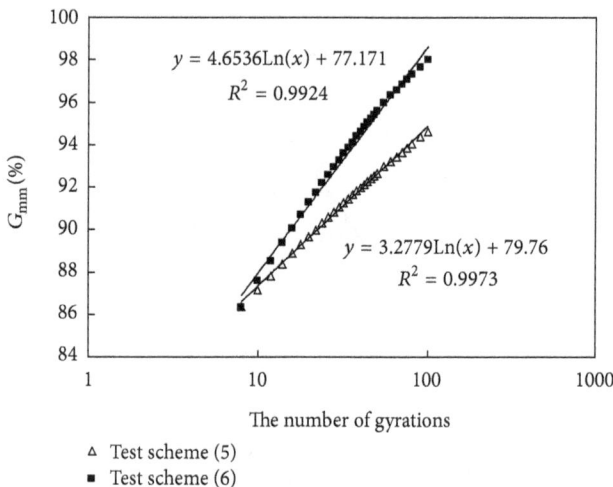

FIGURE 11: Densification curves of schemes (5) and (6).

mixture having more of asphalt has a faster density-increasing speed; it is more easy to compact and has a smaller air void when their gradations are the same.

Test schemes (1), (2), (3), and (4) have the same amount of asphalt and mineral powder, but the content of coarse aggregates increases gradually and that of fine aggregates decreases gradually; the variety rule of compaction characteristic with the content of fine aggregates is depicted in Figure 4. It can be seen from Figure 4 that when the content of fine aggregates is small and that of coarse aggregates is large the mixture is difficult to compact; with the gradual increase of fine aggregates compaction performance of mixtures improves gradually; but with the continual increase of fine aggregates compaction performance of mixtures starts to fall. So there is an optimal amount of fine aggregates.

When the contents of coarse aggregates, mineral powder, and asphalt are all kept constant, the thicker ones in fine aggregates has a prominent effect on compaction performance; more of thicker ones will make compaction more

difficult. When the contents of fine aggregates, mineral powder, and asphalt are all kept constant, the proportion of each size section among coarse aggregates above 4.75 mm has no obvious influence on compaction characteristic; maybe more of thicker coarse aggregates will contribute to compaction; this is probably because there are less of contact points between coarse aggregate particles. When the contents of coarse aggregates and asphalt are kept fixed, the air void of the mixture with more mineral powder is smaller and more easy to compact, which shows that an appropriate amount of mineral powder will improve workability of mixtures and contribute to field compaction.

Mineral powder continuing to add after it has added to a certain amount will not do good to compaction any longer; this is because overmuch mineral powder will make mixtures dry and hard and do harm to field construction compaction. When the content of coarse aggregates is kept constant, the mixture having more of mineral powder and asphalt has a smaller air void and its density-increasing speed is faster. The mixture having more of coarse aggregates and mineral powder but less of fine aggregates has a smaller final air void, and its density-increasing speed is fast; although it is more easy to compact, its final air void will be too small. The mixture having overmany fine aggregates but lacking enough mineral powder is more difficult to compact and the final air void is usually large.

References

[1] R. Delgadillo and H. U. Bahia, "Effects of temperature and pressure on hot mixed asphalt compaction: field and laboratory study," *Journal of Materials in Civil Engineering*, vol. 20, no. 6, pp. 440–448, 2008.

[2] Z. Q. Zhang, N. L. Li, and H. X. Chen, "Determining method of mixing and compaction temperatures for modified asphalt mixture," *Journal of Traffic and Transportation Engineering*, vol. 7, no. 2, pp. 36–40, 2007 (Chinese).

[3] W. Xiao and X. L. Wang, "Research on rolled-compactness performance of multilevel interlocked-dense type asphalt mixture," *Maintenance Machinery & Construction Technology*, no. 3, pp. 25–27, 2007 (Chinese).

[4] A. D. Stakston, H. U. Bahia, and J. J. Bushek, "Effect of fine aggregate angularity on compaction and shearing resistance of asphalt mixtures," *Transportation Research Record*, no. 1789, pp. 14–24, 2002.

[5] B. D. Aho, W. R. Vavrik, and S. H. Carpenter, "Effect of flat and elongated coarse aggregate on field compaction of hot-mix asphalt," *Transportation Research Record*, no. 1761, pp. 26–31, 2001.

[6] U. B. Hussain and P. F. Timothy, "Optimization of constructibility and resistance to traffic: a new design approach for HMA using the superpave compactor," *Journal of the Association of Asphalt Paving Technologists*, vol. 67, no. 2, pp. 189–232, 1998.

[7] L. H. Li, X. J. Li, and Z. X. Zhong, "Analysis of factors effecting on densification characteristics of hot mix asphalt," *China Journal of Highway and Transport*, vol. 14, supplement, pp. 31–34, 2001 (Chinese).

[8] F. Leiva and R. C. West, "Analysis of hot-mix asphalt lab compactability using lab compaction parameters and mix

characteristics," *Transportation Research Record*, no. 2057, pp. 89–98, 2008.

[9] R. B. Mallick, "Use of Superpave gyratory compactor to characterize hot-mix asphalt," *Transportation Research Record*, no. 1681, pp. 86–96, 1999.

[10] R. J. Cominsky, G. A. Huber, and T. W. Kennedy, *The Superpave Mix Design Manual for New Construction and Overlays*, National Research Council, Washington, DC, USA, 1994.

[11] G. Fitts, "Compaction principles for heavy-duty HMA," *Asphalt*, vol. 16, no. 2, pp. 17–19, 2001.

[12] Z. Q. Zhang, Y. J. Yuan, and B. G. Wang, "Information of gyratory compaction densification curve of asphalt mixture and its application," *China Journal of Highway and Transport*, vol. 18, no. 3, pp. 1–6, 2005 (Chinese).

16

Characteristics of Traffic Flow at Nonsignalized T-Shaped Intersection with U-Turn Movements

Hong-Qiang Fan, Bin Jia, Xin-Gang Li, Jun-Fang Tian, and Xue-Dong Yan

MOE Key Laboratory for Urban Transportation Complex Systems Theory and Technology, Beijing Jiaotong University, Beijing 100044, China

Correspondence should be addressed to Bin Jia; bjia@bjtu.edu.cn

Academic Editors: S. Chen and B. Uy

Most nonsignalized T-shaped intersections permit U-turn movements, which make the traffic conditions of intersection complex. In this paper, a new cellular automaton (CA) model is proposed to characterize the traffic flow at the intersection of this type. In present CA model, new rules are designed to avoid the conflicts among different directional vehicles and eliminate the gridlock. Two kinds of performance measures (i.e., flux and average control delay) for intersection are compared. The impacts of U-turn movements are analyzed under different initial conditions. Simulation results demonstrate that (i) the average control delay is more practical than flux in measuring the performance of intersection, (ii) U-turn movements increase the range and degree of high congestion, and (iii) U-turn movements on the different direction of main road have asymmetrical influences on the traffic conditions of intersection.

1. Introduction

U-turn movements increase the complexity of urban intersections. However, it is applied more and more widely. The main reasons are forbidding left-turn movements at the intersection and no median opening on the road. Figure 1 displays the process of U-turn movements.

Recently, The effects of U-turn movements on the safety and operation of intersections have attracted great research interests. These studies can be substantially categorized into using the analytical models [1–8] and simulation techniques [9–23].

Some researchers investigated the capacity of signalized intersections permitting U-turn movements [1–4]. The related conclusions were used in Highway Capacity Manual (HCM 2010). Several studies [5, 6] have analyzed the characteristics of U-turn movements at the unsignalized intersections from four aspects (i.e., headway acceptance, impedance effects of minor movements, conflicting traffic volume, and shared-lane capacity of the major street exclusive left-turn lane). Liu et al. [7] evaluated the impacts of indirect driveway left-turn treatments on traffic operations

at signalized intersections. Guo et al. [8] have developed a negative-binomial model to predict U-turn volume on a left-turn approach at a signalized intersection during weekday peak periods.

Analytical models adapt to analyze a single intersection. However, it is not suitable for more complex cases, for example, two or more intersections. Traffic simulation techniques can compensate the disability of analytical models. One kind of simulation approaches are based on some commercial softwares, including CORSIM, AIMSUN, and VISSIM [9–12]. CA models were also used to investigate the characteristics of traffic flow at intersections of different types [13–23]. The intersection types include roundabout, cross intersection, and T-shaped intersection. At a T-shaped intersection, Li et al. [19] investigated the influence of the left-turning car on the whole traffic situation by introducing the priority probability of the through vehicle; Wu et al. [20] analyzed the interactions between vehicles on different lanes and effects of traffic flow states of different roads on capacity of nonsignalized system; Li et al. [21] considered three input flows and two left turnings to study the traffic behaviors under two crash avoiding rules; Ding et al. [22] investigated and compared the phase diagram,

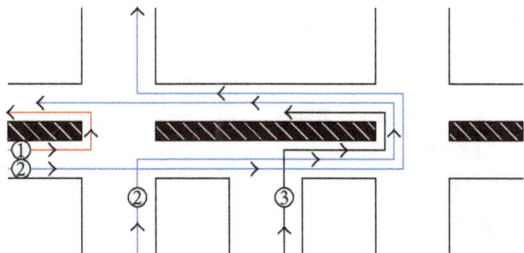

FIGURE 1: Origination of U-turn movement at the intersection. Three types vehicles need make U-turn movements at the nearest intersections: (1) vehicles which move the opposite direction; (2) vehicles which are forbidden to make direct left-turn movement at the intersection; (3) vehicles which turn left from a minor driveway.

the capacity and the average travel time of two different signal controlling systems. Fan et al. [23] explored the characteristic of traffic flow at the nonsignalized T-shaped intersection with all directional vehicles.

However, these studies did not measure the performance of intersections by the average control delay but by the flux for convenience. This is not practical. In addition, the effect of U-turn movements at nonsignalized intersections was also ignored. For these reasons and based on the former work in [23], this paper proposes a new CA model to characterize the nonsignalized T-shaped intersection with U-turns. For this, new avoiding conflicts and gridlock avoiding rules are developed, and the average control delay (while not flux) is introduced as the performance index. The remainder of this paper is organized as follows. In Section 2, the model is illustrated in detail. In Section 3, the numerical and analytical results are analyzed. Conclusions are summarized in Section 4.

2. Model

The geometric design of analyzed T-shaped intersection is illustrated in Figure 2. That is the same as in the literature [19–21]. Each of the approaches has one lane. The major street is consisted of lane A and lane B. The minor street is composed by lane C and lane D. Vehicles move in the right lane. The length of major street is L, and minor street is $L/2 - 1$. The intersection is made up of four cells, T1, T2, T3, and T4. There are seven types of vehicles: (a) straight-driving vehicles on lane A and B, $s = 1$; (b) left-turning vehicles on lane A, $s = 2$; (c) U-turning vehicles on lane A, $s = 3$; (d) right-turning vehicles on lane B, $s = 4$; (e) U-turning vehicles on lane B, $s = 5$; (f) left-turning vehicles on lane C, $s = 6$; (g) right-turning vehicles on lane C, $s = 7$. Here, s denotes the type of vehicles.

2.1. The Rules of Vehicle Moving on the Road.
NaSch model [24] is used to simulate the vehicle movement. Although simply, it can reproduce many basic phenomena in realistic traffic, such as the start-stop waves. In NaSch model, time and space are discrete. The road is divided into L cells. Each cell has two conditions, occupied by one vehicle or empty.

The vehicle speed can be $0, 1, 2, \ldots, v_{\max}$; here $v_{\max}$ is the maximum speed. The update rules of NaSch model are as follows:

step 1: acceleration, $v_n \rightarrow \min(v_n + 1, v_{\max})$;

step 2: deceleration, $v_n \rightarrow \min(v_n, d_n)$;

step 3: randomization, $v_n \rightarrow \max(v_n - 1, 0)$ with probability p;

step 4: position update, $x_n \rightarrow x_n + v_n$.

Here, v_n and x_n denote the speed and position of vehicle n, respectively; $d_n = x_{n-1} - x_n - l$ is the number of empty cells in front of vehicle $n - 1$, and l is the length of vehicle n; p is the randomization probability.

2.2. The Rules of Avoiding Conflict at the Intersection.
Due to the intersection cell that can be occupied only by one vehicle, many potential conflicts may occur when vehicles approach to the intersection. Four types of conflicts are classified.

(a) At T1, the straight-driving or left-turning vehicles on lane A or the left-turning vehicles on lane C may conflict with the U-turning vehicles on lane B.

(b) At T2, the vehicles on lane A may conflict with left-turning on lane C.

(c) At T3, the vehicles on lane B may conflict with left-turning on lane A.

(d) At T4, the U-turning vehicles on lane A, the straight-driving vehicles on lane B, and the vehicles on lane C may conflict with each other.

In order to prevent accidents, some control rules should be applied when the potential conflicts occur. The velocity, position, and type of the first vehicle upstream cell T1 (include T1) on lane A are denoted as x_A, v_A, and s_A, respectively; those of the first vehicle upstream cell T4 on lane B are x_B, v_B, and s_B, and those of the first vehicle on lane C (T4 as the $L/2$ cell on lane C) are x_C, v_C, and s_C. When the potential conflict occurs, the time, which the first vehicle that upstreams the intersection on each lane needs to reach the conflict cell, is calculated. The times are denoted as t_A, t_B, and t_C, respectively. The conflict cell will be occupied by the vehicle which uses less time to reach the conflict cell. If the times are equal, the priority vehicle will occupy the conflict cell. According to the Highway Capacity Manual (HCM 2000), the priority of right of way given to each traffic stream can be identified as follows. Movements of rank 1 include through traffic stream on the major street and right-turning traffic stream from the major street. Movements of rank 2 include left-turning traffic stream from the major street and right-turning traffic stream onto the major street. Movements of rank 3 include left-turning traffic stream from the minor street. Movements of rank 4 include U-turning traffic stream from the major street.

Three types of gridlock in the system are identified (a) the cell T1 is occupied by a left-turning vehicle on lane A and the cell T3 is occupied by a U-turning vehicle on lane B at the same time; (b) the cell T2 is occupied by a U-turning vehicle

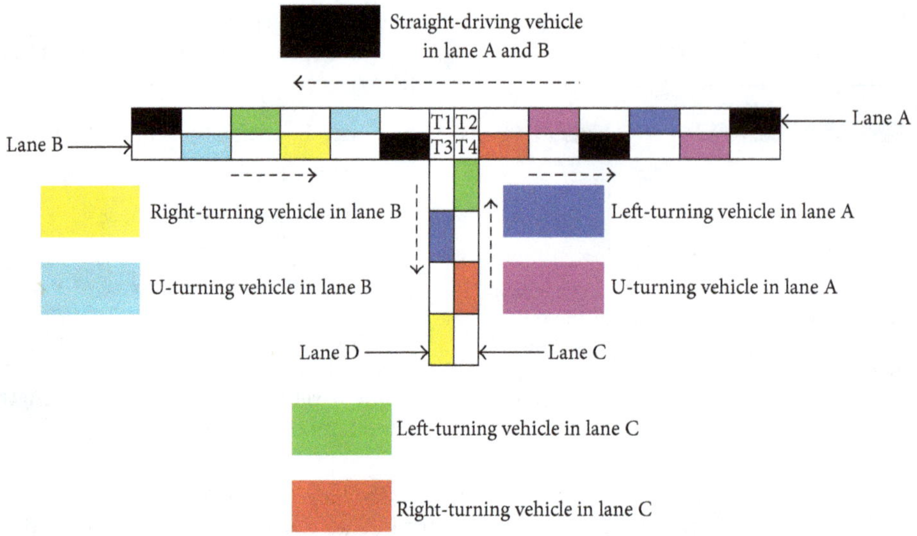

FIGURE 2: Type of nonsignalized T-shaped intersection.

on lane A and the cell T4 is occupied by a left-turning vehicle on lane C at the same time; (c) the cell T1 or the cell T2 is occupied by a left-turning vehicle on lane A, the cell T3 is occupied by a straight-driving vehicle on lane B, and the cell T4 is taken up by a left-turning vehicle at the same time. In order to avoid the gridlock, the following rules corresponding to above types are used, respectively: (a) if the cell T1 or T2 has been or will be occupied in the next time step by a left-turning vehicle on lane A, the U-turning vehicle on lane B is not allowed to enter into the cell T3; if the cell T3 has been occupied by the U-turning vehicle on lane B, the left-turning vehicle on lane A is forbidden to enter into the cell T1; (b) if the cell T4 has been or will be occupied in the next time step by a left-turning vehicle on lane C, the U-turning vehicle on lane A is not allowed to enter into the cell T2; if the cell T2 has been occupied by the U-turning vehicle on lane A, the left-turning vehicle on lane C is forbidden to enter into cell T4; (c) if the cell T1 or T2 has been or will be occupied in the next time step by a left-turning vehicle on lane A, the left-turning vehicle on lane C is not allowed to enter into the cell T4; if the cell T4 has been occupied by the left-turning vehicle on lane C, the left-turning vehicle on lane A is forbidden to enter into the cell T1 and T2.

The system operates as follows. Firstly, the velocities of all vehicles are updated. Then, the conflicts and gridlock are identified and disposed. At last, the positions of all vehicles are updated. The simulations are carried out under open boundary condition. At each time step, we check the position of the last vehicle on each lane, which is represented as x_λ^{last}. If $x_\lambda^{last} > v_{max} + l$, a new vehicle with the maximum velocity v_{max} is injected with inflow rate p_λ at the position $\min(v_{max}, x_\lambda^{last} - v_{max})$. Here, $\lambda = $ A, B, C. The vehicle on lane A is set as a left-turning vehicle with probability p_{LA} and a U-turning vehicle with probability p_{UA}. The vehicle on lane B is set as a right-turning vehicle with probability p_{RB} and a U-turning vehicle with probability p_{UB}. The vehicle on lane C is set as a left-turning vehicle with probability p_{LC}. If the position of first

vehicle on lane A, B, and D is larger than the length of lane, the vehicle will be removed, and the following vehicle will be the new leading vehicle.

3. Simulation and Discussion

In the simulation, $L = 1000$, $l = 2$, $p = 0.3$, $p_{LA} = 0.1$, $p_{RB} = 0.1$, $p_C = 0.1$, $p_{LC} = 0.5$, and $v_{max} = 6$. Each cell corresponds to 3.75 m; thus, the length of a vehicle is 7.5 m. The first 50000 time steps are discarded to avoid the transient behaviors. The detector is set at the 490th cell on each lane upstream the intersection to obtain the flux by counting the number of vehicles in 20000 time steps. The flux of lane A, B, and C are denoted as $flux_A$, $flux_B$, and $flux_C$ which represent the average flux of each time step. The total flux of T-shaped intersection is the sum of $flux_A$, $flux_B$, and $flux_C$, which is denoted as $flux_{All}$. The travel time detectors are placed relatively far upstream and downstream of the intersection to better capture the delays of each movement at the intersection. Delay data are extracted for the 20000 time step. The $averageDelay_A$, $averageDelay_B$, $averageDelay_C$ and $averageDelay_{All}$ denote the average control delay of each vehicle on lane A, B, C and the intersection.

3.1. Performance of Flux and Average Control Delay. The flux and average control delay can reflect the traffic condition of intersection. But their performance is different. Figure 3 shows (a) the flux and (b) the average control delay of intersection in space (p_A, p_B). The simulation system possesses a certain ability of self-organization. So, it can suffer a limited fluctuate of inflow rates and remain a small change. Figure 3(a) contains four plane regions and three transitional regions which are very narrow and steep. The plane region means the average control delay keeps a stable value with the vary of inflow rates. The transitional region means the average control delay increases dramatically with the increase of inflow rates. If the inflow rates exceed the critical value,

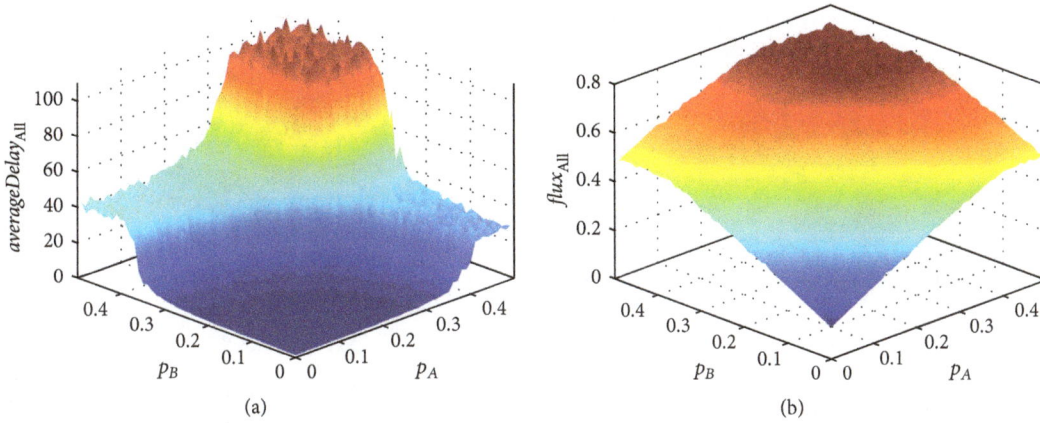

FIGURE 3: Performance of (a) average control delay and (b) flux of intersection in space (p_A, p_B). Assumed the parameter is $p_{UB} = 0$.

the average control delay will increase dramatically and then reach a new stable value. It means the simulate system can exchange from one equilibrium state to another equilibrium state quickly. When the inflow rates are very high, the simulate system can remain the equilibrium state, but it appears a little obvious fluctuation. A curved surface is seen in Figure 3(b). According to the change of color, it can be known that the flux is increasing steadily with the increase of inflow rates and then becomes to be constant value. It means the intersection capacity has a maximum value under a certain condition. It could not be changed by increasing the inflow rates and reflected the flexible equilibrium states. Compared the Figures 3(a) and 3(b), both of them can reflect the performance of intersection. The flux can reflect the intersection state in macrolevel such as free flow and jam flow. But the degree of traffic jam could not be known. The average control delay can reflect the condition of a vehicle movement at the intersection in microlevel. According to the average control delay, the free flow and degree of traffic jam can be classified. One plane region represents one particular state of intersection. And the average control delay can tell the driver how long he will waste to pass the intersection. It is very useful for a driver decide whether pass the intersection. The traffic management department can decide whether implements some traffic measures according to the average control delay. So, the average control delay can reflect the performance of intersection more reasonable than flux. Figure 4 shows the spatiotemporal diagram of lane A (a), (c), (e), and (f) and lane B (b), (d), (f), and (h) with different inflow rates. Each subgraph represents the traffic condition of each plane region in Figure 3(a). The free flow or congestion flow of each lane can be observed in Figure 4.

3.2. Influence of U-Turn Movements.

The influence of U-turn movements on the intersection is investigated in this part. The average control delay is used to reflect the performance of intersection.

Figure 5 shows the influence of U-turn movements on the performance of intersection under different inflow rates in the cases of (a) $p_{UA} = p_{UB} = 0$, (b) $p_{UA} = p_{UB} = 0.05$. Compared Figures 5(a) and 5(b), the traffic conditions become very sensitive to U-turn movements with the increase of inflow rates. A little increase of U-turn movements causes a large increase of average control delay. So, this plane region range, when the inflow rates are large enough, enlarged due to the appearance of U-turn movements. That is because the appearance of U-turn movements makes the traffic condition of intersection more worse, and more potential conflicts with other traffic streams are occurred. The vehicles must stop more frequently to avoid collisions. So, they spend more time to pass the intersection. That means it caused more control delay. It is not obvious when the inflow rates are small. But it is very obvious when the inflow rates are large. Therefore, reducing the U-turn movements can improve the traffic condition when the inflow rates are large.

Although the lane A and lane B are the major streets, the influence of U-turn movements on the intersection is different. Figure 6 shows the influence of U-turn movements at lane A and lane B on the performance of intersection in the cases of (a) $p_{UA} = 0.05$, $p_{UB} = 0$, (b) $p_{UA} = 0$, $p_{UB} = 0.05$. Compared Figures 6(a) and 6(b), the traffic conditions become more sensitive to U-turn movements at lane A than lane B with the increase of inflow rates. When the inflow rates are large enough, the average control delay caused by U-turn movements at lane A is larger than that at lane B. This plane region range with U-turn movements at lane A is also larger than at lane B. The reason is that the U-turning vehicles at lane A pass the intersection may conflict with vehicles on lane B and C. But the the U-turning vehicles at lane B just may conflict with vehicles on lane A. So, the U-turns at lane A make more conflict with other traffic streams than lane B. It suggests that control the U-turn movements at lane A can improve the traffic condition when the inflow rates are large.

4. Conclusion

In this paper, a new CA model is proposed to characterize the nonsignalized T-shaped intersection with U-turn movements. For this, new avoiding conflicts and gridlock avoiding rules are defined, and the average control delay (while not

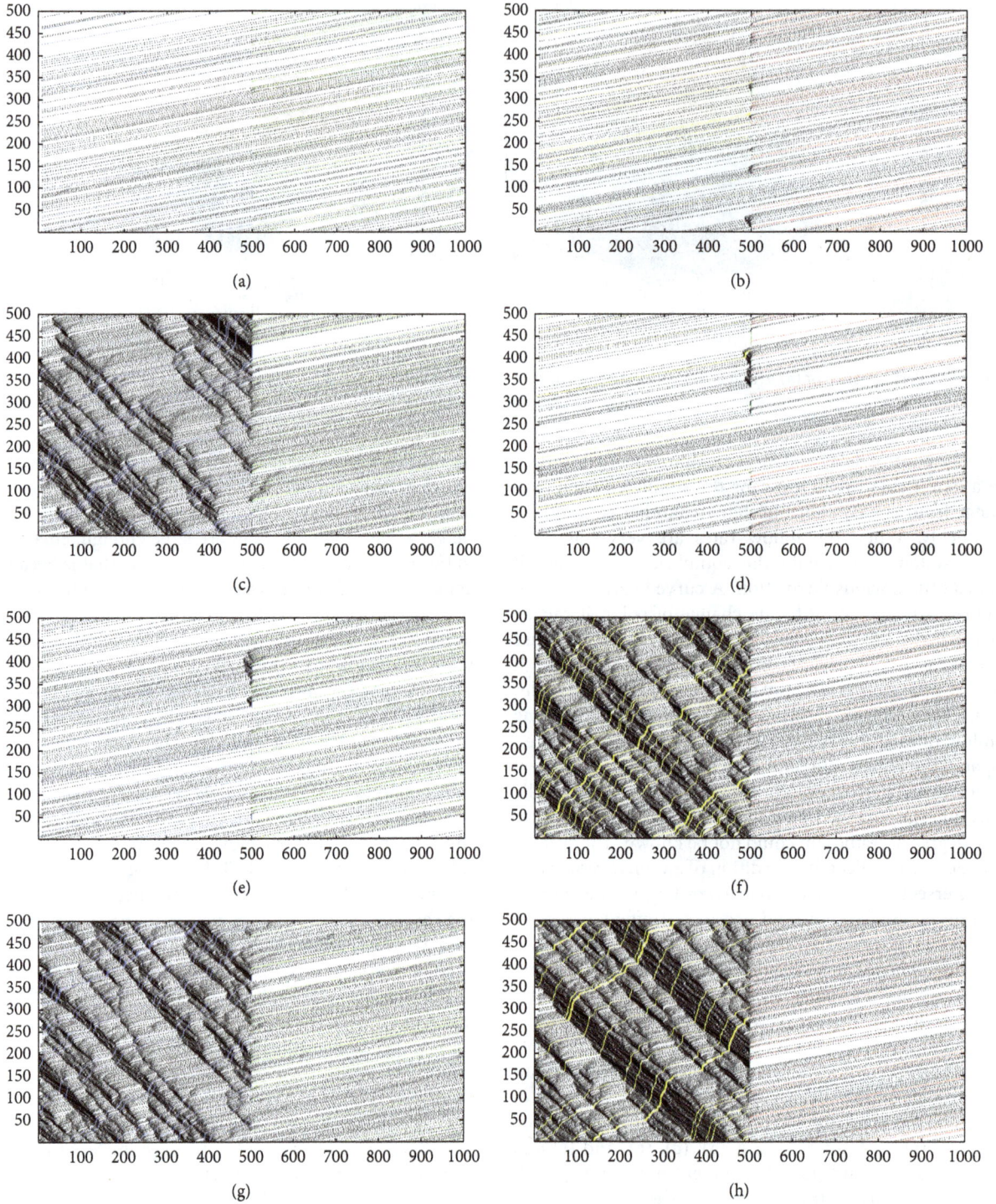

FIGURE 4: Spatiotemporal diagram of lane A (a), (c), (e), and (f) and lane B (b), (d), (f), and (h) with different inflow rates.

flux) is introduced as the performance measure. Simulations based on the present new CA model are executed. Three findings can be concluded from the simulation results: firstly, compare with flux, the average control delay is more practical to measure the performance of intersection; secondly, when the inflow rates are large, U-turn movements can worsen the traffic condition of intersection, that is, increasing both range and degree of high congestion; finally, U-turn movements on the different direction of main-road have asymmetrical influences on the traffic condition of intersection, for example, in the present example, the U-turn movements on lane A has more influence than that on lane B. Consequently, to

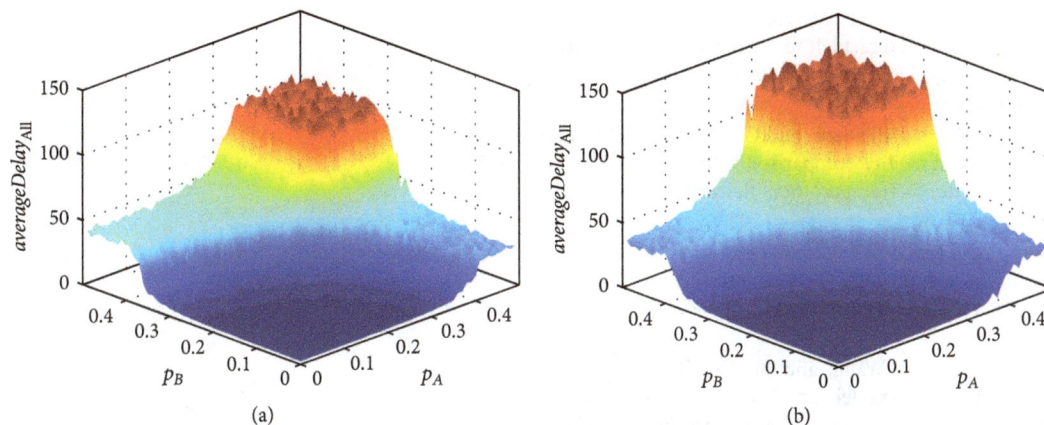

FIGURE 5: Influence of U-turns on the performance of intersection in space (p_A, p_B). In the cases of (a) $p_{UA} = p_{UB} = 0$ and (b) $p_{UA} = p_{UB} = 0.05$.

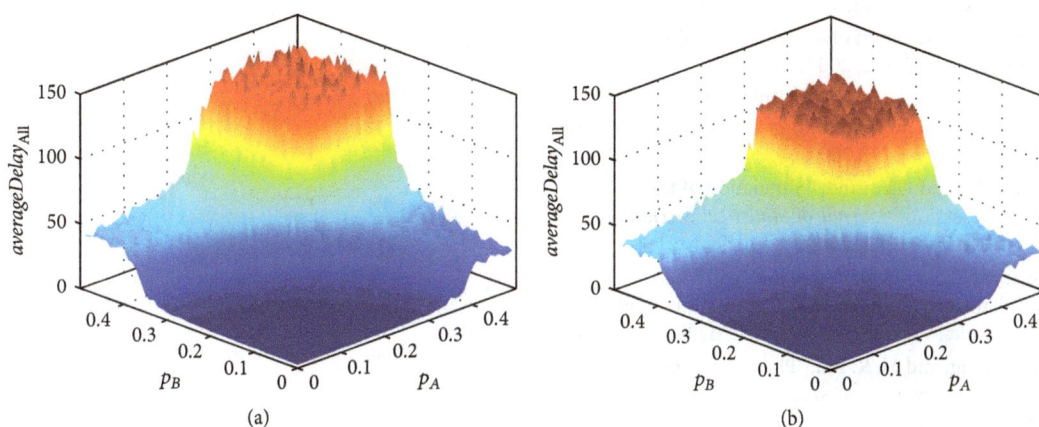

FIGURE 6: Influence of U-turns at different lane on the performance of intersection in space (p_A, p_B). In the cases of (a) $p_{UA} = 0.05$, $p_{UB} = 0$, (b) $p_{UA} = 0$ and $p_{UB} = 0.05$.

improve the traffic condition of intersection, U-turn movexments should be restricted when the inflow rates are large enough.

Acknowledgments

This work is financially supported by the National Basic Research Program of China (no. 2012CB725400), the National High Technology Research and Development Program (2011AA110303), the China National Science Funds for Excellent Young Researchers (71222101), the National Natural Science Foundation of China (71071013), the Major Projects of International Cooperation (71210001), and the Fundamental Research Funds for the Central Universities (2013YJS053).

References

[1] J. C. Adams and J. E. Hummer, "Effects of U-turns on left-turn saturation flow rates," *Transportation Research Record*, no. 1398, pp. 90–100, 1993.

[2] S. M. Tsao and S. W. Chu, "Study on adjustment factors for U-turns in left-turn lanes at signalized intersections," *Journal of Advanced Transportation*, vol. 29, no. 2, pp. 183–192, 1995.

[3] D. Carter, J. E. Hummer, R. S. Foyle, and S. Phillips, "Operational and safety effects of U-turns at signalized intersections," *Transportation Research Record*, no. 1912, pp. 11–18, 2005.

[4] P. Liu, J. J. Lu, J. Fan, J. C. Pernia, and G. Sokolow, "Effects of U-turns on capacities of signalized intersections," *Transportation Research Record*, no. 1920, pp. 74–80, 2005.

[5] P. Liu, X. Wang, J. Lu, and G. Sokolow, "Headway acceptance characteristics of U-turning vehicles at unsignalized intersections," *Transportation Research Record*, no. 2027, pp. 52–57, 2007.

[6] P. Liu, T. Pan, J. J. Lu, and B. Cao, "Estimating capacity of U-turns at unsignalized intersections: conflicting traffic volume, impedance effects, and left-turn lane capacity," *Transportation Research Record*, no. 2071, pp. 44–51, 2008.

[7] P. Liu, J. J. Lu, H. Zhou, and G. Sokolow, "Operational effects of U-turns as alternatives to direct left-turns," *Journal of Transportation Engineering*, vol. 133, no. 5, pp. 327–334, 2007.

[8] T. Guo, P. Liu, J. J. Lu, L. Lu, and B. Cao, "Procedure for evaluating the impacts of indirect driveway left-turn treatments on traffic operations at signalized intersections," *Journal of Transportation Engineering*, vol. 137, no. 11, pp. 760–766, 2011.

[9] X. K. Yang and H. G. Zhou, "CORSIM-based simulation approach to evaluation of direct left turn versus right turn plus U-turn from driveways," *Journal of Transportation Engineering*, vol. 130, no. 1, pp. 68–75, 2004.

[10] A. Pirdavani, T. Brijs, T. Bellemans, and G. Wets, "Travel time evaluation of a U-turn facility: comparison with a conventional signalized intersection," *Transportation Research Record*, no. 2223, pp. 26–33, 2011.

[11] M. El Esawey and T. Sayed, "Operational performance analysis of the unconventional median U-turn intersection design," *Canadian Journal of Civil Engineering*, vol. 38, no. 11, pp. 1249–1261, 2011.

[12] P. Liu, X. Qu, H. Yu, W. Wang, and B. Cao, "Development of a VISSIM simulation model for U-turns at unsignalized intersections," *Journal of Transportation Engineering*, vol. 138, pp. 1333–1339, 2012.

[13] M. E. Fouladvand, Z. Sadjadi, and M. R. Shaebani, "Characteristics of vehicular traffic flow at a roundabout," *Physical Review E*, vol. 70, no. 4, Article ID 46132, 8 pages, 2004.

[14] D. W. Huang, "Phase diagram of a traffic roundabout," *Physica A*, vol. 383, no. 2, pp. 603–612, 2007.

[15] M. E. Foulaadvand and S. Belbasi, "Vehicular traffic flow at a non-signalized intersection," *Journal of Physics A*, vol. 40, no. 29, article 006, pp. 8289–8297, 2007.

[16] S. Belbasi and M. E. Foulaadvand, "Simulation of traffic flow at a signalized intersection," *Journal of Statistical Mechanics*, vol. 2008, no. 7, Article ID P07021, 2008.

[17] M. E. Foulaadvand, M. Fukui, and S. Belbasi, "Phase structure of a single urban intersection: a simulation study," *Journal of Statistical Mechanics*, vol. 2010, no. 7, Article ID P07012, 2010.

[18] Q. L. Li, B. H. Wang, and M. R. Liu, "Phase diagrams properties of the mixed traffic flow on a crossroad," *Physica A*, vol. 389, no. 21, pp. 5045–5052, 2010.

[19] X. B. Li, R. Jiang, and Q. S. Wu, "Cellular automaton model simulating traffic flow at an uncontrolled T-shaped intersection," *International Journal of Modern Physics B*, vol. 18, no. 17–19, pp. 2703–2707, 2004.

[20] Q. S. Wu, X. B. Li, M. B. Hu, and R. Jiang, "Study of traffic flow at an unsignalized T-shaped intersection by cellular automata model," *European Physical Journal B*, vol. 48, no. 2, pp. 265–269, 2005.

[21] X. G. Li, Z. Y. Gao, B. Jia, and X. M. Zhao, "Cellular automata model for unsignalized T-shaped intersection," *International Journal of Modern Physics C*, vol. 20, no. 4, pp. 501–512, 2009.

[22] Z. J. Ding, X. Y. Sun, R. R. Liu, Q. M. Wang, and B. H. Wang, "Traffic flow at a signal controlled t-shaped intersection," *International Journal of Modern Physics C*, vol. 21, no. 3, pp. 443–455, 2010.

[23] H.-Q. Fan, B. Jia, X.-G. Li, and J.-F. Tian, "Characteristics of traffic flow at non-signalized T-shaped intersection," *Journal of Transportation Systems Engineering and Information Technology*, vol. 12, no. 1, pp. 185–192, 2012.

[24] K. Nagel and M. Schreckenberg, "A cellular automaton model for freeway traffic," *Journal de Physique*, vol. 2, pp. 2221–2229, 1992.

17

Fracture Property of Y-Shaped Cracks of Brittle Materials under Compression

Xiaoyan Zhang, Zheming Zhu, and Hongjie Liu

Department of Engineering Mechanics, Sichuan University, Chengdu 610065, China

Correspondence should be addressed to Zheming Zhu; zhuzm1965@163.com

Academic Editor: Xiao-wei Gao

In order to investigate the properties of Y-shaped cracks of brittle materials under compression, compression tests by using square cement mortar specimens with Y-shaped crack were conducted. A true triaxial loading device was applied in the tests, and the major principle stresses or the critical stresses were measured. The results show that as the branch angle θ between the branch crack and the stem crack is 75°, the cracked specimen has the lowest strength. In order to explain the test results, numerical models of Y-shaped cracks by using ABAQUS code were established, and the J-integral method was applied in calculating crack tip stress intensity factor (SIF). The results show that when the branch angle θ increases, the SIF K_I of the branch crack increases from negative to positive and the absolute value K_{II} of the branch crack first increases, and as θ is 50°, it is the maximum, and then it decreases. Finally, in order to further investigate the stress distribution around Y-shaped cracks, photoelastic tests were conducted, and the test results generally agree with the compressive test results.

1. Introduction

Cracks are frequently encountered in many engineering structures, such as rock and concrete, and such cracks usually play a dominative role in structure stability. In order to predict and prevent engineering disasters induced by such cracks, it is necessary to investigate the properties of crack propagation so as to obtain a better understanding of the dominant parameters that control material fracturing. Therefore, a great deal of efforts from theoretical and experimental points of view has been devoted to the study of the physics of crack fracture, and accordingly many significant results have been presented in the literatures [1–5].

Cracks are not always straight, and they may have many different shapes, such as curved cracks, intersected cracks, and branched cracks. Y-shaped crack is one kind of branched cracks and exists widely in structures because they are easily developed as two cracks intersect or coalesce. For the problem of multiple interacting cracks in an infinite plate, Yavuz et al. [6] formulated the stress intensity factor (SIF) by using integral equations expressed in terms of unknown edge dislocation along crack lines, and finally some new and

challenging crack interaction problems including branched Y-cracks, two-kinked V-cracks were analyzed. For symmetry Y-shaped cracks under compression, Li and Zhu [7] reported that there is a particular relation between the crack angle and the SIF, and the results showed that the SIF at crack tips is strongly dependent both on crack angle and crack length. By using finite element method, Isida and Noguchi [8] solved the basic problems of branched cracks in an infinite body through diverse numerical methods. Moreover, also by using finite element method, several researchers [9–15] calculated the growing path and the SIF values of branched cracks. For nonstraight cracks, Tilbrook and Hoffman [16] employed a simple analytical model to predict mechanical energy release rate and deflection angle for a range of crack shapes under mixed-mode loading. It was found that the crack length and orientation of the crack tip with respect to loading direction are the key influences on fracture parameters. The experimental study of the coalescence property of two cracks under compression conducted by Bobet and Einstein [17] showed that as the vertical load increases, new cracks emanate from the flaws and eventually coalesce; flaw slippage, wing-crack initiation, secondary crack initiation, crack coalescence, and

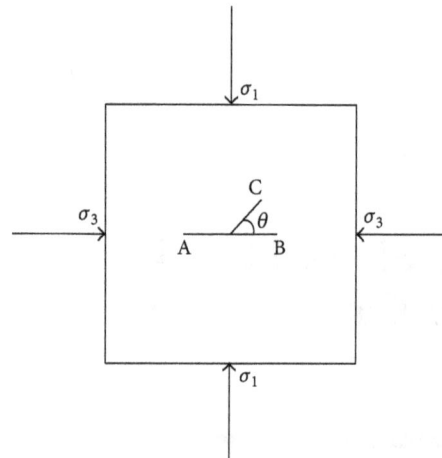

FIGURE 1: Y-shaped crack under compression.

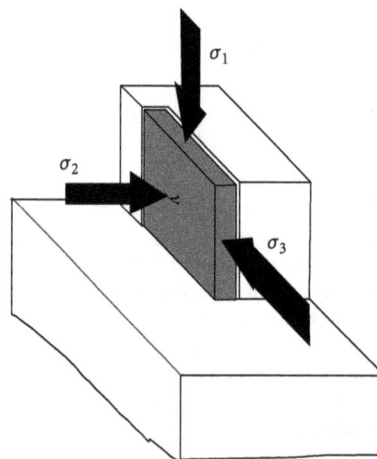

FIGURE 2: Sketch of a true triaxial loading device.

failure were observed. Two types of cracks arised: wing-cracks, which are tensile cracks, and secondary cracks which initiate as shear cracks in a plane roughly coplanar with the flaw.

In order to investigate the fracture behavior of Y-shaped cracks, experimental and numerical studies are implemented in this paper. The experimental studies include true triaxial compression test and photoelastic test. A true triaxial loading device is applied in the measurement of the compressive critical stress of the cement mortar specimens with Y-shaped cracks, and meanwhile in order to qualitatively investigate the crack tip SIFs for the specimens with various branch angles, photoelastic tests are conducted. In the numerical simulation, the finite element code ABAQUS is applied in the calculation of SIF values of Y-shaped crack tips.

2. Compression Tests

The specimens were square plates, 150 mm ∗ 150 mm ∗ 50 mm, with a Y-shaped artificial and penetrated crack, as

shown in Figure 1. The main crack AB measures 40 mm in length, while the branch crack measures 20 mm, and the branch angle θ varies from 15° to 90°. The material is cement mortar, and the ratio of cement : water : sand is 1 : 0.5 : 3 by weight. The cracks were made by using a very thin, 0.1 mm, film. The films were placed inside the samples during the process of casting in a mold until they were loaded. The curing period of the samples is 28 days. It is found that after the specimens have been stored in a heating apparatus with a controlled temperature for more than 2 hrs, the films can be easily pulled out from the specimens.

The specimens were loaded by a true triaxial loading device, as shown in Figure 2. The vertical loading is the major principal stress σ_1, and the two horizontal confining stresses are kept as constants during the process of vertical loading. One of the horizontal stress is σ_3, and the other one is σ_2, In order to avoid the specimen damage before testing, the specimens, at beginning, are loaded simultaneously along three directions, and after σ_2 and σ_3 are fixed to the values predesigned, we continue to increase σ_1 gradually until

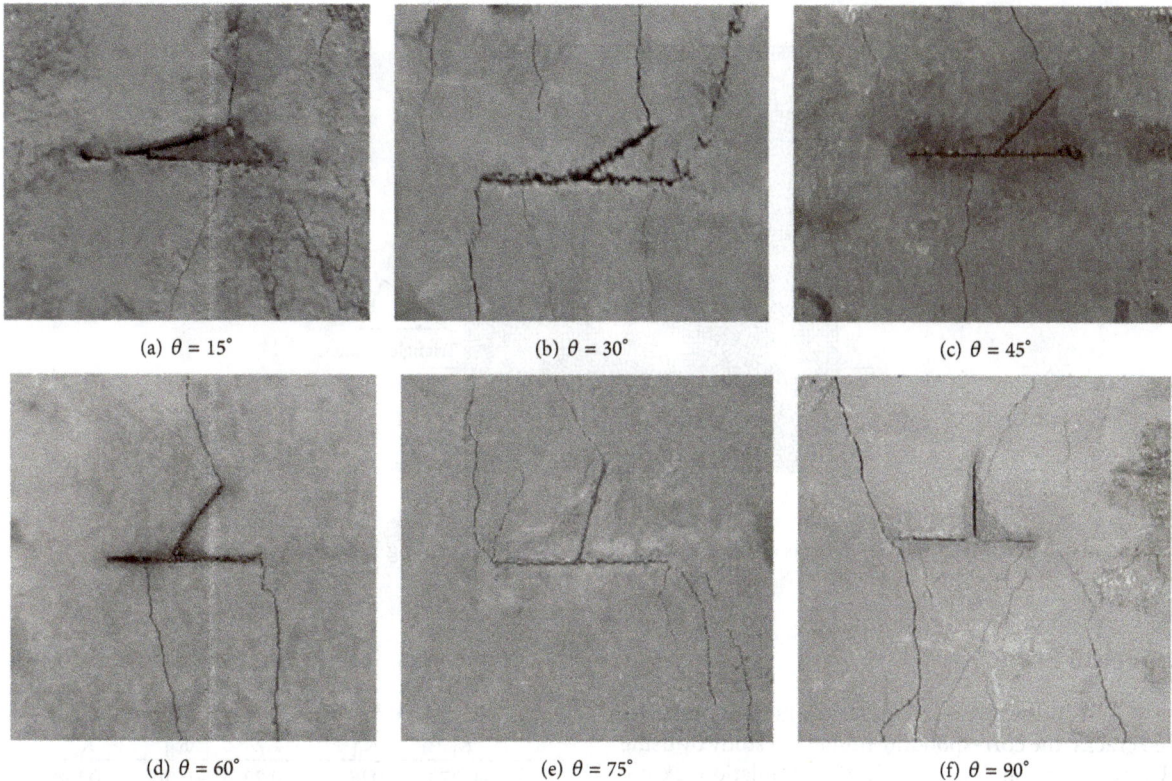

(a) $\theta = 15°$ (b) $\theta = 30°$ (c) $\theta = 45°$

(d) $\theta = 60°$ (e) $\theta = 75°$ (f) $\theta = 90°$

FIGURE 3: Test result of the fracture patterns for the specimens with different branch angles.

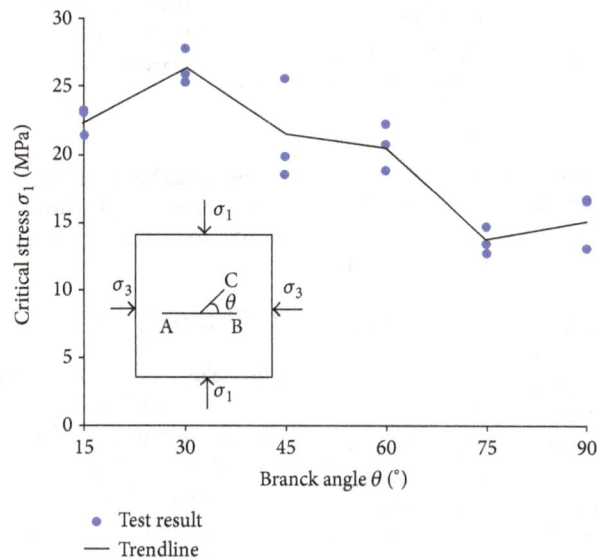

FIGURE 4: Test result for the relation between the critical stress and branch crack angles θ.

the specimen fails. The maximum vertical stress σ_1 is selected as the critical stress of the specimens. In order to avoid the effect of the friction between the specimen and the loading device, the specimen surfaces were smeared with oil before testing.

Figure 3 shows the fracture patterns for the specimens with branch angles 15°, 30°, 45°, 60°, 75°, and 90°. It can be seen that wing-cracks often initiate at branch crack tip C and then they grow and curve to the uploading boundary of σ_1.

Figure 4 shows the test result of the critical stress σ_1 versus branch angle θ. As the branch angle is 30°, the average test result of the critical stress is the largest, and as it increases from 30° to 90°, the critical stress generally decreases, and as the branch angle is 75°, the average test result is the lowest.

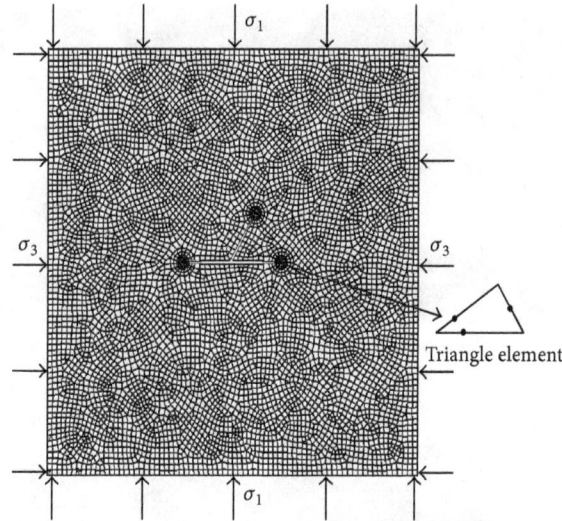

FIGURE 5: Elements applied in dividing the domain containing Y-shaped cracks.

3. Numerical Study

In order to obtain the stress intensity factors of asymmetric Y-shaped cracks, the corresponding numerical study by using ABAQUS code was implemented. In this simulation, except for the zone near crack tips where triangular elements CPS6 were applied, quadrilateral elements CPS8 were employed, as shown in Figure 5. The J-integral or contour integral method was applied in calculating crack tip stress intensity factor (SIF), and the material parameters applied in this numerical study are as follows. Young's modulus is 29.5 GPa; Poisson's ratio is 0.2; the major principal stress σ_1 is 1.0 MPa; and σ_3 is 0.05 MPa. In Total 20 models with different branch angles were established, and the corresponding calculation results are plotted in Figures 6 and 7. For the branch angle θ is 15°, 30°, 45°, 60°, 75°, and 90°; the corresponding calculation results of SIF values at crack tips A, B, and C are presented in Table 1.

It can be seen that the values of K_I at crack tips A and B are always negative. This is because under compression, there is no stress concentration at crack tips, and accordingly the mode I SIF is negative which cannot cause crack propagation. The values of K_I at crack tip C increase as the branch angles θ increase, and as θ is larger than 65°, the K_I changes from negative to positive which could cause crack propagation.

The mode II SIFs K_{II} changes with the branch angle θ, and the absolute value of K_{II} at crack tip A increases slightly with the branch angle θ. This is due to the effect from tip C which increases as θ increases. As $\theta = 90°$, the absolute values of K_{II} at crack tips A and B are roughly equal, but their signs are inverse. The absolute value of K_{II} at crack tip C increases as θ increases from 0° to 50°, and as $\theta = 50°$, the K_I absolute value is the maximum and then it starts to decrease as θ increases from 50° to 90°.

From Figures 5 and 6, one can find that for the branch crack C, as θ is larger than 65°, the values of K_I is positive, and meanwhile the absolute values of K_{II} is large. Combing the values of K_I and K_{II}, one can find that as the branch angle θ is

TABLE 1: Calculation results of SIF (MPa∗m$^{1/2}$) from ABAQUS code.

Angle	Crack A		Crack B		Crack C	
	K_I	K_{II}	K_I	K_{II}	K_I	K_{II}
15°	−0.270	−0.002	−0.22	−0.018	−0.168	−0.010
30°	−0.271	−0.008	−0.247	0.001	−0.124	−0.051
45°	−0.272	−0.017	−0.265	0.017	−0.067	−0.068
60°	−0.273	−0.026	−0.275	0.029	−0.013	−0.067
75°	−0.276	−0.034	−0.279	0.036	0.026	−0.039
90°	−0.279	−0.038	−0.279	0.038	0.041	0

in the range between 70° and 75°, the compressive strength of the cracked sample should be low, which agrees with the test results shown in Figure 4.

4. Photoelastic Tests

Photoelasticity is an experimental technique for stress and strain analysis that is particularly useful for the samples with a complex geometry and loading conditions. It is a nondestructive, whole-field, graphic stress-analysis technique based on an optical mechanical property called birefringence [18, 19].

According to the standard of marking photoelastic specimen method, a group of specimens as shown in Figure 1 with the inclined angles 0°, 15°, 30°, 45°, 60°, 75°, and 90° are conducted, respectively. The material was polycarbonate (PC) which possesses high transparency. The size of the specimen was 100 mm ∗ 100 mm ∗ 8 mm, and the width of cracks in the specimens was 1 mm.

According to stress-optical law of plane photoelastic experiment, the maximum shear stress τ_m can be calculated based on the following equation:

$$\tau_m = \frac{1}{2}\left(\sigma_1 - \sigma_3\right) = \frac{Nf_\sigma}{2h}, \tag{1}$$

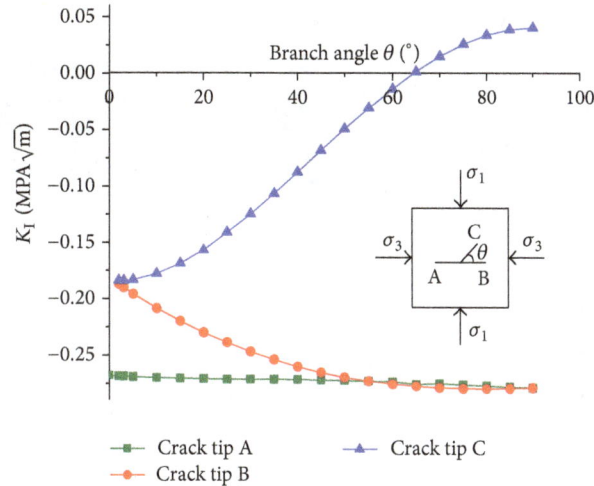

FIGURE 6: Curves of SIF K_I versus the branch angle θ.

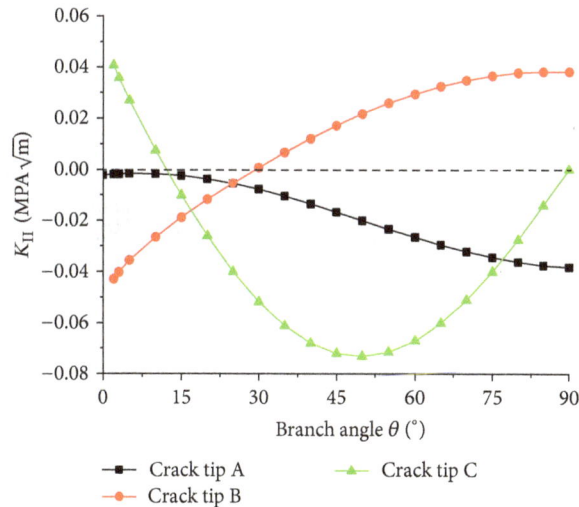

FIGURE 7: Curves of SIF K_{II} versus the branch angle θ.

where N is the number of isochromatic fringes, f_σ is fringe value, and h is the thickness of specimen.

Figure 8 shows the test results of isochromatic fringes for the specimens with different branch angles θ. It can be seen that as θ increases, the fringes number near crack tips increases gradually, which means that crack SIF values increase, and as the branch angle is in the range between 60° and 75°, the fringes number is more. This means that the compressive strength is low, which can be proved from the test results of the compressive strength shown in Figure 4.

5. Conclusion

In order to study the fracture property of Y-shaped cracks under compression, compression tests by using a true triaxial loading device and the corresponding photoelastic experiments have been conducted. The stress intensity factors of Y-shaped cracks have been calculated by using ABAQUS code. In the numerical study, J-integration method has been applied in calculating stress intensity factors. Through the experimental and numerical studies, the following conclusions can be obtained.

(1) For a Y-shaped crack under compression, at tips A and B, the mode I SIFs are always negative which will not induce fracturing, and their maximum mode II SIFs are much less than those at tip C.

(2) As the branch angle is larger than 65°, the mode I SIF at tip C is positive, and as the branch angle is 50°, tip C has the largest K_{II} value.

(3) The critical stress changes with the branch angles, and as the branch angle increases from 15° to 75°, the critical stress decreases, and as the branch angle is 30°, the principle stress is the largest, and as it is 75°, the principle stress is the lowest.

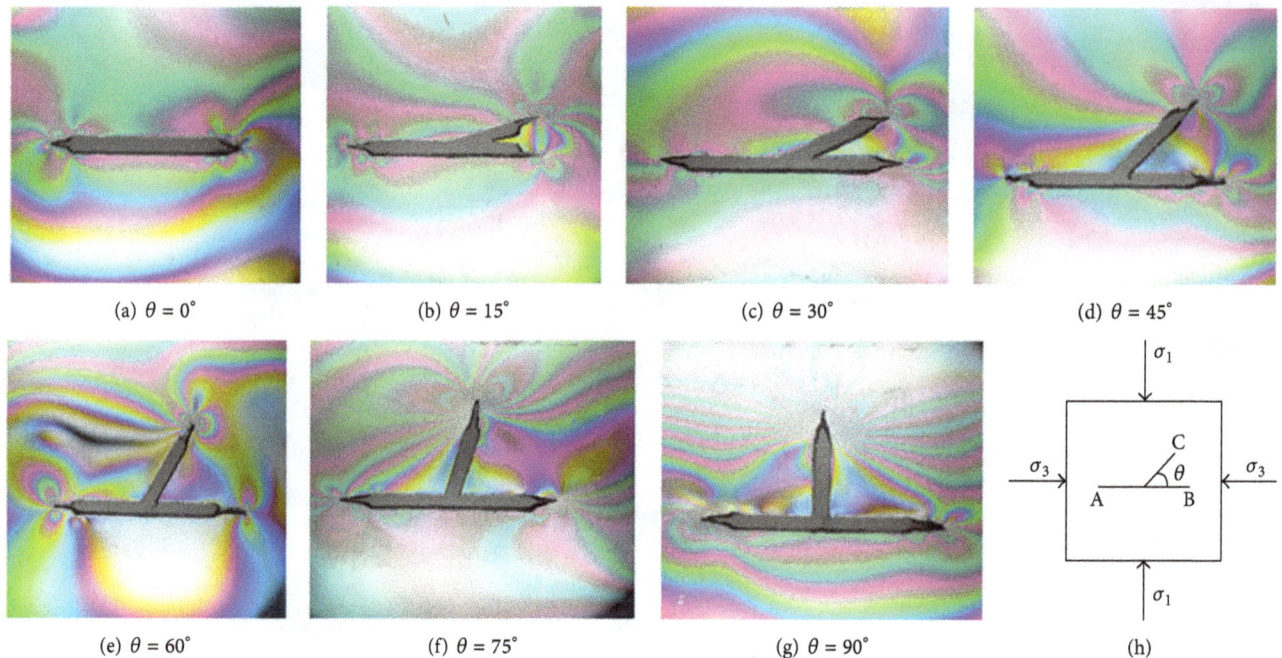

(a) $\theta = 0°$ (b) $\theta = 15°$ (c) $\theta = 30°$ (d) $\theta = 45°$

(e) $\theta = 60°$ (f) $\theta = 75°$ (g) $\theta = 90°$ (h)

FIGURE 8: Isochromatic fringes of Y-shaped cracks under biaxial compression.

Conflict of Interests

The authors declare that there is no conflict of interests regarding the publication of this paper.

Acknowledgments

This work was financially supported by the Open Fund of State Key Laboratory of Oil and Gas Reservoir Geology and Exploitation (PLN1202), by the project of Science and Technology of Sichuan province (2014JY0002), and by the Major State Basic Research Project (2010CB732005).

References

[1] S. Nemat-Nasser and H. Horii, "Compression-induced nonplanar crack extension with application to splitting, exfoliation, and rockburst," *Journal of Geophysical Research*, vol. 87, no. 8, pp. 6805–6821, 1982.

[2] Z. M. Zhu, L. Wang, B. Mohanty, and C. Huang, "Stress intensity factor for a cracked specimen under compression," *Engineering Fracture Mechanics*, vol. 73, no. 4, pp. 482–489, 2006.

[3] Z. M. Zhu, "An alternative form of propagation criterion for two collinear cracks under compression," *Mathematics and Mechanics of Solids*, vol. 14, no. 8, pp. 727–746, 2009.

[4] Z. M. Zhu, "New biaxial failure criterion for brittle materials in compression," *Journal of Engineering Mechanics, ASCE*, vol. 125, no. 11, pp. 1251–1258, 1999.

[5] P. Baud, T. Reuschlé, and P. Charlez, "An improved wing crack model for the deformation and failure of rock in compression," *International Journal of Rock Mechanics and Mining Sciences and Geomechanics*, vol. 33, no. 5, pp. 539–542, 1996.

[6] A. K. Yavuz, S. L. Phoenix, and S. C. Termaath, "An accurate and fast analysis for strongly interacting multiple crack configurations including kinked (V) and branched (Y) cracks," *International Journal of Solids and Structures*, vol. 43, no. 22-23, pp. 6727–6750, 2006.

[7] B.-Y. Li and Z.-M. Zhu, "Numerical and experimental research on the fracture and propagation of the branch crack under compression," *Journal of China Coal Society*, vol. 38, no. 7, pp. 1207–1214, 2013.

[8] M. Isida and H. Noguchi, "Stress intensity factors at tips of branched cracks under various loadings," *International Journal of Fracture*, vol. 54, no. 4, pp. 293–316, 1992.

[9] Y. K. Cheung, C. W. Woo, and Y. H. Wang, "A general method for multiple crack problems in a finite plate," *Computational Mechanics*, vol. 10, no. 5, pp. 335–343, 1992.

[10] C. Colombo and L. Vergani, "A numerical and experimental study of crack tip shielding in presence of overloads," *Engineering Fracture Mechanics*, vol. 77, no. 11, pp. 1644–1655, 2010.

[11] C. W. Smith and O. Olaosebikan, "Use of mixed-mode stress-intensity algorithms for photoelastic data," *Experimental Mechanics*, vol. 24, no. 4, pp. 300–307, 1984.

[12] C. W. Smith, J. J. McGowan, and M. Jolles, "Effects of artificial cracks and poisson's ratio upon photoelastic stress-intensity determination," *Experimental Mechanics*, vol. 16, no. 5, pp. 188–193, 1976.

[13] D.-C. Shin, J.-H. Nam, J.-S. Hawong, and J.-H. Lee, "Influences of equal biaxial tensile loads on the stress fields near the mixed mode crack," *Journal of Mechanical Science and Technology*, vol. 23, no. 8, pp. 2320–2329, 2009.

[14] I. Katsuhiko, "New model materials for photo," *Experimental Mechanics*, vol. 2, pp. 373–376, 1962.

[15] G. R. Irwin, "Analysis of stresses and strains near the end of a crack traversing a plate," *Journal of Applied Mechanics*, vol. 24, pp. 361–364, 1957.

[16] M. Tilbrook and M. Hoffman, "Approximation of curved cracks under mixed-mode loading," *Engineering Fracture Mechanics*, vol. 74, no. 7, pp. 1026–1040, 2007.

[17] A. Bobet and H. H. Einstein, "Fracture coalescence in rock-type materials under uniaxial and biaxial compression," *International Journal of Rock Mechanics and Mining Sciences*, vol. 35, no. 7, pp. 863–888, 1998.

[18] R. Patyṅska and J. Kabiesz, "Scale of seismic and rock burst hazard in the Silesian companies in Poland," *Mining Science and Technology*, vol. 19, no. 5, pp. 604–608, 2009.

[19] Z. P. Zhang, "Mixed mode stress intensity factors from photoelastic five-parameter method," *Chinese Journal of Applied Mechanics*, vol. 17, pp. 80–86, 2000.

18

Structural Damage Identification Based on Rough Sets and Artificial Neural Network

Chengyin Liu,[1,2] Xiang Wu,[3] Ning Wu,[1] and Chunyu Liu[1]

[1] *Harbin Institute of Technology, Shenzhen Graduate School, Shenzhen 518055, China*
[2] *Key Laboratory of C&PC Structures, Southeast University, Nanjing 211189, China*
[3] *State Key Laboratory of Robotics and System, Harbin Institute of Technology, Dazhi Street, Nangang District, Harbin 150001, China*

Correspondence should be addressed to Ning Wu; aning.wu@gmail.com

Academic Editor: Hua-Peng Chen

This paper investigates potential applications of the rough sets (RS) theory and artificial neural network (ANN) method on structural damage detection. An information entropy based discretization algorithm in RS is applied for dimension reduction of the original damage database obtained from finite element analysis (FEA). The proposed approach is tested with a 14-bay steel truss model for structural damage detection. The experimental results show that the damage features can be extracted efficiently from the combined utilization of RS and ANN methods even the volume of measurement data is enormous and with uncertainties.

1. Introduction

Structures are very vulnerable to influence like impact, earthquake and hurricanes. Therefore it is crucial for the decision maker to know the damage and health status of the structure in time, so that necessary maintenance can be taken. Recently, more and more innovative structural damage detection techniques have been applied to the existing structures for Structural Health Monitoring (SHM), especially large-scale structures, and many testing methods are nondestructive [1–3]. Attention has been drawn to how to use the current measurement data to produce a result with less uncertainty regardless of measurement noises and environmental variation, such as changing temperature, moisture, and load condition [4]. Many different approaches have been applied to solve the inaccurate measurement problem, for example, Sohn et al. proposed a probabilistic damage detection methodology to reduce measurement noises [5]. Worden and Dulieu-Barton investigated the influence of uncertainties both in practical measurement and in finite element model of damage detection [6], and they proposed a statistical method to resolve the inaccuracy that resulted from the modeling and measurement errors [7]. In recent studies, intelligent information processing techniques such as the autoregressive integrated moving average model, linear regression technique, ANN methods, and grey models are introduced to SHM applications.

ANN methods have been used extensively in structural damage identification. In practice, damage indexes in structures are firstly extracted by using signal processing techniques such as wavelet transform and Fourier analysis; then ANN models are built to detect structural damages from those indexes. It has been widely accepted that the ANN methods have helped to achieve a greater accuracy in structural damage detection. However, ANN has two obvious drawbacks when applying to a large number of data [8, 9]. The first one is that training an ANN model with big amount of data is time consuming, and the second one is that ANN cannot reach an analytical solution. In consequence, a reliable ANN model that can select the relevant factors automatically from the historical data is required.

As a useful mathematical tool, RS theory applies the unclear relation and data pattern comparison based on the concept of an information system with indiscernible data, where the data is uncertain or inconsistent. The characteristics of RS theory are to create approximate descriptions of objects for data analysis, optimization, and recognition, and it does not need the prior knowledge. Therefore using RS theory can evaluate the importance of various attributes and retain some key attributes with no additional knowledge except for

TABLE 1: Decision table general form.

Target	Condition attribute			Decision attribute
X	C_1	$\cdots$	C_n	d
x_1	$u_{1,1}$	$\cdots$	$u_{1,n}$	v_1
x_2	$u_{2,1}$	$\cdots$	$u_{2,n}$	v_2
$\cdots$	$\cdots$	$\cdots$	$\cdots$	$\cdots$
x_m	$u_{m,1}$	$\cdots$	$u_{m,n}$	v_m

the supplied data required [10]. To date, the RS approach has been applied in many domains, such as machine fault diagnosis, stock market forecast, decision support systems, medical diagnosis, data filtration, and software engineering [11–14].

The classical RS model can only be used to process categorical features with discrete values. For the RS based damage index selection in structural damage identification, a discretizing algorithm is required to partition the value domains of real-valued variables into several intervals as categorical features. Many discretization methods of numerical attributes have been proposed in recent years, including equal distance method, equal frequency method, and maximum entropy method [9]. However, discretization of numerical attributes may cause information loss because the degrees of membership of numerical values to discretized values are not considered [15, 16]. Recently, a discretization algorithm based on information entropy has been reported to be a potential mechanism for the measurement of uncertainty in RS. The information entropy has been widely employed in RS, and different information entropy models have been proposed. In particular, Düntsch and Gediga presented a well-justified information entropy model for the measurement of uncertainty in RS [17].

A novel application of integrating RS theory and ANN is presented in this paper for structural health monitoring and damage detection particularly for problems with large measurement data with uncertainties. The objective of the paper is to study how the RS and ANN techniques can be combined to detect structural damages. This method consists of three stages. First, RS will be applied to find relevant factors for structural modal parameters derived from structural vibration responses. Then, relevant information will be fed to the ANN as input. Finally, a synthesizing RS-ANN model based on the data-fusion technique will be used to assess the structural damage.

This paper is organized as follows. In Section 2, a brief introduction of fundamental theories on RS with information entropy is presented, and an overview of the ANN methods is given in Section 3. A three-stage damage detection model using combined RS and ANN technique is presented in Section 4. Laboratory experiment of a 14-bay truss model will be carried out to test and validate the proposed method in Section 5. Finally, concluding remarks are summarized in Section 6.

2. Information Entropy Based RS Theory

RS theory was proposed by Pawlak [18] as a new mathematical tool for reasoning about vagueness, uncertainty, and imprecise information. In this section, we introduce the concepts of decision table, discretization algorithm, and information entropy in RS theory and explain their relationships.

2.1. RS Theory. We have the following.

Definition 1. Decision table is a knowledge representation system in the application of RS theory with a quaternary (X, R, V, f) set, where X is a set of targets, and R is a set of attributes, $R = C \cup D$. C and D are condition attribute set and decision attribute set, respectively. $V = \cup V_r$ is a set of attributes' data range. V_r is the range of attribute r. $f : X \times R \to V$; f is an information function, which assigns the range of each attribute. Table 1 is a typical decision table.

Definition 2. X is a domain of discourse. P and Q are equivalence relations of universe X; then the P-positive region of Q is defined by the union of all the objects of U which can be classified as the equivalence class of U/Q by the knowledge U/P; that is,

$$\text{POS}_P(Q) = \bigcup_{Z \in X/Q} P(Z). \tag{1}$$

Definition 3. Let P and Q be equivalence relations of U. If (2) is satisfied, then $r \in P$ is said to be Q-dispensable in P; otherwise, $r \in P$ is Q-indispensable in P. If all r are Q-indispensable in P, P is said to be independent with respect to Q. Consider

$$\text{POS}_P(Q) = \text{POS}_{P-\{r\}}(Q). \tag{2}$$

Definition 4. If $S \subseteq P$ is P-independent and $\text{POS}_S(Q) = \text{POS}_P(Q)$ is satisfied, then S is said to be the Q-reduct of P, that is, $\text{RED}_Q(P)$, and the union of all the Q-indispensable attributes is said to be the Q-core of P, that is, $\text{CORE}_Q(P)$. The relation of these two notions is expressed as

$$\text{CORE}_Q(P) = \cap \text{RED}_Q(P). \tag{3}$$

2.2. Discretization Algorithm Based on Information Entropy. Let $U \subseteq X$ be a subset, and the number of instances is $|U|$. The number of jth ($j = 1, 2, \ldots, r$) decision attribute is k_j. Let the information entropy of this subset be

$$H(U) = -\sum_{j=1}^{r(d)} p_j \log_2 p_j, \quad p_j = \frac{k_j}{|U|}. \tag{4}$$

In general, $H(U) \geq 0$. If the information entropy is small, it reveals that several decision attributes are predominant, and the complexity is small. All the decision attributes especially are the same, and $H(U) = 0$. For the breakpoint c_i^a in the example, its decision attribute is j ($j = 1, 2, \ldots, r$); the

TABLE 2: Structural damage database.

Damage case	Damage condition			Natural frequency			$\cdots$	Mode curvature		
	Bay	Position	Degree	1	2	3	$\cdots$	1	$\cdots$	13
1	2	1	5%	8.76	32.34	61.57	$\cdots$	0.006	$\cdots$	0.006
2	2	1	10%	8.73	32.31	61.49	$\cdots$	0.006	$\cdots$	0.006
$\cdots$	$\cdots$	$\cdots$	$\cdots$	$\cdots$	$\cdots$	$\cdots$	$\cdots$	$\cdots$	$\cdots$	$\cdots$
684	13	3	95%	7.84	24.53	52.32	$\cdots$	0.006	$\cdots$	0.018

TABLE 3: Minimum property set 1 after reduction.

Damage case	Damage condition			Natural frequency			First order strain mode		
	Span	Position	Degree	1	2	3	1	$\cdots$	12
1	2	1	5%	8.76	32.34	61.57	0.003	$\cdots$	0.003
2	2	1	10%	8.73	32.31	61.49	0.004	$\cdots$	0.003
$\cdots$	$\cdots$	$\cdots$	$\cdots$	$\cdots$	$\cdots$	$\cdots$	$\cdots$	$\cdots$	$\cdots$
684	13	3	95%	7.84	24.53	52.32	0.003	$\cdots$	0.018

TABLE 4: Minimum property set 2 after reduction.

Damage case	Damage condition			Natural frequency			Second order strain mode		
	Span	Position	Degree	1	2	3	1	$\cdots$	12
1	2	1	5%	8.76	32.34	61.57	0.009	$\cdots$	-0.006
2	2	1	10%	8.73	32.31	61.49	0.012	$\cdots$	-0.006
$\cdots$	$\cdots$	$\cdots$	$\cdots$	$\cdots$	$\cdots$	$\cdots$	$\cdots$	$\cdots$	$\cdots$
684	13	3	95%	7.84	24.53	52.32	0.006	$\cdots$	-0.036

TABLE 5: Minimum property set 3 after reduction.

Damage case	Damage condition			Natural frequency			Third order strain mode		
	Span	Position	Degree	1	2	3	1	$\cdots$	12
1	2	1	5%	8.76	32.34	61.57	0.009	$\cdots$	0.009
2	2	1	10%	8.73	32.31	61.49	0.015	$\cdots$	0.009
$\cdots$	$\cdots$	$\cdots$	$\cdots$	$\cdots$	$\cdots$	$\cdots$	$\cdots$	$\cdots$	$\cdots$
684	13	3	95%	7.84	24.53	52.32	0.009	$\cdots$	0.054

TABLE 6: Sections of each attribute set and condition attributes.

Damage case	Natural frequency			Strain mode											
	1	2	3	1	2	3	4	5	6	7	8	9	10	11	12
Set 1, DB	3	2	5	5	6	5	7	7	5	5	7	7	5	6	5
Set 2, DB	3	2	5	9	7	8	9	10	11	11	9	9	8	7	9
Set 3, DB	3	2	5	8	7	7	8	10	10	10	10	8	7	7	8
Set 1, DP	4	10	10	3	5	2	4	4	4	4	4	4	2	5	3
Set 2, DP	4	10	10	1	2	4	4	3	5	5	3	4	4	2	1
Set 3, DP	4	10	10	1	1	1	5	4	3	3	4	5	1	1	1

TABLE 7: Attribute set 1 rules generation for damage bay.

Damage case	Natural frequency			First order strain mode											
	1	2	3	1	2	3	4	5	6	7	8	9	10	11	12
1	3	2	3	4	3	3	2	3	3	3	4	2	3	3	3	2
2	3	2	3	5	3	3	2	2	2	3	2	2	2	3	2	2
$\cdots$	$\cdots$	$\cdots$	$\cdots$	$\cdots$	$\cdots$	$\cdots$	$\cdots$	$\cdots$	$\cdots$	$\cdots$	$\cdots$	$\cdots$	$\cdots$	$\cdots$	$\cdots$
139	3	2	2	2	3	2	2	2	3	2	2	2	3	3	5	12
140	2	1	1	2	2	2	2	2	2	2	2	2	2	2	5	12

TABLE 8: Attribute set 2 rules generation for damage bay.

Damage case	Natural frequency			Second order strain mode											
	1	2	3	1	2	3	4	5	6	7	8	9	10	11	12
1	3	2	3	1	1	2	3	2	3	8	6	6	6	5	6	2
2	3	2	3	1	2	2	3	2	3	8	6	6	6	5	6	2
...	...	...	...	...	...	...	...	...	...	...	...	...	...	...	...
194	3	1	1	6	5	6	6	7	7	3	3	4	3	3	1	12
195	2	1	1	6	5	6	6	7	7	3	3	5	3	4	1	12

TABLE 9: Attribute set 3 rules generation for damage bay.

Damage case	Natural frequency			Third order strain mode											
	1	2	3	1	2	3	4	5	6	7	8	9	10	11	12
1	3	2	3	7	5	5	6	7	9	5	3	1	1	1	1	2
2	3	2	3	7	5	5	6	7	10	5	3	1	1	1	1	2
...	...	...	...	...	...	...	...	...	...	...	...	...	...	...	...
229	3	1	1	2	1	1	1	3	5	10	7	6	5	5	8	12
230	2	1	1	2	1	1	1	3	6	10	7	6	5	5	8	12

number of decision attributes less than c_i^a in the set U is $l_j^U(c_i^a)$, and the number of decision attributes greater than c_i^a in the set U is $r_j^U(c_i^a)$. Let

$$l^U(c_i^a) = \sum_{j=1}^{r(d)} l_j^U(c_i^a),$$

$$r^U(c_i^a) = \sum_{j=1}^{r(d)} r_j^U(c_i^a). \tag{5}$$

Therefore the breakpoint c_i^a could divide the set U into two subsets X_l and X_r. Let

$$H(X_l) = -\sum_{j=1}^{r(d)} p_j \log_2 p_j, \quad p_j = \frac{l_j^U(c_i^a)}{l^U(c_i^a)},$$

$$H(X_r) = -\sum_{j=1}^{r(d)} q_j \log_2 q_j, \quad q_j = \frac{r_j^U(c_i^a)}{r^U(c_i^a)}. \tag{6}$$

The information entropy of the breakpoint c_i^a to the set U is rewritten as

$$H^U(c_i^a) = \frac{|X_l|}{|X|} H(X_l) + \frac{|X_r|}{|X|} H(X_r). \tag{7}$$

Assume that $L = \{Y_1, Y_2, \ldots, Y_m\}$ is the equivalence selected by decision table; the new information entropy of the new breakpoint $c \notin P$ can be written as

$$H(c, L) = H^{Y_1}(c) + H^{Y_2}(c) + \cdots + H^{Y_m}(c). \tag{8}$$

Let P be the set of the chosen breakpoints, L is an equivalent set divided by breakpoint set P, S is the set of the initial breakpoint, and H is the information entropy of decision table; our discretization algorithm can be expressed as follows.

Step 1. $P = \emptyset$; $L = \{X\}$; $H = H(X)$.

Step 2. To any $c \in S$, calculate $H(c, L)$.

Step 3. If $H \leq \min\{H(c, L)\}$, go to the end.

Step 4. Select $c_{\max}$ into P to make $H(c, L)$ be minimum, $H = H(c, L)S = S - \{c\}$.

Step 5. To all $U \in L$, if $c_{\max}$ divide the equivalence U into X_1 and X_2, then delete U from L and join the equivalence X_1 and X_2 into L.

Step 6. If any equivalence in L has the same decision, go to the end. Otherwise go to Step 2.

3. Artificial Neural Network (ANN)

An artificial neural network (ANN) is an information processing paradigm inspired by biological nervous systems like brains. Although ANNs model the mechanism of brain, they do not have analytical function form, and therefore ANNs are data based instead of model based. An ANN is usually composed of a large number of highly interconnected processing elements (neurons) working in unison to solve specific problems.

The ANN used in this study is arranged in three layers of neurons, namely, the input, hidden, and output layers. The input layers introduce the model inputs, and the middle layer of hidden units feeds into an output layer through variable weight connections. The ANN learns by adjusting the values of these weights through a back-propagation algorithm that permits error corrections to be fed through the layers. Output layer provides the estimations of the network. An ANN is renowned for their ability to learn and generalize from example data, even when the data is noisy and incomplete. This ability has led to an investigation into the application

TABLE 10: Attribute set 1 rules generation for damage position.

Damage case	Natural frequency			First order strain mode											
	1	2	3	1	2	3	4	5	6	7	8	9	10	11	12
1	3	10	8	2	3	1	2	2	2	2	2	2	1	3	1	1
2	3	9	7	3	1	1	2	2	2	2	2	2	1	1	1	1
...	...	...	...	...	...	...	...	...	...	...	...	...	...	...	...	...
251	1	2	1	1	1	1	1	1	1	1	1	1	1	1	3	3
252	1	1	1	1	1	1	1	1	1	1	1	1	1	1	3	3

TABLE 11: Attribute set 2 rules generation for damage position.

Damage case	Natural frequency			Second order strain mode											
	1	2	3	1	2	3	4	5	6	7	8	9	10	11	12
1	3	10	8	1	1	1	2	1	1	3	1	3	3	1	1	1
2	3	9	7	1	1	1	2	1	1	3	1	3	3	1	1	1
...	...	...	...	...	...	...	...	...	...	...	...	...	...	...	...	...
253	1	1	1	1	1	3	3	1	3	1	1	2	2	1	1	3
254	1	1	1	1	1	3	3	1	3	1	1	3	3	1	1	3

of ANNs to automated knowledge acquisition. They also help to discern patterns among input data, require fewer assumptions, and achieve a higher degree of prediction accuracy.

4. The Hybrid Method

The common advantage of RS and ANN is that they do not need any additional information about data like probability in statistics or grade of membership in fuzzy-set theory [19]. RS has proved to be very effective in many practical applications. However, in RS theory, the deterministic mechanism for the description of error is too straightforward [20], and therefore the rules generated by RS are often unstable and have low classification accuracies. In consequence, RS cannot identify structural damage with a high accuracy. ANN is generally considered to be the most powerful classifier for low classification-error rates and robustness to noise. The knowledge of ANN is buried in their structures and weights [21, 22]. It is often difficult to extract rules from a trained ANN. The combination of RS and ANN is very natural for their complementary features.

One typical approach is to use the RS approach as a preprocessing tool for the ANN [12, 23]. RS theory provides useful techniques to reduce irrelevant and redundant attributes from a large database with various attributes. ANN has the ability to approach any complex functions and possess a good robustness to noise. In practice, there are often vast amounts of sensor data that are typically updated every few minutes in SHM system. One of the most important issues of RS theory is the reduction in dimension of the decision table in terms of both attributes and objects, thereby reducing the redundancy.

This paper will develop the structural damage model by using the RS methodology to reduce the dimension of the structural damage database before applying the ANN

method. Firstly, the following reductions can be derived based on the RS theory: attribute reduction, object reduction, and rule generation. Object reduction involves reducing the rows of the database in terms of redundant objects (rows). Rule generation involves the generation of If-Then rules from the database. Then the ANN is trained to learn in order to predict the damage conditions.

5. Experimental Validation

5.1. Test Structure. The test structure is a steel truss with 14 bays, shown in Figure 1. Each bay is 585 mm long, 490 mm wide, and 350 mm high. Totally, the steel truss has 52 longitudinal rods, 50 crosswise rods, and 54 diagonal rods. Each rod is forged with steel pipe. The section of the rods is hollow circular with an outer diameter of 18 mm, and inner diameter of 12 mm. Node board uses equilateral angle steel. Rods are bolted on the node board. Damages of the structure are simulated by two kinds of reduced thickness rods. One is 2 mm thick, and the other is 1 mm thick.

Accelerometers are mounted on each node of the structure as shown in Figure 2. The sampling interval of measurements retrieved from the data acquisition system is 5 min.

5.2. Establishment of Damage Database. A FE model was built to simulate the test structure as shown in Figure 3. In this study, three types of damage conditions are investigated, respectively, including damage bay, damage position, and damage degree. Since the end bays have no upper rod, the damage bay starts from the second span. Thus 12 bays are assumed to be damaged. In these bays, damage positions in upper rod, diagonal rod, and bottom rod are all known. For damage degree, we simulate the stiffness from 95% to 5% with the interval of 5%. In total there are 19 different kinds of damage degrees. Combining these three damage conditions, we have 684 damage conditions in total.

TABLE 12: Attribute set 3 rules generation for damage position.

Damage case	Natural frequency			Third order strain mode											
	1	2	3	1	2	3	4	5	6	7	8	9	10	11	12
1	3	10	8	1	1	1	4	4	3	3	1	1	1	1	1	1
2	3	9	7	1	1	1	4	4	3	3	1	1	1	1	1	1
...	...	...	...	...	...	...	...	...	...	...	...	...	...	...	...
273	3	2	3	1	1	1	1	1	3	3	4	4	1	1	1	3
274	2	2	3	1	1	1	1	2	3	3	4	4	1	1	1	3

TABLE 13: Damage identification by using attribute set 1.

Expectation	Bay	Position	Degree	Recognition	Bay	Position	Degree
1	7	upper	28.8%	1	7.32	1.28	15.96%
2	7	upper	62.2%	2	6.93	1.02	48.53%
3	7	diagonal	28.8%	3	7.23	1.97	30.68%
4	7	diagonal	62.2%	4	7.08	2.01	57.75%
5	7	bottom	28.8%	5	7.11	3.33	10.84%
6	7	bottom	62.2%	6	7.07	3.13	20.68%
7	5	upper	28.8%	7	5.12	1.17	40.52%
8	5	upper	62.2%	8	4.88	0.74	63.84%
9	5	diagonal	28.8%	9	4.53	1.45	82.56%
10	5	diagonal	62.2%	10	4.54	1.15	35.53%
11	5	bottom	28.8%	11	5.16	2.63	68.45%
12	5	bottom	62.2%	12	5.22	3.04	72.34%

FIGURE 1: Test structure.

FIGURE 2: Accelerometer.

According to the FEA results, 13 structural damage indexes are extracted, including the first three natural frequencies, the first three strain modes, the first three vibration mode shapes, modal assurance criterion (MAC), coordinate modal assurance criterion (COMAC), curvature mode, and natural frequency square. These indexes, together with damage conditions, form a 684 rows and 124 columns structural damage database (decision table) in this study. Table 2 lists part of the database. Note that in the damage position column, number 1, 2, and 3 represent the upper rod, diagonal rod, and bottom rod, respectively.

5.3. Attribute Reduction. In this section, application of RS to data reduction involves three steps (see below).

5.3.1. Step 1: Reduction of Decision Table. The damage database is reduced in batches as shown in Tables 3, 4, and 5. From the reduced database, it can be seen that the data volume has been greatly reduced. The core of the database is the first three natural frequencies. In order to ensure the integrity of the damage indexes, less reduced condition attributes are remained. There are 3 minimum properties in total. They are the first three frequencies with the first order strain mode (set 1), the first three frequencies with the second

TABLE 14: Damage identification by using attribute set 2.

Expectation	Bay	Position	Degree	Recognition	Bay	Position	Degree
1	7	upper	28.8%	1	6.24	1.34	13.41%
2	7	upper	62.2%	2	7.42	1.34	35.42%
3	7	diagonal	28.8%	3	7.14	1.84	42.52%
4	7	diagonal	62.2%	4	6.73	1.93	45.14%
5	7	bottom	28.8%	5	7.24	2.04	85.31%
6	7	bottom	62.2%	6	7.21	3.54	51.96%
7	5	upper	28.8%	7	4.76	1.42	45.15%
8	5	upper	62.2%	8	4.62	0.67	13.56%
9	5	diagonal	28.8%	9	5.25	2.02	41.08%
10	5	diagonal	62.2%	10	5.11	1.44	68.28%
11	5	bottom	28.8%	11	5.62	3.52	49.31%
12	5	bottom	62.2%	12	6.21	3.13	25.19%

TABLE 15: Damage identification by using attribute set 3.

Expectation	Bay	Position	Degree	Recognition	Bay	Position	Degree
1	7	upper	28.8%	1	7.22	1.17	58.74%
2	7	upper	62.2%	2	5.82	0.74	20.84%
3	7	diagonal	28.8%	3	6.81	1.88	66.68%
4	7	diagonal	62.2%	4	7.03	2.35	59.52%
5	7	bottom	28.8%	5	6.81	2.63	60.39%
6	7	bottom	62.2%	6	7.59	3.04	63.84%
7	5	upper	28.8%	7	4.84	1.14	62.56%
8	5	upper	62.2%	8	5.04	1.36	83.15%
9	5	diagonal	28.8%	9	5.14	2.44	39.16%
10	5	diagonal	62.2%	10	5.15	1.97	43.18%
11	5	bottom	28.8%	11	4.91	3.74	25.44%
12	5	bottom	62.2%	12	4.70	3.02	72.82%

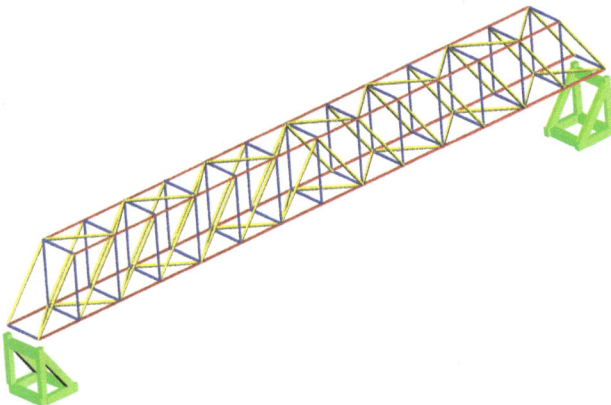

FIGURE 3: Test structure FE model.

order strain mode (set 2), and the first three frequencies with the third order strain mode (set 3), respectively.

5.3.2. Step 2: Discretization of Reduced Decision Table. Through the discretization of the three attribute sets, a set of reduced decision tables can be obtained. The attribute sets (1, 2, and 3) are discretized according to the decision attributes, the damage bay (DB), and the damage position (DP), respectively.

Table 6 summarizes the intervals of each decision attribute resulted from the discretization of the three attribute sets. It is found that, for the decision attribute of damage bay, the intervals are much more in the strain mode condition attributes than those in natural frequency condition attributes. While for the decision attribute of damage position, the intervals are much more at the natural frequency condition attributes than those in strain mode condition attributes. The result demonstrates that the strain mode has more weights in identification of structural damage bay, while the natural frequency has more weights in identification of structural damage position.

5.3.3. Step 3: Rules Generation. Rules generation is a key step in the RS analysis. In this study, the rules are generated from the discretized decision table in the form of knowledge. According to the exclusive rule extraction method, the same condition and decision attributes are removed. Therefore, simplified decision tables are obtained as shown in Tables 7, 8, 9, 10, 11, and 12. These decision tables demonstrate that every single damage case is unique.

From Table 7 to Table 12, it can be seen that the rows of each table are decreased to less than half of the original ones after rules generation. Each attribute set has its own rule of damage identification. The values of rule generation result for damage bay are less than those for damage position on average. It illustrates that the identification of damage bay is easier than that of damage position.

5.4. Identification of Structure Damage Using ANN. In this section, back-propagation ANN is applied to the reduced database for further identification of structural damages. The reduced database in terms of attributes can be described as the best subset of variables which describe the structural damage database completely. This reduction in number of attributes decreases the time of decision-making process and consequently reduces the cost of efficiency analysis. As mentioned above, three attribute sets are chosen as the input, and three damage conditions are chosen as the output to train the ANN model. The back-propagation network computes the weights in a recurrence mode from the last layer backward to the first layer.

Using real data obtained from the experimental testing, we put the experimental measurements into the trained ANN input layer to identify the structural damage. The results in Tables 13, 14, and 15 show that the RS method determines the group of input variables and generates the structural damage rule sets before using ANN. While the performance of the ANN model on identification of damaged degree is not very good, the hybrid method proposed in the paper is helpful to construct a good identification model for structural damage, offering an excellent performance of identifying the damaged bay and damaged position of the test structure.

6. Conclusions

In this paper, a novel method of combining RS and ANN methods is applied to the identification of structural damages. This study uses RS theory and integrates the inductive reduction algorithm and discretization algorithm based on information entropy to improve the ANN model for structural damage identification. Through a detailed experimental analysis of a 14-bay truss structure, this paper presents and discusses the conversion of damage index to RS object, predicting variables selection, removal of redundant from information table, and rules generation. The experiments data is preprocessed and reduced by RS before using ANN for identifying the damages of truss structure. The identification accuracy is mainly attributed to RS since it can remove redundant attributes without any classification information loss. Furthermore, the improvement in tolerance and accuracy with the proposed method shows that there is a great potential for integration of various techniques to improve the performance of an individual technique.

Conflict of Interests

The authors declare that there is no conflict of interests regarding the publication of this paper.

Acknowledgments

This work is supported by the Programme of Introducing Talents of Discipline to Universities (Grant no. B07018), Natural Science Foundation of China (Grant no. 60772072, 51108129), Key Laboratory of C&PC Structures Ministry of Education, Guangdong Major Science and Technology Plan (Grant no. 2007AA03Z117), Shenzhen Foundmental Research Project (Grant no. JC201105160538A) and Shenzhen Overseas Talents Project (Grant no. KQCX20120802140634893).

References

[1] C. R. Farrar, S. W. Doebling, and D. A. Nix, "Vibration-based structural damage identification," *Philosophical Transactions of the Royal Society A: Mathematical, Physical and Engineering Sciences*, vol. 359, no. 1778, pp. 131–149, 2001.

[2] S. W. Doebling and C. R. Farrar, "The state of the art in structural identification of constructed facilities," Tech. Rep., Los Alamos National Laboratory, Los Alamos, Mexico, 1999.

[3] S. W. Doebling, C. R. Farrar, and M. B. Prime, "A summary review of vibration-based damage identification methods," *Shock and Vibration Digest*, vol. 30, no. 2, pp. 91–105, 1998.

[4] C. R. Farrar and K. Worden, "An introduction to structural health monitoring," *Philosophical Transactions of the Royal Society A: Mathematical, Physical and Engineering Sciences*, vol. 365, no. 1851, pp. 303–315, 2007.

[5] H. Sohn, C. R. Farrar, F. M. Hemez, D. D. Shunk, D. W. Stinemates, and R. N. Brett, "A review of structural health monitoring literature: 1996–2001," Tech. Rep. LA-13976-MS, Los Alamos National Laboratory Report, 2003.

[6] K. Worden and J. M. Dulieu-Barton, "An overview of intelligent fault detection in systems and structures," *Structural Health Monitoring*, vol. 3, no. 1, pp. 85–98, 2004.

[7] C. R. Farrar, S. W. Doebling, P. J. Cornwell, and E. G. Straser, "Variability of modal parameters measured on the Alamosa Canyon Bridge," in *Proceedings of the 15th International Modal Analysis Conference*, pp. 257–263, February 1996.

[8] P. Lingras, "Comparison of neofuzzy and rough neural networks," *Information Sciences*, vol. 110, no. 3-4, pp. 207–215, 1998.

[9] B. S. Ahn, S. S. Cho, and C. Y. Kim, "Integrated methodology of rough set theory and artificial neural network for business failure prediction," *Expert Systems with Applications*, vol. 18, no. 2, pp. 65–74, 2000.

[10] X. Xiang, J. Zhou, C. Li, Q. Li, and Z. Luo, "Fault diagnosis based on Walsh transform and rough sets," *Mechanical Systems and Signal Processing*, vol. 23, no. 4, pp. 1313–1326, 2009.

[11] F. E. H. Tay and L. Shen, "Fault diagnosis based on rough set theory," *Engineering Applications of Artificial Intelligence*, vol. 16, no. 1, pp. 39–43, 2003.

[12] R. Zhou and J. G. Yang, "The research of engine fault diagnosis based on rough sets and support vector machine," *Transactions of CSICE*, vol. 24, no. 4, pp. 379–383, 2006 (Chinese).

[13] J. R. Li, L. P. Khoo, and S. B. Tor, "RMINE: a rough set based data mining prototype for the reasoning of incomplete data in condition-based fault diagnosis," *Journal of Intelligent Manufacturing*, vol. 17, no. 1, pp. 163–176, 2006.

[14] Z. Geng and Q. Zhu, "Rough set-based heuristic hybrid recognizer and its application in fault diagnosis," *Expert Systems with Applications*, vol. 36, no. 2, pp. 2711–2718, 2009.

[15] R. Jensen and Q. Shen, "Semantics-preserving dimensionality reduction: rough and fuzzy-rough-based approaches," *IEEE Transactions on Knowledge and Data Engineering*, vol. 16, no. 12, pp. 1457–1471, 2004.

[16] Q. Hu, D. Yu, J. Liu, and C. Wu, "Neighborhood rough set based heterogeneous feature subset selection," *Information Sciences*, vol. 178, no. 18, pp. 3577–3594, 2008.

[17] I. Düntsch and G. Gediga, "Uncertainty measures of rough set prediction," *Artificial Intelligence*, vol. 106, no. 1, pp. 109–137, 1998.

[18] Z. Pawlak, *Rough Sets: Theoretical Aspects of Reasoning about Data*, Kluwer Academic, 1991.

[19] R. Li and Z.-O. Wang, "Mining classification rules using rough sets and neural networks," *European Journal of Operational Research*, vol. 157, no. 2, pp. 439–448, 2004.

[20] J. Bazan, A. Skowron, and P. Synak, "Dynamic reducts as a tool for extracting laws from decisions tables," in *Proceedings of the Symposium on Methodologies for Intelligent Systems*, pp. 346–355, 1994.

[21] M. W. Craven and J. W. Shavlik, "Using neural networks for data mining," *Future Generation Computer Systems*, vol. 13, no. 2-3, pp. 211–229, 1994.

[22] H. Lu, R. Setiono, and H. Liu, "Effective data mining using neural networks," *IEEE Transactions on Knowledge and Data Engineering*, vol. 8, no. 6, pp. 957–961, 1996.

[23] R. W. Swiniarski and L. Hargis, "Rough sets as a front end of neural-networks texture classifiers," *Neurocomputing*, vol. 36, pp. 85–102, 2001.

19

Study on Typhoon Characteristic Based on Bridge Health Monitoring System

Xu Wang,[1] **Bin Chen,**[2] **Dezhang Sun,**[3] **and Yinqiang Wu**[4]

[1] *State Key Laboratory Breeding Base of Mountain Bridge and Tunnel Engineering, Chongqing Jiaotong University, Chongqing 400074, China*
[2] *College of Civil Engineering and Architecture, Zhejiang University, Hangzhou 310058, China*
[3] *Institute of Engineering Mechanics, China Earthquake Administration, Harbin 150080, China*
[4] *Bureau of Public Works of Shenzhen Municipality, Shenzhen 518006, China*

Correspondence should be addressed to Dezhang Sun; sundz2008@163.com

Academic Editor: Ying Lei

Through the wind velocity and direction monitoring system installed on Jiubao Bridge of Qiantang River, Hangzhou city, Zhejiang province, China, a full range of wind velocity and direction data was collected during typhoon HAIKUI in 2012. Based on these data, it was found that, at higher observed elevation, turbulence intensity is lower, and the variation tendency of longitudinal and lateral turbulence intensities with mean wind speeds is basically the same. Gust factor goes higher with increasing mean wind speed, and the change rate obviously decreases as wind speed goes down and an inconspicuous increase occurs when wind speed is high. The change of peak factor is inconspicuous with increasing time and mean wind speed. The probability density function (PDF) of fluctuating wind speed follows Gaussian distribution. Turbulence integral scale increases with mean wind speed, and its PDF does not follow Gaussian distribution. The power spectrum of observation fluctuating velocity is in accordance with Von Karman spectrum.

1. Introduction

Typhoon disaster is a major disaster in China, which has caused serious property and casualty losses annually and threatened sustainable development along the east coast of China. To have a better understanding of bridge destruction reasons from typhoon, wind characteristics near ground during typhoon attack must be studied firstly. However, experimental simulation of typhoon is very difficult because of its particularity. This has led to field measurement that has been recognized as the most effective research method and long-term direction in wind-resistant research of structure [1]. A huge mass of data was accumulated and parts of research result had been used in model of conduct in developed countries, where the field-measured research in strong wind characteristics developed earlier. Database for wind characteristics had been built in some countries, where the research of wind engineering is developed [2–6]. China's field-measured research of wind characteristics developed later but very fast; some valuable research results had been achieved in recent years [7–12]. Currently, most researches on near ground wind characteristics during typhoon attack focus on buildings; few have been performed on bridges, but it is important for wind engineering research.

In China, health monitoring systems have been installed on many long span bridges [13–16] to capture health states of bridges. Among all these systems, the health monitoring of wind load is of primary importance to bridge health monitoring systems because the health monitoring information gathered could be used in bridge design and construction. However, current practices of strong wind characteristics using wind load monitoring system on bridge site are still encountering shortage.

Based on the wind data measured from the anemometer positioned on Jiubao Bridge at 6 m height, Hangzhou city, near-ground wind characteristics under typhoon HAIKUI in

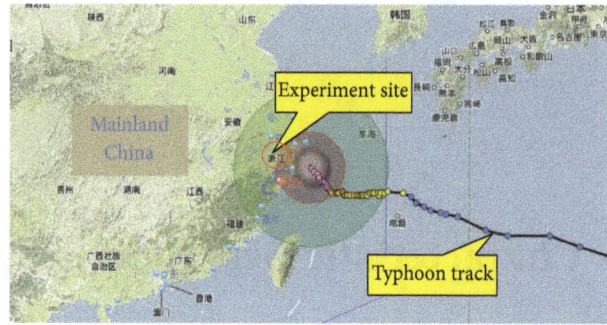

FIGURE 1: Experiment site and track of typhoon HAIKUI.

FIGURE 2: The location and photo of anemometers.

2012 were discussed in this paper, which include wind speed and direction, gust factor, turbulence intensity, peak factor, turbulence integral scale, and power spectrum of wind speed. Obviously, such studies are useful for promotions of wind-resistant design of bridge in the future.

2. Site Description and Measurement Characterization

2.1. Introduction of Typhoon HAIKUI. In 2012, the 11th tropical storm "HAIKUI" (named by HAIKUI, number 1211 tropical cyclone), named from a Chinese marine animal, headed to the coast of East China Sea at 17:00 on August 17, 2012, and landed at Zhejiang province in the morning of August 8. The location of the eye was 640 km from southwest of Zhejiang province, which was very close to the test bridge. The maximum 10 min mean wind speed at 10 m height was found to be 28 m/s, and the lowest pressure was 980 hPa. Figure 1 shows a satellite view of the path of typhoon HAIKUI and the experiment site. According to the statistical results of meteorological data from Chinese Ministry of Civil Affairs and Disaster Reduction Office, typhoon HAIKUI caused 6 casualties in Zhejiang, Shanghai, Jiangsu, and Anhui provinces and forced 2.173 million people to evacuate. The most damaged area was Zhejiang province, resulting in 7.001 million people involved, 1.546 million people were forced to move in emergency, and almost 5100 houses were destroyed.

Jiubao Bridge was located on Qiantang River, Hangzhou. The construction work began on December 18, 2009, and was completed in 2012. The bridge was designed to a standard bidirectional and six traffic lanes freeway, combining with 80 km/h of speed limit and about 1855 meters in span length. The whole bridge health monitoring systems which were comprised of vehicle system, wind speed and direction measurement system, and fatigue detection system were installed and distributed on the bridge. The meteorological data of typhoon was detected by a two-dimensional sonic anemometer in the wind speed and direction measurement system (see Figure 2). The anemometer, produced by British Gill Company, was Windsonic and has a sampling frequency of 4 Hz. Wind direction was defined north as 0° along a clockwise direction varying from 0° to 360°. Specifications and parameters of anemometer were considered with the records indicated in Table 1.

3. Study of the Field-Measured Wind Data

3.1. Mean Wind Speeds and Wind Directions. Figures 3 and 4 show the time history of wind speeds and directions of typhoon HAIKUI from 0:00 on August 8, 2012, to 0:00 August 9, 2012. Because typhoon HAIKUI did not directly impact the experimental site in lateral direction, an uphill and a downhill of wind speed were observed in the time history, indicating that a peak value was recorded.

Before analyzing wind characteristics, the overall sample needs to be separated into several segments by a specified time interval. They are different from regulations of time interval of mean wind speed among standards in many countries, such as 3 s used in America [17] and India [18] and 10 min used in Japan [19], Europe [20], and China [21]. In this paper, according to Chinese standard, wind data was separated into 144 samples, and the maximum 10 min mean

TABLE 1: Specifications of anemometer.

Brand (type)	Wind speed		Wind direction		Measurement and structure	
	Range	0–60 m/s (116 Knots)	Range	0–359°	Frequency	0.25 Hz, 0.5 Hz, 1 Hz, 2 Hz or 4 Hz
Gill Windsonic	Accuracy	±2%; 12 m/s	Accuracy	±3° at 12 m/s	Parameter	U and V
	Resolution	0.01 m/s (0.02 Knots)	Resolution	1°	Measurement unit	m/s, knots, mph, kph, ft/min
	Reaction time	0.25 s	Reaction time	0.25 s	Size	142 mm × 160 mm
	Lowest	0.01 m/s	/	/	Wight	0.5 Kg

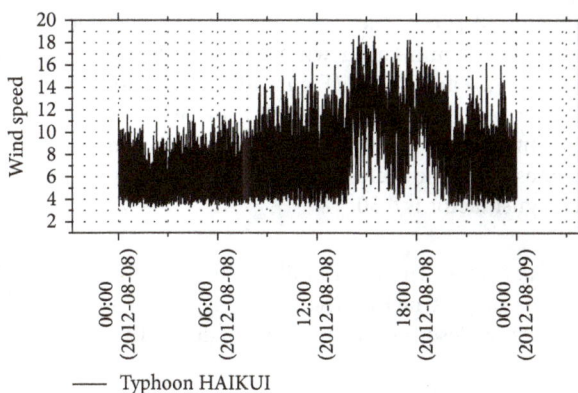

FIGURE 3: Instantaneous wind speed during typhoon HAIKUI.

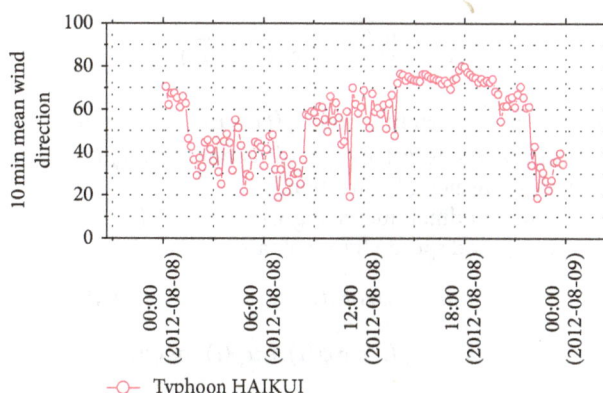

FIGURE 4: Instantaneous wind direction during typhoon HAIKUI.

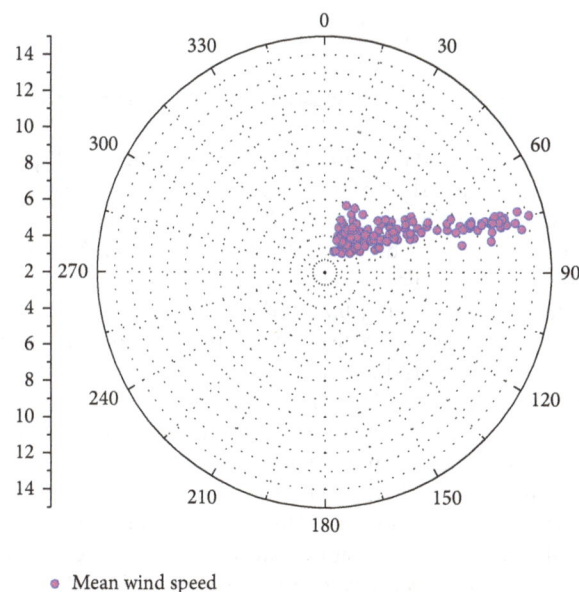

FIGURE 5: 10 min mean wind speed during typhoon HAIKUI.

FIGURE 6: 10 min mean wind direction during typhoon HAIKUI.

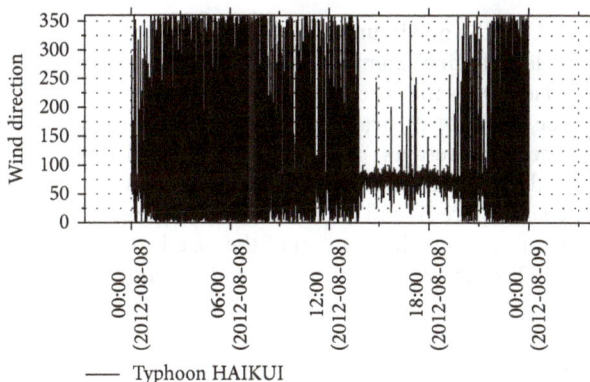

FIGURE 7: 10 min mean wind speed versus mean wind direction.

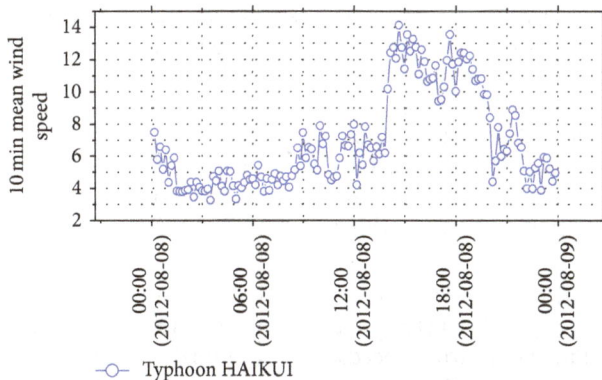

wind speed was 14.12 m/s. Figures 5 and 6 show time histories of 10 min mean wind speed and horizontal wind direction. Variations of 10 min mean wind speed with horizontal wind direction were shown in Figure 7, which made the variation trend of mean wind speed with horizontal wind direction intuitive.

After breaking down to 10 min time interval, the wind speed vector was resolved into three components [22], two of which are orthogonal, namely, $u_x(t)$ and $u_y(t)$, and can be measured synchronously by a two-dimensional sonic anemometer, denoting directions on x-axis and y-axis, respectively. The horizontal wind speed U and main wind direction ϕ are calculated by the following formulas:

$$U = \sqrt{\overline{u_x(t)}^2 + \overline{u_y(t)}^2},$$

$$\cos(\phi) = \frac{\overline{u_x(t)}}{U}, \qquad \sin(\phi) = \frac{\overline{u_y(t)}}{U}, \qquad (1)$$

$$\phi = \arccos\frac{\overline{u_x(t)}}{U} + \text{step}\left(-\overline{u_y(t)}\right)\cdot 180°,$$

where step($\cdot$) is step function, $\overline{u_x(t)}$ is the mean wind speed in 10 min time interval along X-direction, and $\overline{u_y(t)}$ has the same meaning along Y-direction.

The longitudinal fluctuating wind speed $u(t)$ and lateral fluctuating wind speed $v(t)$ are obtained from

$$u(t) = u_x(t)\cos\phi(t) + u_y(t)\sin\phi(t) - U,$$
$$v(t) = -u_x(t)\sin\phi(t) + u_y(t)\cos\phi(t). \qquad (2)$$

3.2. Turbulence Intensity. Turbulence intensity is considered the relative index for intensity of turbulence and then determined as

$$I_i = \frac{\sigma_i}{U} \quad (i = u, v, w), \qquad (3)$$

where σ_i is the standard deviation of fluctuating wind speed for component i ($i = u, v, w$).

Similar diurnal patterns of 10 min fluctuating wind speed distribution are shown in longitudinal (Figure 8(a)) and lateral (Figure 8(b)) directions. The changes in turbulence intensities are slow between 00:00 and 12:00 on August 8, with average turbulence intensities being 0.47 and 0.40 and maximum values being 0.71 and 0.67 in longitudinal and lateral directions, respectively. In both directions, a sharp decrease is found in the period between 12:00 and 20:00 (the lowest average intensities being 0.16 and 0.13, resp.) before an increase in the period between 20:00 and 00:00 on August 9 (variation ranging from 0.1 to 0.7).

Figure 9 shows both longitudinal and lateral turbulence intensity as a function of 10 min mean wind speed. The overall turbulence intensities reduce with decreased mean wind speed. It is clear from Figure 9 that an obvious decrease is found in turbulence intensities before mean wind speed approaching 8 m/s, but the changes become small when mean wind speed is greater than 8 m/s.

Table 2 shows the ratio of average longitudinal turbulence intensity to average lateral turbulence intensity in both domestic and international field research results. It can be seen from Table 2 that the ratio of the case is slightly larger than that in Peng et al. [12] and Fu et al.'s [23] results but quite close to the results from normal storm wind and typhoon by Li et al. [9], Cao et al. [6], and Shiau and Chen [24–26].

3.3. Gust Factor. Gust factor is defined as the ratio of maximum gust wind speed over average gust wind speed. It can be expressed as

$$G_u(t) = 1 + \frac{\max\left(\overline{u(t)}\right)}{U},$$
$$G_v(t) = \frac{\max\left(\overline{v(t)}\right)}{U}, \qquad (4)$$

where $\max(\overline{u(t)})$ and $\max(\overline{v(t)})$ are the maximum gust wind speed in the period of t for longitudinal and lateral fluctuates, respectively.

Figure 10 shows the distribution of 3 s gust factor as a function of 10 min mean wind speed both in longitudinal (Figure 10(a)) and lateral (Figure 10(b)) directions. An obvious decrease of gust factor under low wind speed can be seen, but the rate of reduction goes smaller when wind speed becomes high. The average ratio of longitudinal gust factor to lateral gust factor is 0.30.

3.4. Peak Factor. There is a similarity in the definition of peak factor as of gust factor. The following expression for peak factor which describes the intensity of fluctuating wind speed is employed:

$$g_u = \frac{\widehat{U}_t - U}{\sigma_u}, \qquad (5)$$

where $\widehat{U}_t$ is the maximum value of t min average wind speed of the longitudinal component of fluctuating wind velocity record and σ_u is the corresponding standard deviation.

The quantity of peak factor was determined from the three-second averages, shown as a function of time during a one-day period coincident with the passage of typhoon HAIKUI (Figure 11) and 10 min mean wind speed (Figure 12). In each figure, the distribution of the data is in the range of 0.6 and 2.5; the average and standard deviations are 1.52 and 0.34, respectively. A noticeable decrease was found from the analysis of the average peak factor compared with Huang's results [12].

3.5. Probability Density Distribution. The probability density function of wind speed fluctuations is customarily with Gauss hypothesis. However, a similar statistical description of the wind speed fluctuations is generally the lack of certainty when applied to strong typhoon. Analysis of the wind speed fluctuations in longitudinal (Figure 13) and lateral (Figure 14) directions confirms a fairly good agreement with a Gaussian distribution through moment estimation method. It is believed the wind speed fluctuations are examined as Gaussian processes in this region of wind speed.

3.6. Turbulence Integral Scale. Turbulence is a three-dimensional spatial structure, and its nine parameters correspond to fluctuating wind speeds in longitudinal u, lateral v, and vertical w directional components. For example, L_u^x, L_u^y, and L_u^z are the average integral scales of longitudinal-dependent

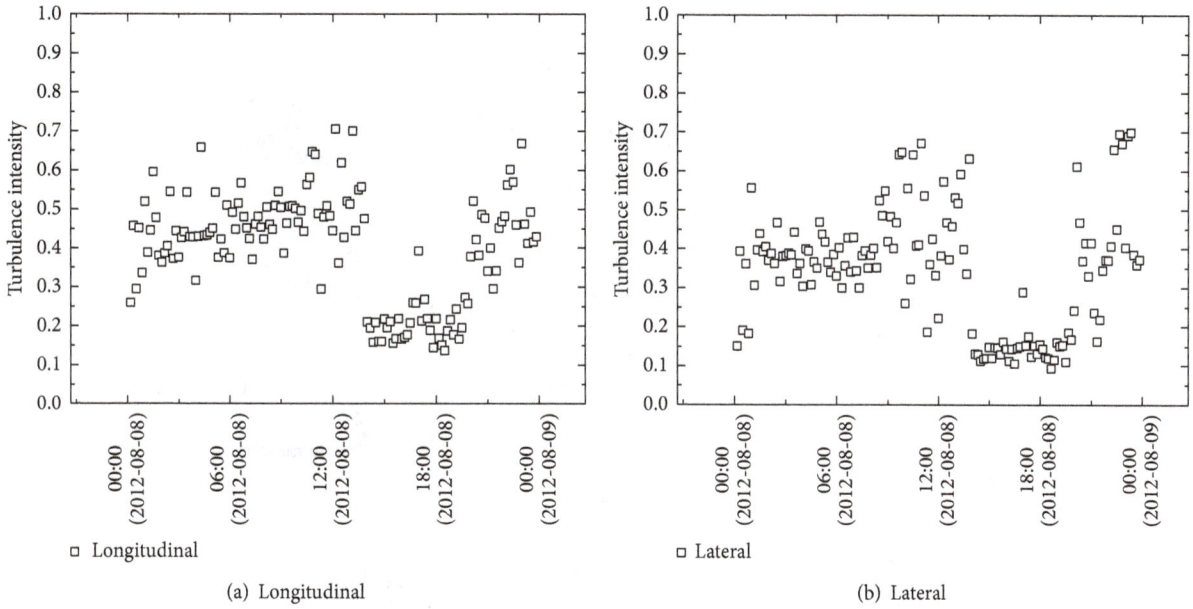

(a) Longitudinal

(b) Lateral

FIGURE 8: Variation of turbulence intensity during typhoon HAIKUI.

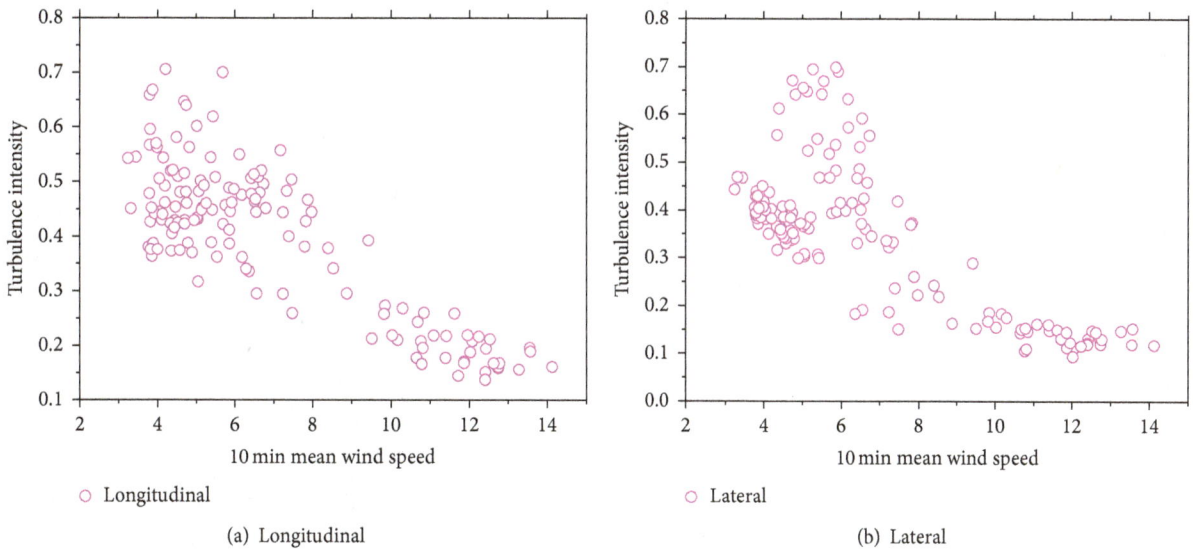

(a) Longitudinal

(b) Lateral

FIGURE 9: Variation of turbulence intensities with wind speed.

TABLE 2: Ratios of turbulence intensity among the turbulence components.

Research man	Wind	Height (m)	$I_u : I_v$	Sites
Tieleman [27]	Normal strong wind	5	1 : 0.80	Holland
Cao et al. [6]	Typhoon Maemi	10	1 : 0.83	Japan
Shiau	Typhoon Zeb [24]	26	1 : 0.98	Taiwan, China
	Typhoon Babs [25]		1 : 0.78	
	Normal strong wind [26]		1 : 0.78	
Fu et al. (2008) [23]	Typhoon Sanvu	10	1 : 0.7	Guangzhou, China
Peng et al. [12]	Typhoon Muifa	10	1 : 0.66	Shanghai, China
		20	1 : 0.65	
		40	1 : 0.74	
Present test	Typhoon HAIKUI	6 (upper bridge)	1 : 0.85	Hangzhou, China

(a) Longitudinal

(b) Lateral

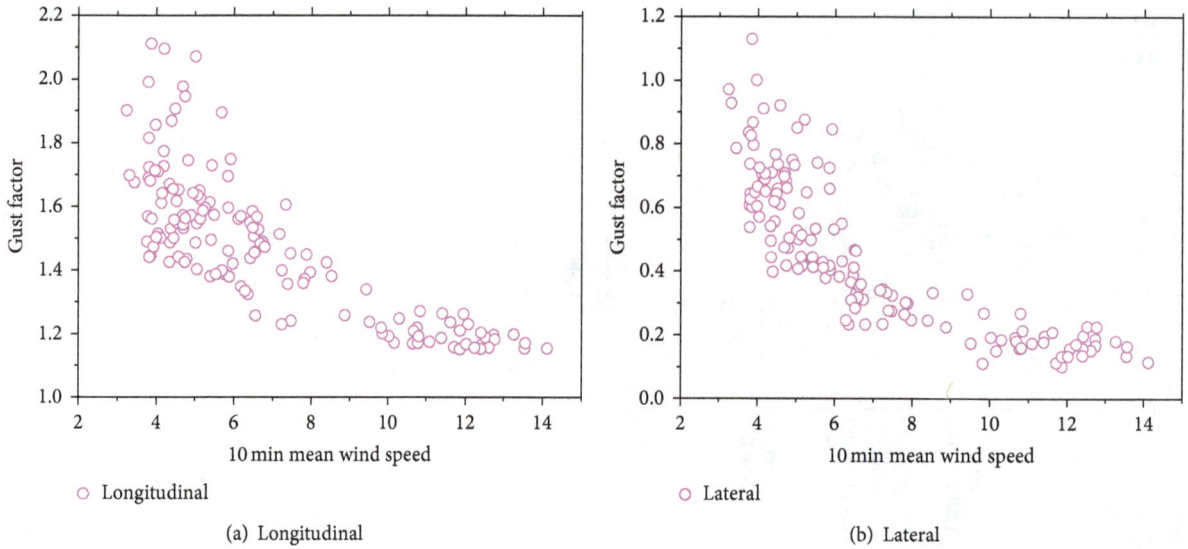

FIGURE 10: Variation of gust factors with wind speed.

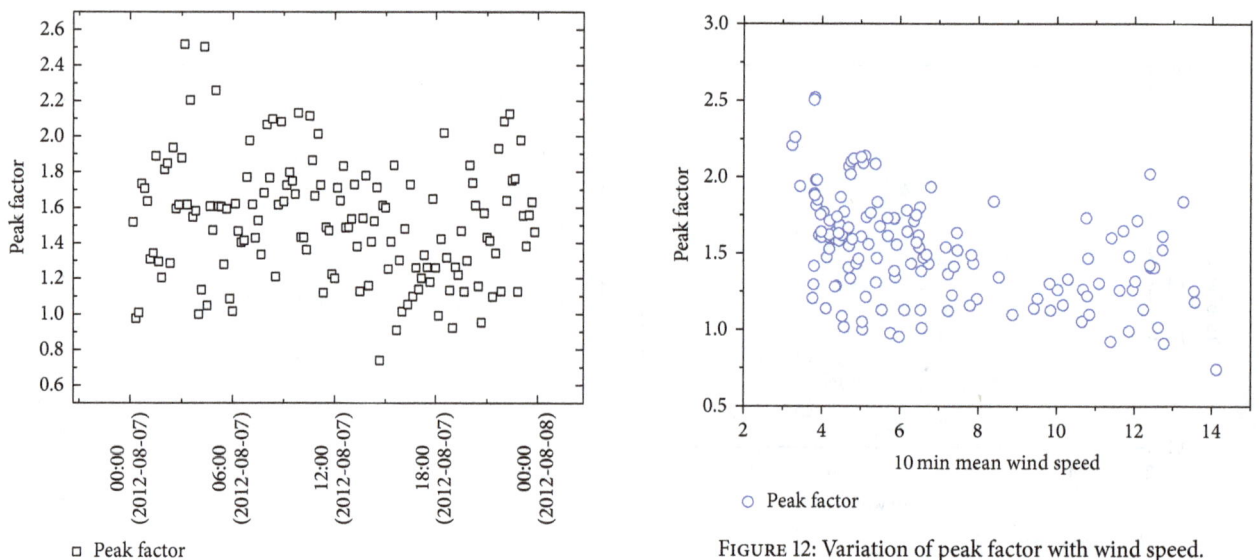

FIGURE 11: Variation of peak factor during typhoon HAIKUI.

FIGURE 12: Variation of peak factor with wind speed.

turbulence component fluctuations in longitudinal, lateral, and vertical directions, respectively. Turbulence integral scale L_i^x is mathematically defined as

$$L_i^x = \frac{1}{\sigma_i^2} \int_0^\infty R_{i_1 i_2}(x)\, dx, \quad i = u, v, w, \tag{6}$$

where $R_{i_1 i_2}(x)$ is the covariance function of fluctuating components in two positions.

In general, spatial correlation needs to be transferred into time correlation by Taylor hypothesis due to the fact that simultaneously measuring recorded points in space are complex and difficult. Thus, turbulence integral scale can

be calculated by autocorrelation function integral method depending on Taylor hypothesis and interpreted as

$$L_i^x = \frac{U}{\sigma_i^2} \int_0^\alpha R(\tau)\, d\tau, \quad i = u, v, w, \tag{7}$$

where $R(\tau)$ is the autocorrelation function and α is the variable when autocorrelation coefficient drops to 0.05 [4].

In Figure 15, turbulence integral scales of longitudinal and lateral components determined from 10 min average show a clear dependency on 10 min mean wind speed, and increasing quantity is recorded by increased 10 min mean wind speed. For 10 min mean wind speed is greater than 8 m/s, turbulence integral scales drastically change and distributed in a large region between 100 and 250. From the statistical analysis of Figure 16 for the turbulence integral scales both in longitudinal (Figure 16(a)) and lateral (Figure 16(b)) directions,

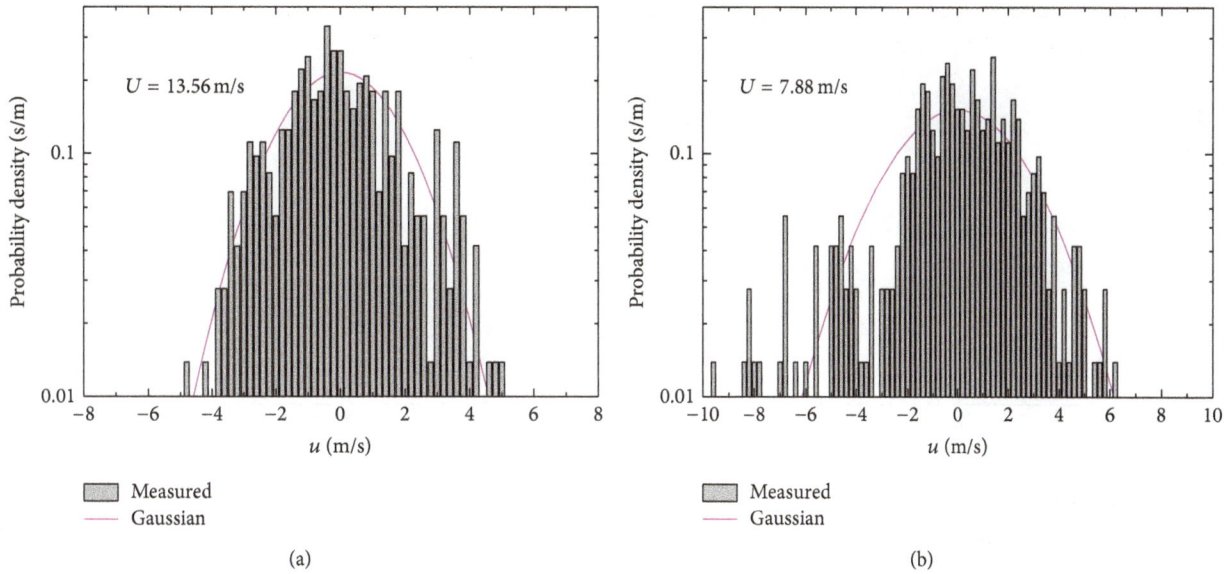

(a)

(b)

FIGURE 13: Probability density function of longitudinal wind speed fluctuations.

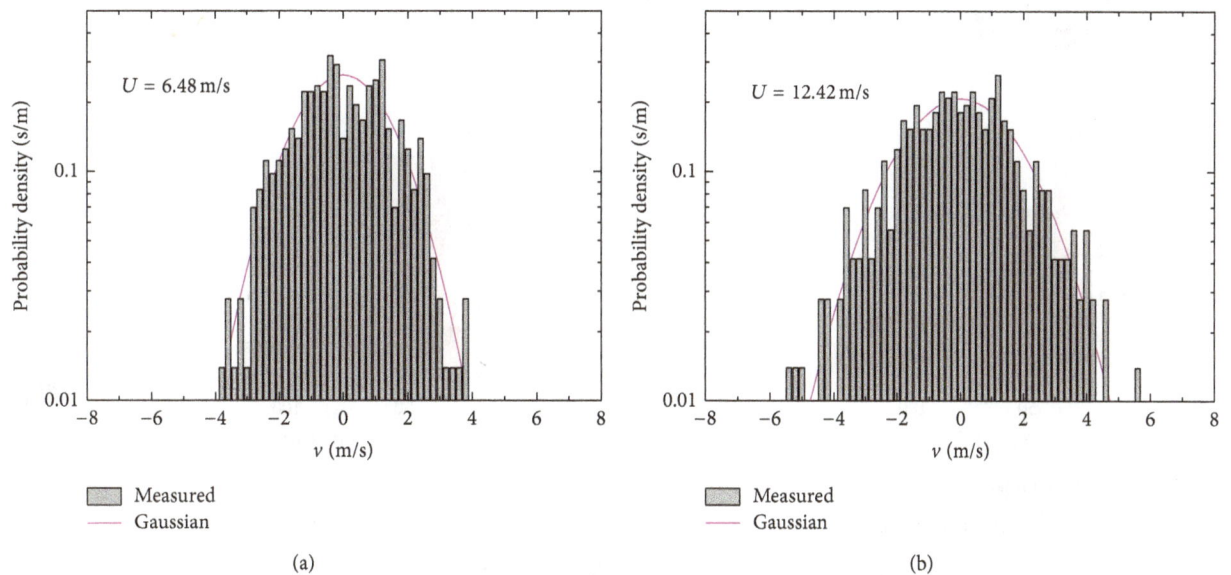

(a)

(b)

FIGURE 14: Probability density function of lateral wind speed fluctuations.

a similarity of probability density distribution is observed in each component; as it is obviously asymmetrical and differs from a standardized Gaussian distribution, in particular at low turbulence integral scales which is between 20 and 40, the probability is the largest with most data locate in this region.

3.7. Correlation. Autocorrelation describes the correlation of two dependent values of a time series at different moments. $X(t)$ is here defined as a time series, and the correlation function can be derived as

$$R_{XX}(t_1, t_2) = E[X(t_1) X(t_2)]. \tag{8}$$

Whereas $X(t)$ denotes a stationary random process, the correlation function can be derived as

$$R_{XX}(\tau) = E[X(t) X(t + \tau)], \tag{9}$$

where R_{XX} is the autocorrelation function and τ is the delaying time.

Indicating correlation intensity of wind speed fluctuations for different directions, cross-correlation coefficient can be expressed as

$$C_{R(ij)} = \frac{R_{ij}(0)}{\sqrt{R_{ii}(0)} \sqrt{R_{jj}(0)}}, \tag{10}$$

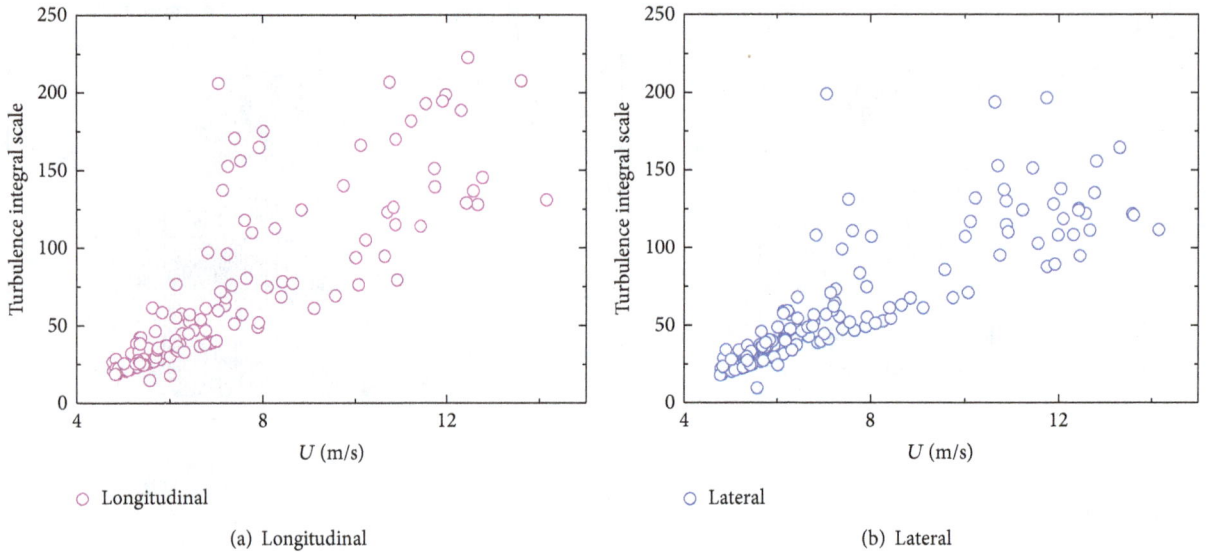

(a) Longitudinal

(b) Lateral

FIGURE 15: Variation of turbulence integral scale with 10 min mean wind speed.

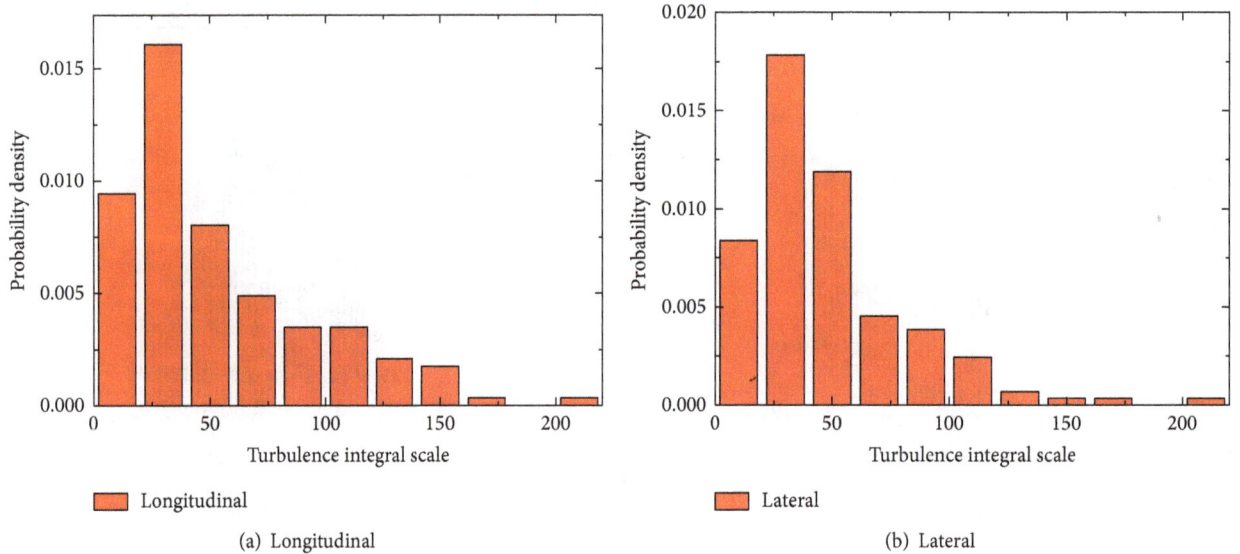

(a) Longitudinal

(b) Lateral

FIGURE 16: Probability density function of turbulence integral scale.

where $C_{R(ij)}$ is cross-correlation coefficient of wind speed fluctuations in i and j directions, R_{ij} is cross-correlation function of wind speed fluctuations in i and j directions, and R_{ii} and R_{jj} are autocorrelation functions of wind speed fluctuations in i and j directions, respectively.

The autocorrelation coefficients for both longitudinal and lateral wind speed fluctuations are plotted in Figure 17 at the average level of 10 min time interval. It is found that a noticeable similarity of tendency is observed from the two plots; in particular with τ increases, the autocorrelation coefficients for both directions decrease, whereas the longitudinal autocorrelation coefficient is shown slightly larger than the lateral autocorrelation coefficient at the same delaying time τ.

3.8. *Power Spectra of Wind Speed Fluctuations.* Turbulent power spectrum is a describer of turbulent energy distribution in frequency domain and characteristics of wind fluctuation. It can be expressed as [28]

$$\frac{S_u(n, z)}{U_0^{*2}} = \frac{A}{(1 + Bf^\beta)^\gamma}, \tag{11}$$

where n is frequency; z is the observed height; f denotes reduced frequency derived as $f = nz/U(z)$; and A, B, β, and γ are undetermined parameters.

Functions of power spectra densities of wind speed fluctuations were expressed in accordance with the Kolmogorov principle as Davenport spectrum, Von Karman spectrum, Simiu spectrum, Kaimal spectrum, and Harris spectrum.

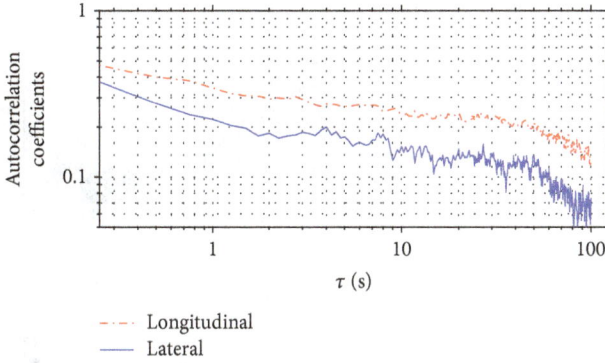

FIGURE 17: Autocorrelation coefficients of longitudinal and lateral wind speed fluctuation component.

Based on the comparisons with field measurement and wind tunnel test, Von Karman spectrum is believed as the best fitted function for wind speed fluctuations. The longitudinal fluctuating component can be expressed as [25, 29]

$$S_u(n) = \frac{2\overline{u'^2}L_u^x}{\overline{U}\left[1 + \left(2cnL_u^x/\overline{U}\right)^2\right]^{5/6}},$$ (12)

where $\overline{U}$ is mean wind speed; L_x^u is longitudinal turbulence integral scale; $\overline{u'^2}$ is standard deviation of the longitudinal fluctuating component; and constant coefficient c is 4.2065.

Power spectral densities of the longitudinal and lateral fluctuating components are derived from the same characteristic of turbulence hypothesis:

$$S_v(n) = S_w(n) = \frac{1}{2}\left[S_u(n) - \frac{ndS_u(n)}{dn}\right].$$ (13)

For isotropic turbulences, $\overline{v'^2} = \overline{w'^2} = \overline{u'^2}$, $L_u^x = L_v^y = L_w^z$, some equations are obtained from plugging (12) into (13):

$$S_v(n) = \frac{\overline{v'^2}L_v^y\left[1 + (8/3)\left(cnL_v^y/\overline{U}\right)^2\right]}{\overline{U}\left[1 + \left(2cnL_v^y/\overline{U}\right)^2\right]^{11/6}},$$
$$S_w(n) = \frac{\overline{w'^2}L_w^z\left[1 + (8/3)\left(cnL_w^z/\overline{U}\right)^2\right]}{\overline{U}\left[1 + \left(2cnL_w^z/\overline{U}\right)^2\right]^{11/6}}.$$ (14)

If $L_v^x = L_w^x = 0.5L_u^x$ for isotropic turbulences, the expressions of Von Karman spectrum are derived from plugging $L_v^x = L_w^x = 0.5L_u^x$ into (14):

$$\frac{nS_u(n)}{\sigma_u^2} = \frac{4f}{\left(1 + 70.8f^2\right)^{5/6}},$$
$$\frac{nS_i(n)}{\sigma_i^2} = \frac{4f\left(1 + 755.2f^2\right)}{\left(1 + 283.2f^2\right)^{11/6}}, \quad i = v, w.$$ (15)

In Figures 18 and 19, field-measured power spectral densities of longitudinal and lateral fluctuate components are shown at wind speed of 13.56 m/s and 7.88 m/s, presenting a good agreement with the Karman empirical spectra for the whole set of data. However, the PSD data estimated by field measurement are slightly larger than that by empirical method when the reduced frequency $nz/U(z)$ is smaller than 0.1 or larger than 0.1.

4. Conclusions

Through a full-scale wind speed and direction monitoring system on Jiubao Bridge in Hangzhou city, this paper presents a reliable study of wind speed and direction characteristics during typhoon HAIKUI, such as time histories of mean wind speed and direction, turbulence intensity, peak factor, gust factor, probability distribution of fluctuating velocity component, correlation among fluctuating velocities, turbulence integral scale, and the power spectrum of fluctuating velocity, and the results are summarized as follows.

(1) Turbulence intensities decrease with increased anemometer elevation for both longitudinal and lateral components of mean wind speed. A remarkable decrease of turbulence intensity is shown with increased mean wind speed, typically at low mean wind speeds ($u < 8$ m/s), but an unknown tendency when mean wind speed exceeds 8 m/s. The ratio of turbulence intensities of longitudinal component to lateral component is 0.85 in this paper, indicating a clear similarity with the results by Tieleman, Cao, and Bao-Shi, whereas a certain deviation from Huang et al.'s results [11].

(2) Gust factors decrease with mean wind speed. An obvious decrease of the change rate is found at low wind speeds, but wind speed seems to have less impact on the change rate when it becomes high. The average ratio for gust factor of longitudinal component to lateral component is 0.30. The upper and lower peak factors are recorded as 0.6 and 2.5; the average and standard deviation are 1.52 and 0.34, respectively. A noticeable decrease is found from the analysis of the average peak factor compared with Peng et al.'s results [12].

(3) Probability density distribution of wind speed fluctuations at different wind speeds agrees well with a Gaussian distribution, which means the full-measured recorded data of wind speed fluctuations are in accordance with Gauss assumption.

(4) Increasing quantity of turbulence integral scales is recorded by increased 10 min mean wind speed. A similarity of probability distributions of turbulence integral scales of the three components is that each distribution is obviously asymmetrical and does not follow a standardized Gaussian distribution.

(5) Power spectral densities of recorded wind speed fluctuations are fairly good in agreement with Von Karman spectrum.

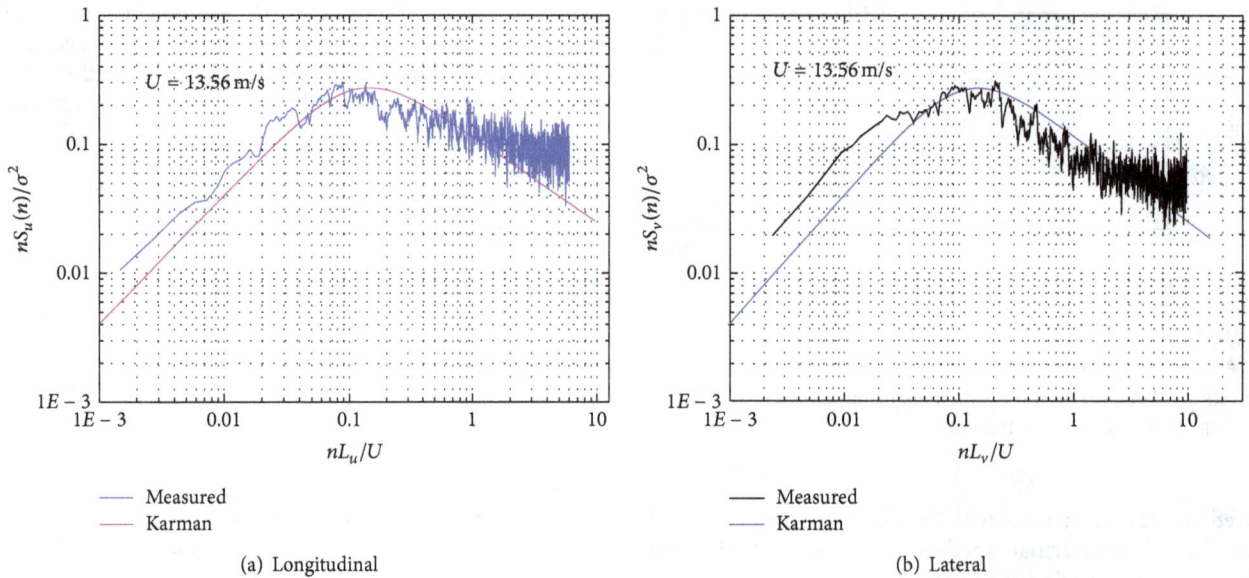

(a) Longitudinal

(b) Lateral

FIGURE 18: Power spectral density of fluctuation wind speed components ($U = 13.56$ m/s).

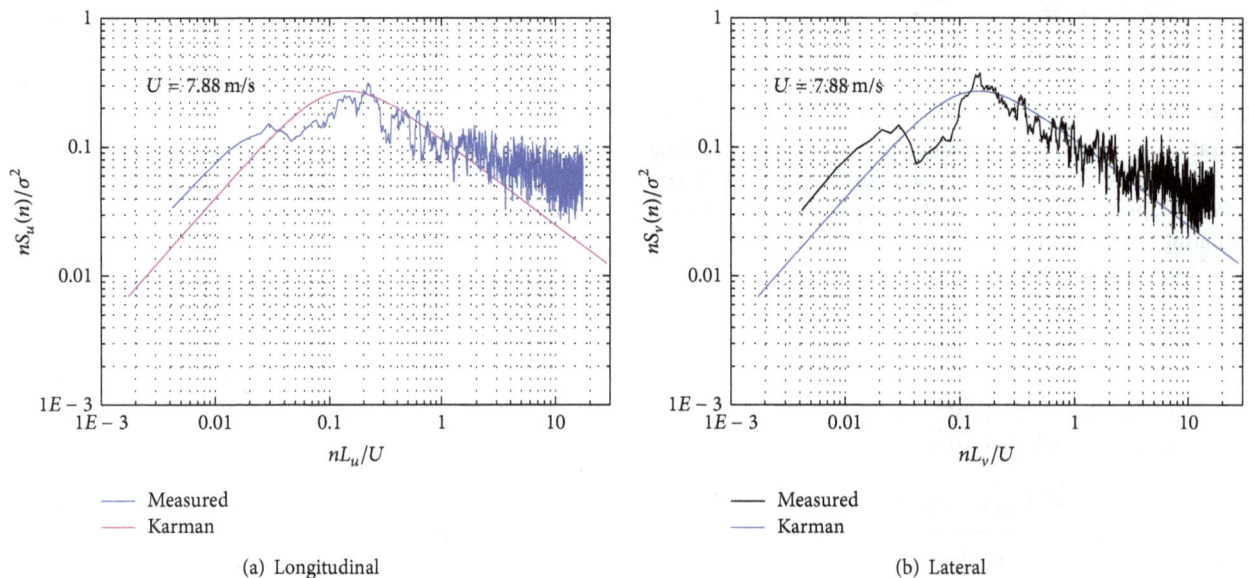

(a) Longitudinal

(b) Lateral

FIGURE 19: Power spectral density of fluctuation wind speed components ($U = 7.88$ m/s).

Conflict of Interests

The authors declare that there is no conflict of interests regarding the publication of this paper.

Acknowledgments

This project is supported by Open Fund of State Key Laboratory Breeding Base of Mountain Bridge and Tunnel Engineering (CQSLBF-Y14-15), the National Program on Key Basic Research Project (973 Program) Grant no. 2012CB723305, the Chinese National Natural Science Foundation (51308510), and Zhejiang Provincial Natural Science Foundation (Q12E080026) which are gratefully acknowledged.

References

[1] M. Gu, *The Research Process and Basic Scientific Issues about Civil Structure*, Science Press, Beijing, China, 2006.

[2] C. S. Durst, "Wind speeds over short periods of time," *The Meteorological Magazine*, vol. 89, pp. 181–186, 1960.

[3] N. Kato, T. Ohkuma, J. R. Kim, H. Marukawa, and Y. Niihori, "Full scale measurements of wind velocity in two urban areas using an ultrasonic anemometer," *Journal of Wind Engineering and Industrial Aerodynamics*, vol. 41, no. 1–3, pp. 67–78, 1992.

[4] R. G. J. Flay and D. C. Stevenson, "Integral length scales in strong winds below 20 m," *Journal of Wind Engineering and Industrial Aerodynamics*, vol. 28, no. 1–3, pp. 21–30, 1988.

[5] E. C. C. Choi and F. A. Hidayat, "Gust factors for thunderstorm and non-thunderstorm winds," *Journal of Wind Engineering and Industrial Aerodynamics*, vol. 90, no. 12–15, pp. 1683–1696, 2002.

[6] S. Y. Cao, Y. Tamura, N. Kikuchi, M. Saito, I. Nakayama, and Y. Matsuzaki, "Wind characteristics of a strong typhoon," *Journal of Wind Engineering and Industrial Aerodynamics*, vol. 97, no. 1, pp. 11–21, 2009.

[7] Q. S. Li, Y. Q. Xiao, J. Y. Fu, and Z. N. Li, "Full-scale measurements of wind effects on the Jin Mao building," *Journal of Wind Engineering and Industrial Aerodynamics*, vol. 95, no. 6, pp. 445–466, 2007.

[8] Q. S. Li, Y. Q. Xiao, and C. K. Wong, "Full-scale monitoring of typhoon effects on super tall buildings," *Journal of Fluids and Structures*, vol. 20, no. 5, pp. 697–717, 2005.

[9] Q. S. Li, Y. Q. Xiao, C. K. Wong, and A. P. Jeary, "Field measurements of typhoon effects on a super tall building," *Engineering Structures*, vol. 26, no. 2, pp. 233–244, 2004.

[10] Q. S. Li, L. Zhi, and F. Hu, "Boundary layer wind structure from observations on a 325 m tower," *Journal of Wind Engineering and Industrial Aerodynamics*, vol. 98, no. 12, pp. 818–832, 2010.

[11] P. Huang, X. Wang, and M. Gu, "Field experiments for wind loads on a low-rise building with adjustable pitch," *International Journal of Distributed Sensor Networks*, vol. 2012, Article ID 451879, 10 pages, 2012.

[12] H. Peng, X. Wang, and M. Gu, "Study on near-ground wind characteristics of a strong typhoon-wind speed, turbulence intensities, gust factors and peak factors," *Disaster Advances*, vol. 6, no. 5, pp. 3–18, 2013.

[13] T. H. Yi, H. N. Li, and M. Gu, "Recent research and applications of GPS based technology for bridge health monitoring," *Science China Technological Sciences*, vol. 53, no. 10, pp. 2597–2610, 2010.

[14] T.-H. Yi, H.-N. Li, and M. Gu, "Optimal sensor placement for structural health monitoring based on multiple optimization strategies," *The Structural Design of Tall and Special Buildings*, vol. 20, no. 7, pp. 881–900, 2011.

[15] T.-H. Yi, H.-N. Li, and M. Gu, "Wavelet based multi-step filtering method for bridge health monitoring using GPS and accelerometer," *Smart Structures and Systems*, vol. 11, no. 4, pp. 331–348, 2013.

[16] T.-H. Yi, H.-N. Li, and M. Gu, "Experimental assessment of high-rate GPS receivers for deformation monitoring of bridge," *Journal of the International Measurement Confederation*, vol. 46, no. 1, pp. 420–432, 2013.

[17] American Society of Civil Engineers, *ASCE/SEI 7-10 Minimum Design Loads for Buildings and Other Structures*, American Society of Civil Engineers, Reston, Va, USA, 2010.

[18] Department of Civil Engineering Indian Institute of Technology Roorkee, *IS:875(Part3): Wind Loads on Buildings and Structures-Proposed Draft & Commentary*, Indian Institute of Technology, Roorkee, India, 2003.

[19] Architectural Institute of Japan, *AIJ 2004 Recommendations for Loads on Buildings*, Architectural Institute of Japan, Tokyo, Japan, 2004.

[20] Technical Committee CEN/TC250 "Structural Eurocodes", *Eurocode 1: Actions on Structures-General Actions-Part 1-4: Wind Actions*, British Standards Institution, London, UK, 2004.

[21] GB50009-2001, *Load Code for the Design of Building Structures*, Architecture and Building Press, Beijing, China, 2002.

[22] E. Simiu and R. H. Scanlan, *Wind Effects on Structures: Fundamentals and Applications to Design*, John Wiley & Sons, New York, NY, USA, 1996.

[23] J. Y. Fu, Q. S. Li, J. R. Wu, Y. Q. Xiao, and L. L. Song, "Field measurements of boundary layer wind characteristics and wind-induced responses of super-tall buildings," *Journal of Wind Engineering and Industrial Aerodynamics*, vol. 96, no. 8-9, pp. 1332–1358, 2008.

[24] B.-S. Shiau, "Velocity spectra and turbulence statistics at the northeastern coast of Taiwan under high-wind conditions," *Journal of Wind Engineering and Industrial Aerodynamics*, vol. 88, no. 2-3, pp. 139–151, 2000.

[25] B.-S. Shiau and Y.-B. Chen, "In situ measurement of strong wind velocity spectra and wind characteristics at Keelung coastal area of Taiwan," *Atmospheric Research*, vol. 57, no. 3, pp. 171–185, 2001.

[26] B.-S. Shiau and Y.-B. Chen, "Observation on wind turbulence characteristics and velocity spectra near the ground at the coastal region," *Journal of Wind Engineering and Industrial Aerodynamics*, vol. 90, no. 12–15, pp. 1671–1681, 2002.

[27] H. W. Tieleman, "Strong wind observations in the atmospheric surface layer," *Journal of Wind Engineering and Industrial Aerodynamics*, vol. 96, no. 1, pp. 41–77, 2008.

[28] P. J. Richards, R. P. Hoxey, and J. L. Short, "Spectral models for the neutral atmospheric surface layer," *Journal of Wind Engineering and Industrial Aerodynamics*, vol. 87, no. 2-3, pp. 167–185, 2000.

[29] T. Karman, "Progress in the statistical theory of turbulence," *Proceedings of the National Academy of Sciences*, vol. 34, pp. 530–539, 1948.

20

Evaluation of Compressive Strength and Stiffness of Grouted Soils by Using Elastic Waves

In-Mo Lee,[1] Jong-Sun Kim,[1] Hyung-Koo Yoon,[2] and Jong-Sub Lee[1]

[1] *School of Civil, Environmental and Architectural Engineering, Korea University, Seoul 136-701, Republic of Korea*
[2] *Department of Geotechnical Disaster Prevention Engineering, Daejeon University, Daejeon 300-716, Republic of Korea*

Correspondence should be addressed to Jong-Sub Lee; jongsub@korea.ac.kr

Academic Editor: Dimitrios G. Aggelis

Cement grouted soils, which consist of particulate soil media and cementation agents, have been widely used for the improvement of the strength and stiffness of weak ground and for the prevention of the leakage of ground water. The strength, elastic modulus, and Poisson's ratio of grouted soils have been determined by classical destructive methods. However, the performance of grouted soils depends on several parameters such as the distribution of particle size of the particulate soil media, grouting pressure, curing time, curing method, and ground water flow. In this study, elastic wave velocities are used to estimate the strength and elastic modulus, which are generally obtained by classical strength tests. Nondestructive tests by using elastic waves at small strain are conducted before and during classical strength tests at large strain. The test results are compared to identify correlations between the elastic wave velocity measured at small strain and strength and stiffness measured at large strain. The test results show that the strength and stiffness have exponential relationship with elastic wave velocities. This study demonstrates that nondestructive methods by using elastic waves may significantly improve the strength and stiffness evaluation processes of grouted soils.

1. Introduction

Cement-based soil-grout mixtures (grouted soils) have been commonly used during the construction of underground spaces for the improvement of the strength and stiffness of soils and rocks and for the prevention of the leakage of ground water. A practical cement-based grouting technique, however, is not available for penetrating into the small-sized pore geomaterials. Furthermore, the curing of grouting takes 2~3 days. Thus, fine to micro-fine cement grains have been developed in order to improve groutability [1]. Quick setting agents have been used to control gel-time or to cure the grouting within a few minutes.

The curing time and initial strength of grouted soils play a crucial role in ground improvement. Ground improvements, however, have been commonly evaluated by the destructive method in both the field (in situ) and the laboratory. In situ methods include the standard penetration test (SPT) and the cone penetration test (CPT). Laboratory methods include the uniaxial compression test, the triaxial compression test,

and the direct shear test. The sensitivity of SPT and CPT, however, is not sufficient for the evaluation of ground improvements. Furthermore, the sampling for the laboratory test yields a disturbance. In the field, the weak parts are not generally obtained during the coring or drilling of soil-grout mixtures. The strength, which is based on the laboratory tests, may therefore be slightly overestimated. To overcome the numerous limitations of the destructive strength test, a nondestructive test using elastic waves may be an alternative option. Nondestructive tests based on elastic waves include the transmission and reflection methods. Note that the penetration types of elastic wave measurement techniques were also suggested [2–4]. In this study, the transmission method was used, which uses two transducers on the two opposing faces of the specimen.

The goal of this study is to estimate the uniaxial compressive strength and elastic modulus of the grouted soils with and without grains (particulate materials) by using elastic waves. This paper includes specimen preparation, the measurement system, stress-strain responses during the uniaxial compression test, compressional and shear wave responses under applied uniaxial stress, and correlations between strength or stiffness and elastic wave velocities.

2. Elastic Wave Velocity versus Strength and Stiffness

The strength and stiffness of concrete may be empirically determined from nondestructive methods by using elastic waves. Although the theoretical relationship between the elastic wave velocity and the strength or elastic modulus cannot be derived in a composite material, attempts have been made over the past sixty or more years to correlate the properties of composite materials with elastic wave velocities [5]. The strength of high strength concrete has been obtained based on nondestructive methods [6]. Similarly, the strength and elastic modulus of concrete have been obtained by using elastic wave velocities. The relationships between the strength or elastic modulus and elastic wave velocity are based on empirical correlations (exponential functions) as follows:

$$f_c = \alpha e^{\beta V_P}, \tag{1}$$

$$E = a e^{b V_P}, \tag{2}$$

where f_c is the uniaxial compressive strength, E is the elastic modulus, V_P is the compressional wave velocity, and α, β, a, and b are coefficients determined by experimental results. Equations (1) and (2) have been used in concrete [5, 7, 8]. The empirical Equation (1) may be rewritten in the form of the bilinear function by using a log scale in the f_c-axis [5]. For soil-grout mixtures, since an empirical correlation has not been suggested, a relationship is required for the estimation of the strength of soil-grout mixtures.

3. Experimental Study

Nondestructive tests by using elastic waves were conducted before and during the axial compression test. The material properties, preparation of specimens, and measurement systems are described as follows.

3.1. Material Properties. The cemented grouted soils comprised of two types of uniform sands, fine particle cements, and quick setting agents. The grain size distribution curves and photographic images of the materials are plotted in Figures 1 and 2. Two types of sands were used: fine and coarse sands. The physical properties of the two sands are summarized in Table 1. The median diameters were 0.82 mm and 3.1 mm for fine and coarse sands, respectively. Table 1 shows that the friction angle obtained from the direct shear tests was smaller in fine sands than in coarse sands. The coefficient of uniformity, coefficient of curvature, and specific gravity in both sands were similar to each other.

The median diameters of fine cement and quick setting agents were 6.6 μm and 17 μm, respectively, as shown in Figure 1. The component of fine particle cements was similar to Portland cements except for the average particle size (35 μm). The quick setting agent is an additive for

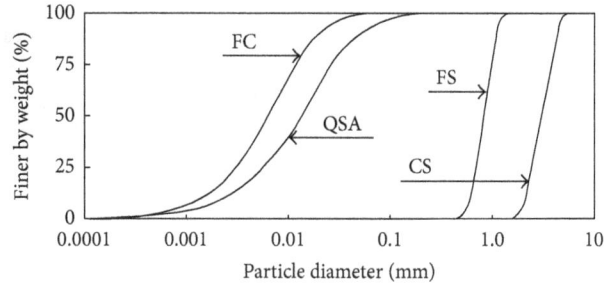

FIGURE 1: Grain-size distribution of coarse and fine grains, cement, and quick setting agent. FC, QSA, FS, and CS denote fine cement, quick setting agent, fine sands, and coarse sands, respectively.

the hydration of cement. The chemical reaction between the fine particle cement and quick setting agent dramatically increases the particle concentration and viscosity of the mixture after mixing due to the generation of ettringite (hydrated calcium aluminum sulfate hydroxide), thaumasite, and alite (tricalcium silicate). Note that ettringite, which was produced during the early hydration of cement [9], was activated by the quick setting agent. The SEM images of mixtures with and without sand grains are shown in Figure 3. Figure 3 shows that sand and cement particles were connected by ettringite. In addition, the mixture with fine cement and quick setting agent produced a relatively larger void ratio and lower density than a regular cement-based soil-grout mixture without a quick setting agent. In addition, because the mixture was hardened before the complete drainage of free water, the strength and elastic wave velocity of the mixture would be lower.

3.2. Preparation of Grout Soils. Five types of the grouted soils without sand grains (P-type), three types of the soil-grout mixtures with fine grain sands (F-type), and three types of the soil-grout mixtures with coarse grain sands (C-type) were prepared to control the uniaxial compressive strength at large strain and elastic wave velocities at small strain. Details of soil-grout mixtures are presented in Table 2 including mixing ratio, water-cement ratio, unit weight of the specimens, and number of specimens. Two to four specimens were prepared in plastic molds for each type, with a total of thirty-four specimens. The diameter and length of each specimen were 45 mm and 90 mm, respectively. The specimens were cured underwater for 7 days.

3.3. Compressive Strength Measurements. The compressive strength was measured by an unconfined axial compression test as shown in Figure 4. The vertical strain rate was 0.001/sec. Three linear variable differential transformers (LVDT) were adopted to measure axial and horizontal deformations. One load cell was used to obtain the axial load. The responses of three LVDTs and one load cell were monitored by the data logger (Agilent 34970A) as shown in Figure 4.

3.4. Elastic Waves Measurement. The compressional waves were measured by compressional wave transducers during

TABLE 1: Properties of fine and coarse sands.

	Median diameter D_{50} (mm)	Coefficient of uniformity C_u	Coefficient of curvature C_c	Specific gravity G_s	Friction angle ϕ (deg)
Fine sands	0.82	1.46	0.93	2.62	39~48
Coarse sands	3.10	1.59	0.91	2.63	45~55

TABLE 2: Properties of soil-grout mixtures.

Soil-grout mixtures	Type	Water : cement : agent : sand (weight ratio)	Water-cement ratio	Unit weight (kN/m^3)	Number of specimens
Mixtures without grains	P1	0.78 : 0.17 : 0.05 : 0.00	4.6	10.3	3
	P2	0.74 : 0.19 : 0.08 : 0.00	3.9	12.0	3
	P3	0.72 : 0.20 : 0.08 : 0.00	3.5	12.1	3
	P4	0.69 : 0.22 : 0.09 : 0.00	3.2	12.8	3
	P5	0.69 : 0.31 : 0.00 : 0.00	2.3	14.5	4
Mixtures with fine grains	F1	0.18 : 0.04 : 0.01 : 0.77	4.6	18.4	2
	F4	0.19 : 0.06 : 0.02 : 0.73	3.2	19.4	3
	F5	0.19 : 0.08 : 0.00 : 0.73	2.3	20.3	3
Mixtures with coarse grains	C1	0.18 : 0.04 : 0.01 : 0.77	4.6	19.3	3
	C4	0.19 : 0.06 : 0.02 : 0.73	3.2	20.7	3
	C5	0.19 : 0.08 : 0.00 : 0.73	2.3	20.8	4

Note: if the typed numbers are the same (e.g., P1, F1, and C1), the mixture ratios of water, cement, and quick setting agent are identical.

FIGURE 2: Photomicrographs: (a) fine cement; (b) quick setting agent; (c) fine sands; (d) coarse sands.

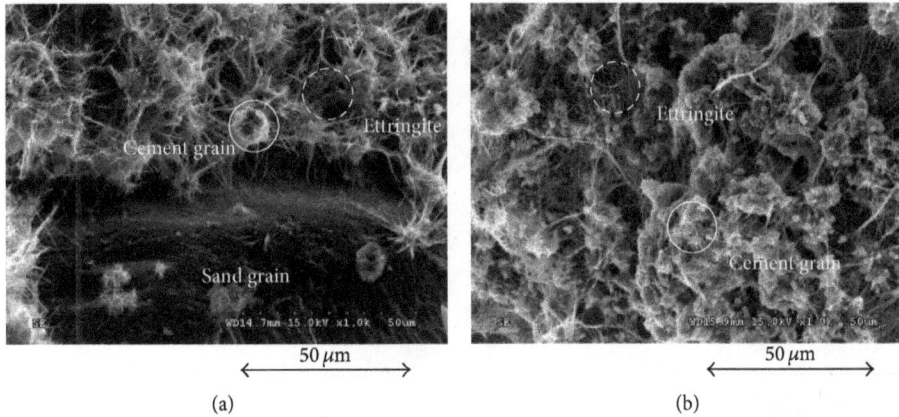

FIGURE 3: SEM images of soil-grout mixtures: (a) specimen with sands; (b) specimen without sands.

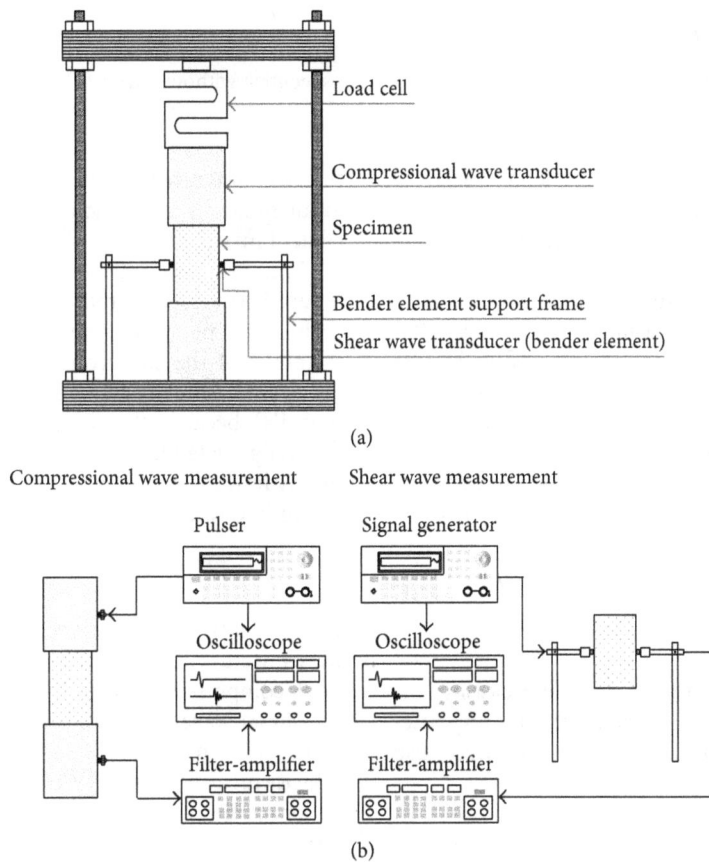

FIGURE 4: Experimental setup: (a) unconfined axial compression test; (b) elastic wave measurements.

an axial compression test as shown in Figure 4. The diameter of the transducers was 50 mm, which corresponds to the diameter of the specimens. The very strong compression wave transducers were installed on the top and bottom of the specimen. Therefore, the axial load was directly applied through the transducers into the specimen. The resonant frequency of the transducers was 50 kHz. The impulse signal, which was generated by the pulser (JSR DPR 300), activated the source transducer. The excitation of the source transducer yielded compressional waves (P-waves). The P-waves were propagated through the specimen, detected by the receiver transducer, and were recorded by the digital oscilloscope (Agilent 54624A). Note that the compressional waves were measured by the pulse velocity method, which was specified in ASTM C597 and C1383 [10, 11]. The details in P-wave measurement were discussed in Lee and Santamarina [12] and Lee et al. [13].

The measurement system of the shear waves was similar to that of the compressional waves. The shear waves were measured by bender elements during an axial compression

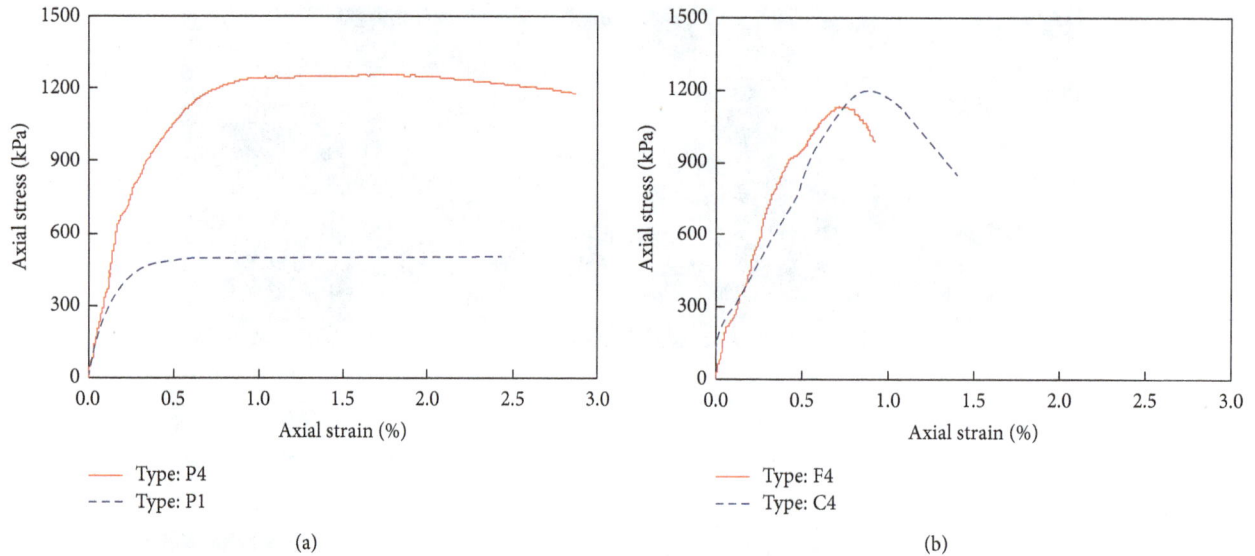

FIGURE 5: Compressive strength of soil-grout mixtures: (a) specimen without grains; (b) specimen with grains.

test as shown in Figure 4. The cantilever length and the resonant frequency of the bender elements were 2 mm and 5 kHz, respectively. The experimental setup for the measurement of shear waves was described in Lee and Santamarina [14, 15] and Lee et al. [16]. The bender elements were installed on the sides of the specimen by using a bender supporter frame as shown in Figure 4. For the excitation of the bender elements, square waves with 10 V in peak-to-peak amplitude were continuously generated by the signal generator (Agilent 33220A). The shear wave responses were measured by the digital oscilloscope (Agilent 54624A) as shown in Figure 4. The details in the bender elements were described in Lee and Santamarina [14].

4. Experimental Results

After all specimens were cured for 7 days after casting, the compressive strength tests were carried out. The responses of elastic waves during axial compression tests were also monitored.

4.1. Stress-Strain Responses at Large Strain. Typical stress-strain responses of the grouted soils with and without grains during the axial compression test are plotted in Figure 5. The grouted soils without grains (P-type) showed a ductile behavior after peak strength even though the peak strength of the specimen was dependent on the paste type as shown in Figure 5(a). The peak strengths of type P4 and type P1 were 1200 kPa and 500 kPa, respectively. The strength of the grouted soils increased with a decrease of the water cement ratio. Furthermore, the strains for peak strength were 0.8% for type P4 and 0.4% for type P1. The quick setting agent activates the generation of early hydration products such as ettringite and thaumasite. A concentration of byproducts of soil-grout mixtures renders a decrease in the strength and stiffness of

the grout mixture [9]. Thus, the strength and stiffness of soil-grout mixtures was much lower than that of the common cement mixtures.

The grouting without grains (P-type) as shown in Figure 5(a) presented more ductile behavior than those with grains (F-type or C-type) as shown in Figure 5(b). The strengths of the soil-grout specimens with similar mixing ratios were similar (1200 kPa for type C4 and 1100 kPa for type F4) because the unconfined compressive strength is generally related to specimen cohesion, which is induced by the hydration of cement. In addition, the strains for peak strength were also similar (0.8% for type C4 and 0.7% for type F4).

4.2. Elastic Waves. Nondestructive tests using compressional waves were conducted during the destructive test (uniaxial compression test). The test procedure for the compressional wave velocity measurements followed the ASTM C597 [10], which described the direct compressional wave velocity measurement. Typical compressional wave responses under applied axial stresses are shown in Figure 6. The first arrival was fairly constant under different axial stresses. The constant first arrival confirmed that the stiffness was less sensitive to the state of stress for the lightly cemented soils [2, 3, 17–19]. As the behavior of the soil-grout mixtures without grains was ductile, the compressional waves were continuously measured after peak strength as shown in Figure 6(a). The grouted soils with grains, however, showed more brittle behavior. Thus, the compressional waves were not measured after peak strength as shown in Figure 6(b). The resonant frequency was 50 kHz for both soil-grout mixtures.

The elastic wave velocities were calculated by using the first arrival and the travel distance of elastic waves. The first arrivals of compressional and shear waves were the first deflection and zero after first bump, respectively [14].

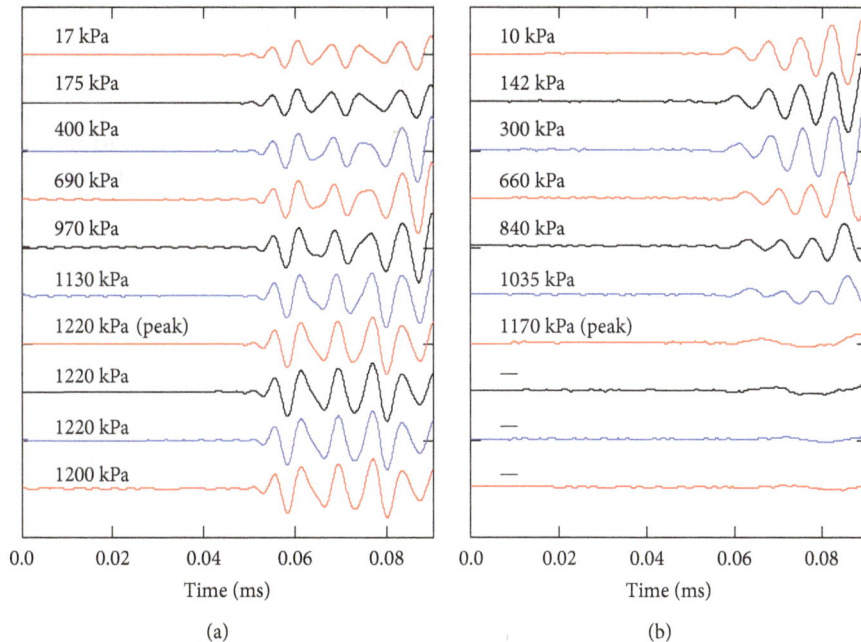

FIGURE 6: Compressional wave signatures during compressive strength tests for soil-grout mixtures: (a) specimen without grains: Type P4; (b) specimen with coarse grains: Type C4. Numbers on left side at each figure indicate applied axial stress.

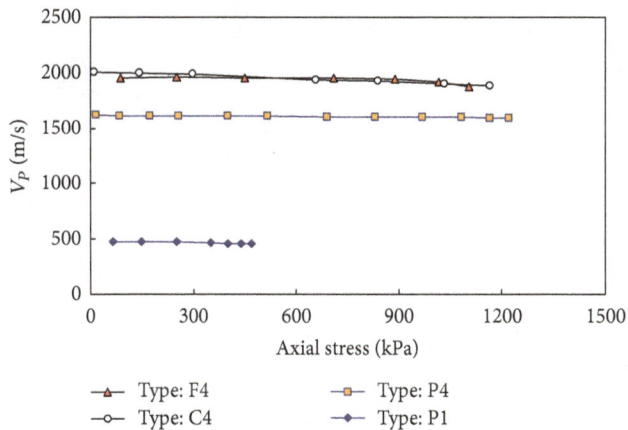

FIGURE 7: Compressional wave velocity (V_P) versus axial stress for different mixture types.

And the travel distances of the elastic waves were the tip-to-tip distance of a pair of transducers [2, 3]. The compressional wave velocities (V_P) versus applied axial stress are plotted in Figure 7. Figure 7 shows that the compressional wave velocities of the soil-grout mixtures without grains were almost constant until the mixtures were broken. For the soil-grout mixtures with fine or coarse grains the compressional wave velocities slightly decreased with an increase in the axial stress. Note that as the compressive stress approached the compressive strength, the specimen volume increased (dilatancy behavior) and the elastic wave velocity slightly decreased.

The resonant frequencies of compressional and shear waves are about 50 kHz and 5 Hz, respectively. And the velocities of compressional and shear waves are 550~2240 m/s and 100~230 m/s, respectively. Thus, the wavelengths are 11~45 mm and 20~46 mm and for compressional and shear waves, respectively. Note that the wavelength of the compressional waves for the mixtures with coarse grains ranges from 22 mm to 43 mm. And the median diameters of coarse and fine grains are 3.1 mm and 0.82 mm, respectively (see Table 1). The ratio of the wavelength to the particle size was at least 7, which is greater than the recommended minimum value of 3 in the ASTM D2845 [20], and thus the dispersion, low-pass filtering, and scattering effects may be minor [17].

5. Analyses

The compressive strength and stiffness are related to the elastic wave velocities [6]. The nondestructive testing method may be an alternative option for the evaluation of soil-grout improvement.

5.1. Compressive Strength versus Compressional Wave Velocity. The compressional wave velocities measured in the air were compared with axial compressive strength for all thirty-four specimens. The compressional wave velocity, which was determined at the initial stage of the stress, versus strength is plotted in Figure 8. Figure 8 shows that the axial compressive strength increased with an increase in the compressional wave velocity. The semiempirical exponential functions for the estimation of the axial compressive strength from the compressional wave velocity are superposed in Figure 8. The compressive strength based on the compressional wave

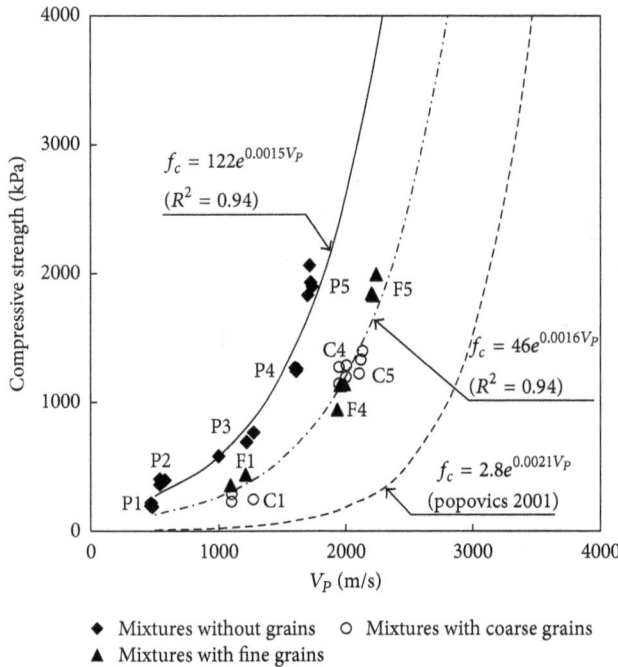

$$f_c = 122e^{0.0015V_P}$$
$$(R^2 = 0.94)$$

$$f_c = 46e^{0.0016V_P}$$
$$(R^2 = 0.94)$$

$$f_c = 2.8e^{0.0021V_P}$$
(popovics 2001)

◆ Mixtures without grains   ○ Mixtures with coarse grains
▲ Mixtures with fine grains

FIGURE 8: Compressive strength versus compressional wave velocity.

velocity for the soil-grout mixtures without grains and the soil-grout mixtures with grains are as follows:

Grouted mixtures without grains (Type P):
$$f_c = 122e^{0.0015V_P}$$
Grouted mixtures with grains (Type F or C):
$$f_c = 46e^{0.0016V_P}.$$

(3)

The correlation coefficients of soil-grout mixtures without grains and with grains were greater than 0.94 as shown in Figure 8. The higher values in the correlation coefficients demonstrate that the elastic wave velocity has a strong relationship with the compressive strength.

The mixture ratio of water, cement, and quick setting agent was the same for types P1, F1, and C1 specimens. Specimens with the same mixture ratio (except grains) produced a similar compressive strength, as shown in Figure 8: ~250 kPa for types P1, F1, and C1 specimens and ~1200 kPa for types P4, F4, and C4 specimens. However, the compressive velocity of type P1 was much smaller than that of types F1 and C1. The difference in the compressional wave velocity for the specimens with the same mixture ratio may result from the different governing factors for the strength and compressional wave velocity. The correlation between strength and compressional velocity in the concrete specimens composed of hardened cement paste, mineral aggregate, and water is also plotted in Figure 8. Figure 8 shows that the α values decreased with an increase in the strength, while the β values were similar regardless of the strength. The bonding material such as cementation agent controls the compressive strength of soil-grout mixtures. The properties of the grains dominate the compressional wave velocity of soil-grout mixture specimen.

5.2. Elastic Modulus versus Compressional Wave Velocity. The elastic modulus at large strain was obtained from the destructive compressive strength test. The elastic modulus, E [MPa], was also estimated from the compressional wave velocity by using a semiempirical exponential relationship. The results are superposed in Figure 9(a) and the functions are as follows:

Grouted mixtures without grains (Type P):
$$E = 43e^{0.0016V_P},$$
Grouted mixtures with grains (Type F or C):
$$E = 22e^{0.0015V_P}.$$

(4)

The trend of the elastic moduli was similar to that of the strength. In addition, the correlation coefficients of soil-grout mixtures without grains and with grains were greater than 0.94. As the slope of the stress-strain curve decreased with an increase in the strain during the axial compression test, the elastic modulus determined at the large strain should be much smaller than the value estimated at the small strain. Note that the strain level, at which the elastic modulus based on the compression wave velocity is determined, is generally less than $10^{-4} \sim 10^{-3}\%$ [21]. The strain level for the elastic modulus from the axial compression test was about 1%. Thus, the difference in strain levels between compressional wave velocity and static tests is 1,000~10,000 times. This strain gap produces the significant modulus difference between wave propagation method and static failure method as shown in Figure 9(b). In addition, as the ratio of the small strain modulus to the large strain modulus increases with an increase in the stiffness of the specimen [22], the modulus for the mixtures with grains was greater than that for the mixtures without grains as shown in Figure 9(b).

5.3. Compressive Strength versus Shear Wave Velocity. Shear waves were also monitored during the axial compression test for six soil-grout mixtures without grains (types P2 and P3). The shear waves were measured on the sides across the diameter (rather than at the top and bottom) of the specimens. For the generation and detection of the shear waves, the coupling between the transducers and medium is critical. As the coupling was not perfect, a few cases of the shear waves were only detected. The shear wave velocity was also fairly constant with axial stresses. The shear wave velocities measured in the air versus axial compressive strength is plotted in Figure 10. Figure 10 shows that the shear wave velocities ranged from 100 m/s to 250 m/s, which are reasonable values in lightly cemented soils [2, 3, 17, 23]. Figure 10 shows that the relationship between the compressive strength and the shear wave velocity also followed the exponential function discussed in the compressional wave section.

5.4. Compressional Velocity versus Shear Wave Velocity. The compressional wave velocity was compared with the shear wave velocity for six soil-grout mixtures without grains. The linear relationship was observed as shown in Figure 11.

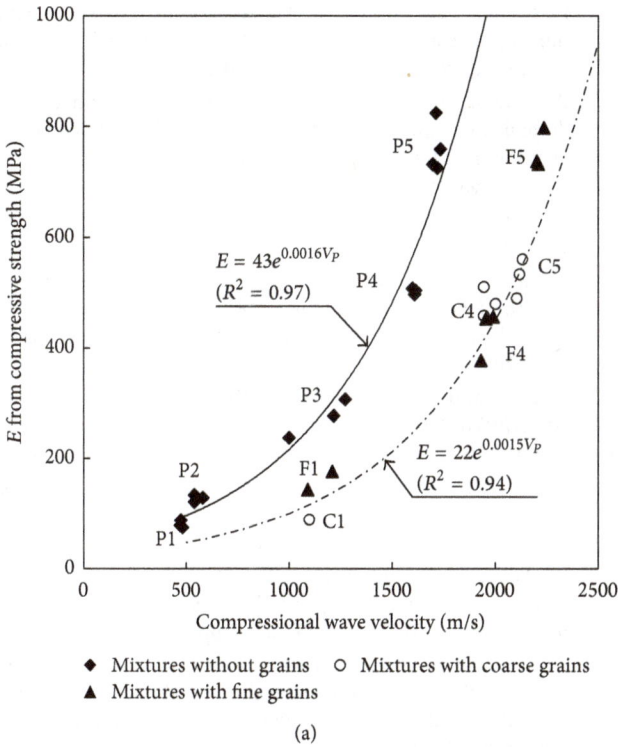

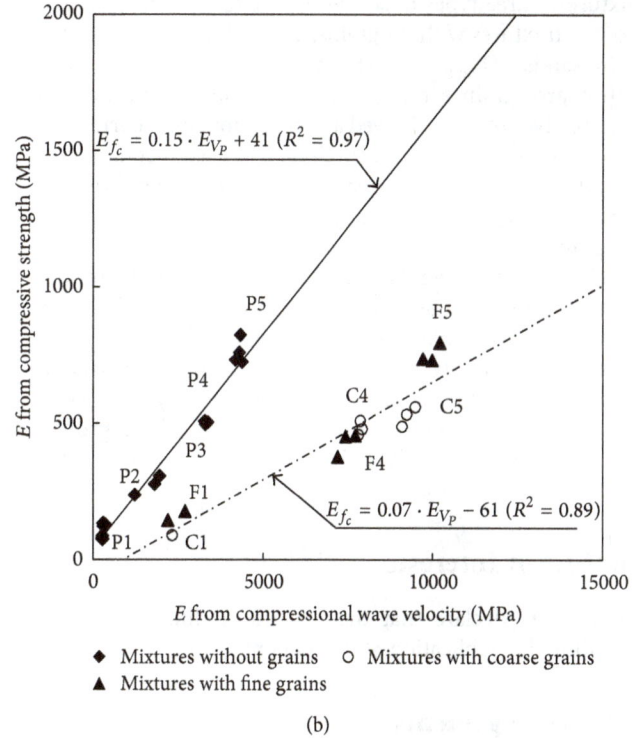

FIGURE 9: Elastic modulus: (a) elastic modulus measured by axial compression test versus compressional wave velocity; (b) elastic moduli measured by axial compression test versus by compressional wave velocity.

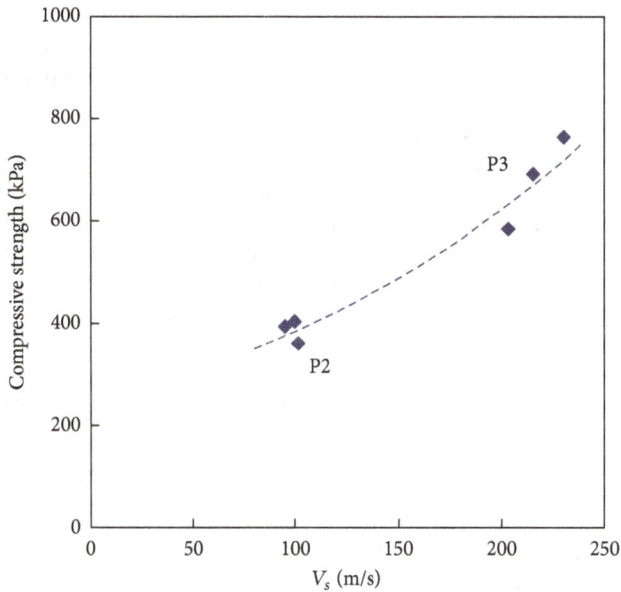

FIGURE 10: Shear wave velocity versus compressive strength for mixtures without grains.

FIGURE 11: Shear wave velocity versus compressional wave velocity for mixtures without grains.

The compressional wave velocity was five to six times greater than the shear wave velocity. Based on elasticity, Poisson's ratio v can be calculated as follows:

$$\text{Poisson's ratio: } v = \frac{1}{2}\left(\frac{V_P^2}{V_s^2}\right) - 1. \qquad (5)$$

Based on (5), the calculated Poisson's ratio v was about 0.48. Note that during the compressive strength test, water was

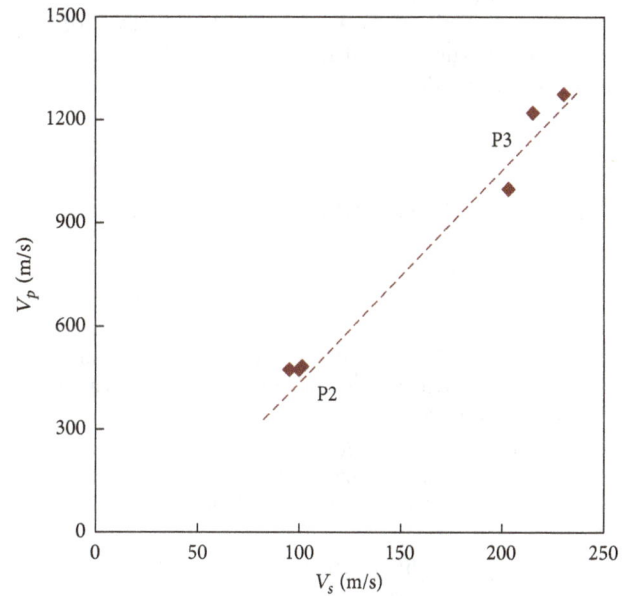

released from the specimen. The leakage during compressive strength tests confirmed the higher Poisson's ratio.

6. Summary and Conclusions

The relationship between the axial compressive strength and elastic wave velocity was investigated by using soil-grout mixtures. Three types of soil-grout mixtures were prepared: grouted mixtures without grains, grouted mixtures with fine grain sands (D_{sand} ≈ 0.82 mm), and grouted mixtures with coarse grain sands (D_{sand} ≈ 3.1 mm). The uniaxial compressive strength showed a good exponential correlation with the compressional wave velocity. The axial compressive strength versus compressional wave velocity showed that the cementation agent governed the strength at large strain and the grains within soil-grout mixtures controlled the compressional wave velocity at small strain. Exponential models for the estimation of strength and elastic modulus were suggested for the mixtures with and without grains by using the compressional wave velocity. The compressive strength was also correlated to the shear wave velocity. This study demonstrates that the strength and stiffness of the grouted mixtures may be simply estimated by the nondestructive elastic wave velocities.

Conflict of Interests

The authors declare that there is no conflict of interests regarding the publication of this paper.

Acknowledgements

This work was supported by the National Research Foundation of Korea (NRF) Grant funded by the Korea government (MSIP) (NRF-2011-0018110) and the Korea Agency of Infrastructure Technology Advancement under the Ministry of Land, Infrastructure, and Transport of the Korean government (Project no. 10 Technology Innovation-E09).

References

[1] J.-C. Dupla, J. Canou, and D. Gouvenot, "Injectability properties of sands by fine cement grouts," in *Proceedings of the 16th International Conference on Soil Mechanics and Geotechnical Engineering (ICSMGE '05)*, pp. 1181–1184, Osaka, Japan, September 2005.

[2] C. Lee, Q. H. Truong, and J. S. Lee, "Cementation and bond degradation of rubber-sand mixtures," *Canadian Geotechnical Journal*, vol. 47, no. 7, pp. 763–774, 2010.

[3] J. S. Lee, C. Lee, H. K. Yoon, and W. Lee, "Penetration type field velocity probe for soft soils," *Journal of Geotechnical and Geoenvironmental Engineering*, vol. 136, no. 1, pp. 199–206, 2010.

[4] H. K. Yoon and J. S. Lee, "Field velocity resistivity probe for estimating stiffness and void ratio," *Soil Dynamics and Earthquake Engineering*, vol. 30, no. 12, pp. 1540–1549, 2010.

[5] S. Popovics, "Analysis of the concrete strength versus ultrasonic pulse velocity relationship," *Materials Evaluation*, vol. 59, no. 2, pp. 123–130, 2001.

[6] G. Pascale, A. Di Leo, and V. Bonora, "Nondestructive assessment of the actual compressive strength of high-strength concrete," *Journal of Materials in Civil Engineering*, vol. 15, no. 5, pp. 452–459, 2003.

[7] S. Pessiki and M. R. Johnson, "Nondestructive evaluation of early-age concrete strength in plate structures by the impact-echo method," *ACI Materials Journal*, vol. 93, no. 3, pp. 260–271, 1996.

[8] H.-K. Lee, H. Yim, and K.-M. Lee, "Velocity-strength relationship of concrete by impact-echo method," *ACI Materials Journal*, vol. 100, no. 1, pp. 49–54, 2003.

[9] R. L. Day, "The effect of secondary ettringite formation on durability of concrete. A literature analysis," PCA Research and Development Bulletin RD108T, 1992.

[10] ASTM C597, *Standard Test Method for Pulse Velocity through Concrete*, ASTM C597-97, ASTM, West Conshohocken, Pa, USA, 1997.

[11] ASTM C1383, *Standard Test Method for Measuring the P-Wave Speed and the Thickness of Concrete Plates Using the Impact-Echo Method*, ASTM C 1383-98a, ASTM, West Conshohocken, Pa, USA, 2002.

[12] J. S. Lee and J. C. Santamarina, "P-wave reflection imaging," *Geotechnical Testing Journal*, vol. 28, no. 2, pp. 197–206, 2005.

[13] I. M. Lee, Q. H. Truong, D. H. Kim, and J. S. Lee, "Discontinuity detection ahead of a tunnel face utilizing ultrasonic reflection: laboratory scale application," *Tunnelling and Underground Space Technology*, vol. 24, no. 2, pp. 155–163, 2009.

[14] J. S. Lee and J. C. Santamarina, "Bender elements: performance and signal interpretation," *Journal of Geotechnical and Geoenvironmental Engineering*, vol. 131, no. 9, pp. 1063–1070, 2005.

[15] J. S. Lee and J. C. Santamarina, "Discussion "measuring shear wave velocity using bender elements" by Leong, E. C., Yeo, S. H., and Rahardjo, H.," *Geotechnical Testing Journal*, vol. 29, no. 5, pp. 439–441, 2006.

[16] C. Lee, J. S. Lee, W. Lee, and T. H. Cho, "Experiment setup for shear wave and electrical resistance measurements in an oedometer," *Geotechnical Testing Journal*, vol. 31, no. 2, pp. 149–156, 2008.

[17] J. C. Santamarina, K. A. Klein, and M. A. Fam, *Soils and Waves: Particulate Materials Behavior, Characterization and Process Monitoring*, John Wiley & Sons, New York, NY, USA, 2001.

[18] Q. H. Truong, Y. H. Eom, Y. H. Byun, and J.-S. Lee, "Characteristics of elastic waves according to cementation of dissolved salt," *Vadose Zone Journal*, vol. 9, no. 3, pp. 662–669, 2010.

[19] Q. H. Truong, C. Lee, Y. U. Kim, and J. S. Lee, "Small strain stiffness of salt-cemented granular media under low confinement," *Geotechnique*, vol. 62, no. 10, pp. 949–953, 2012.

[20] ASTM D2845, *Standard Test Method for Laboratory Determination of Pulse Velocities and Ultrasonic Elastic Constants of Rock*, ASTM D2845, ASTM, West Conshohocken, Pa, USA, 2000.

[21] K. Ishihara, *Soil Behaviour in Earthquake Geotechnics*, Clarendon Press, Oxford, UK, 1996.

[22] A. H. Hammam and M. Eliwa, "Comparison between results of dynamic & static moduli of soil determined by different methods," *HBRC Journal*, vol. 9, no. 2, pp. 144–149, 2013.

[23] T. S. Yun and J. C. Santamarina, "Decementation, softening and collapse: changes in small-strain shear stiffness in k_0-loading," *Journal of Geotechnical and Geoenvironmental Engineering*, vol. 131, no. 3, pp. 350–358, 2005.

21

Parametric Study on Responses of a Self-Anchored Suspension Bridge to Sudden Breakage of a Hanger

Wenliang Qiu, Meng Jiang, and Cailiang Huang

School of Civil Engineering, Dalian University of Technology, Dalian 116024, China

Correspondence should be addressed to Meng Jiang; jiangm@dlut.edu.cn

Academic Editor: Vincenzo Gattulli

The girder of self-anchored suspension bridge is subjected to large compression force applied by main cables. So, serious damage of the girder due to breakage of hangers may cause the collapse of the whole bridge. With the time increasing, the hangers may break suddenly for their resistance capacities decrease due to corrosion. Using nonlinear static and dynamic analysis methods and adopting 3D finite element model, the responses of an actual self-anchored suspension bridge to sudden breakage of hangers are studied in this paper. The results show that the sudden breakage of a hanger causes violent vibration and large changes in internal forces of the bridge. In the process of the vibration, the maximum tension of hanger produced by breakage of a hanger exceeds 2.22 times its initial value, and the reaction forces of the bearings increase by more than 1.86 times the tension of the broken hanger. Based on the actual bridge, the influences of some factors including flexural stiffness of girder, torsion stiffness of girder, flexural stiffness of main cable, weight of girder, weight of main cable, span to sag ratio of main cable, distance of hangers, span length, and breakage time of hanger on the dynamic responses are studied in detail, and the influencing extent of the factors is presented.

1. Introduction

Large-span bridges, including cable-stayed bridge, suspension bridge, and arch bridge, need cables to be the stay cables, main cables, or hangers. With the time increasing, the steel elements of bridge are exposed to corrosion, and their resistance capacities decrease. The stay cables and hangers especially are more prone to corrosion than other elements for the diameters of their steel wires are small. Under the combination action of live loads and corrosion, stay cables and hangers may break suddenly. In china, serious corrosion and breakage of cables occurred in many bridges [1]. In 2001, eight hangers of Yibin Southgate Bridge in China broke, and the deck supported by these hangers fell into the river. Two men died in this accident. In 1995, one stay cable of Guangzhou Haiyin Bridge that was only used for 7 years broke, and it hit an oil tank truck. To avoid the breakage of cables due to corrosion, cables in many bridges were replaced in China. For example, all the stays in Jiujiang Bridge that was used for 10 years were replaced, because 70% of the stays were seriously corroded, and 1/3 of wires in some stays were broken.

Because the breakage of cable usually occurs suddenly, and it can cause strong vibration and large change of internal forces of the structure, the sudden breakage may endanger the safety of the bridge. Because one hanger is broken, Mahakam II suspension Bridge in Indonesia collapsed in 2011, as shown in Figure 1. It happened very fast, only about 30 seconds. This bridge was completed in 2002, and it was less than 10 years old. At least 11 people were dead and 30 people were missing in this accident.

The effects of breakage of stay cables on cable-stayed bridge have been studied by many researchers [2–8]. Considering the influences of different layout of stays, number of planes of stays, and stiffness of deck, Mozos and Aparicio [5, 6] studied the effects of breakage of stays on deck, tower, and stays in detail. The study showed that it was safe for stays when adopting the dynamic amplification factor (DAF) of 2.0 recommended by PTI [9], but it was unsafe for deck and tower. Because the time that the breakage occurs in (breakage time) has significant effects on the responses of structure, Mozos and Aparicio [10] studied the breakage time through experiments and found that the breakage time was 0.00375 s for damaged cables, and 0.0085 s for undamaged cables. Zhou

<table>
<tr><td>(a) Before collapsing</td><td>(b) After collapsing</td></tr>
</table>

FIGURE 1: Mahakam II Bridge in Indonesia.

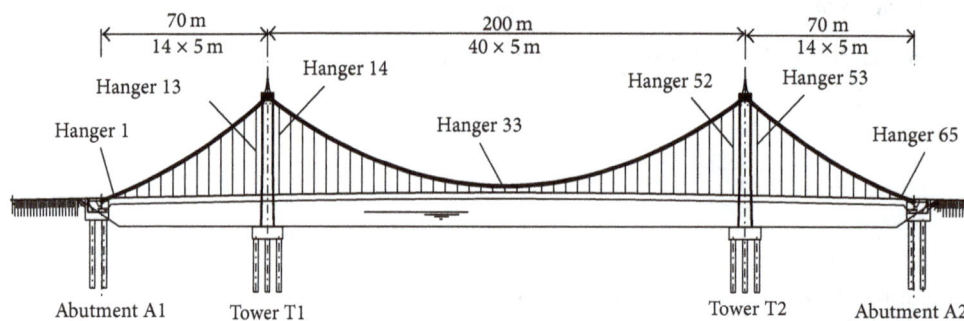

FIGURE 2: Layout of Zhuanghe Jianshe Bridge.

and Chen [8] studied dynamic responses of a cable stayed bridge caused by abrupt cable-breakage considering dynamic bridge-vehicle interactions.

Considering the geometric and material nonlinearity, Cai et al. [11] studied the nonlinear responses and progressive collapses of cable-stayed bridge due to sudden breakage of stays. When a single stay broke, the tension of other stays changed little. When two adjacent stays broke simultaneously, the bridge collapse did not occur, but the deck had large plastic deformation and the stays yielded. Qu et al. [12] studied DAFs of Sutong Bridge in China and found that the DAFs did not exceed 3%.

Ruiz-Teran and Aparicio [13] studied the effects of breakage of stays on underdeck cable-stayed bridge. The influences of the breakage time and the dynamic loads of breakage were investigated. When the breakage time was less than 10% of the fundamental period of the damaged bridge, it had no influences on the results. The study results showed that the responses produced by breakage of stays must be calculated using dynamic analysis method.

Considering the geometric and material nonlinearity, Qiu et al. [14] and Kao et al. [15] studied the static load-bearing capacity of self-anchored suspension bridge. The studies showed that the bridge did not collapse when five hangers broke. Because the dynamic effects of the breakage of hangers were not considered, the load-bearing capacities obtained by the studies are the upper limit value.

In recent years, more than 20 self-anchored suspension bridges have been built in China [16]. Their hangers are all made of parallel high strength galvanized steel wires. In Chinese codes [17], the safety factor of hanger is 3.0 during service, and it is 1.8 during replacement of hanger. No specifications are presented for sudden loss of hangers. Because the hangers are connected with the main cables, the breakage of hangers will induce the main cable to vibrate strongly, which will further induce strong vibration and large changes of internal forces of the whole bridge. Using nonlinear static and dynamic analysis methods, the responses of a self-anchored suspension bridge due to the breakage of hangers are studied in this paper. The results can be used for reference in design and maintenance of this kind of bridge.

2. Structure of a Self-Anchored Suspension Bridge and Analysis Methods

2.1. Structure of a Self-Anchored Suspension Bridge. Zhuanghe Jianshe Bridge built in China is a concrete self-anchored suspension bridge with mid-span of 200 m and side-span of 70 m, as shown in Figure 2. Its stiffening girder with box cross-section is reinforced concrete, as shown in Figure 3. Because the girder is subjected to a large compressive force applied by the main cables and the tension stresses do not occur in the girder under design loads, the tendons are not

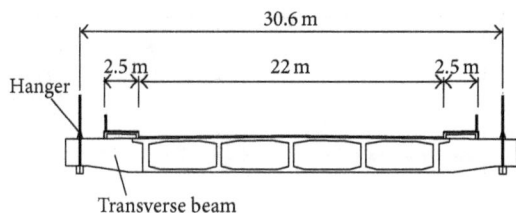

FIGURE 3: Cross-section of girder.

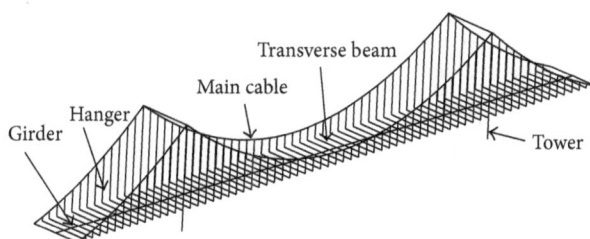

FIGURE 4: FE model of the bridge.

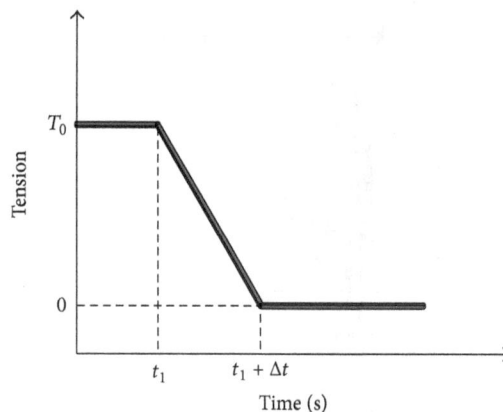

FIGURE 5: Time history of tension of broken hanger.

used in the girder. The tower is reinforced concrete with box cross-section. The bridge has two main cables and each cable is made of 3937 paralleled, 5 mm in diameter, high strength galvanized steel wires. The span to sag ratio of main cable in mid-span is 5.5. There are 65 hangers on each side of the bridge, and the distance of the hangers along the girder is 5 m. The hangers are made of 97 paralleled, 7 mm in diameter, steel wires. The hangers are numbered from 1 to 65.

2.2. Materials. The concrete of the girder and the towers is high strength concrete with a compressive strength of 50 MPa and a modulus of elasticity of 34000 MPa. The density of the reinforced concrete is 25.5 kN/m^3. The steel of the main cables and the hangers has an ultimate tensile strength of 1570 MPa and a modulus of elasticity of 190000 MPa. The density of the steel is 78.5 kN/m^3.

2.3. Loads. Because this study is only used to investigate the responses of bridge caused by sudden breakage of hangers, the analysis of the structure under dead loads is carried out. The self-weight of girder, towers, main cables, and hangers is obtained from multiplying the areas of their cross-sections by their densities. The distributed dead load on the deck except self-weight is 108 kN/m. The distributed dead load on the main cables except self-weight is 0.37 kN/m. Because the weight of clamps affects the vibration of the main cable, it is also considered in dynamic analysis. The weight of each clamp from 1 to 9 is 12 kN, the weight of each clamp from 10 to 17 is 16 kN, and the weight of each clamp from 18 to 33 is 10 kN.

2.4. Analysis Model. The bridge is modeled with three-dimensional (3D) finite element (FE) model, as shown in Figure 4. In the FE model, the girder, towers, and transverse beams are modeled using 3D beam elements, and their torsion stiffness and torsion mass are considered. In order

to study the influence of flexural stiffness, the main cables are modeled using beam elements. The hangers are modeled using truss elements. The influences of initial internal forces under self-weight on the stiffness of the structure are considered. The main cables between two hangers are modeled using 5 elements to consider geometrical configuration and local vibration of the main cables more precisely. The masses of clamps are concentrated on the nodes where the clamps are located. Considering the piles of the bridge had little effects on the analysis of sudden breakage of hanger, they are not modeled in the FE model. The girder is supported by two bearings at each tower as well as at each abutment. Using this FE model, the modal properties of Zhuanghe Jianshe Bridge are calculated. The lowest natural frequency of the vertical vibration mode of the bridge is 0.41 Hz, and the lowest natural frequency of the torsional vibration mode of the girder coupled with transversal vibration is 1.00 Hz.

2.5. Analysis Methods. Responses of the bridge due to sudden breakage of hangers on the bridge are analyzed by means of nonlinear static and dynamic analysis using the finite element software ABAQUS V.6.8. For either static analysis or dynamic analysis, the geometric nonlinearity of the structure and effects of axial forces of the structure on the stiffness are considered, and the nonlinear procedures are carried out using iteration method. In dynamic analysis, the direct time integration method is used. The time history of the tension of the broken cable is shown in Figure 5. Before the broken hanger breaks, the value of its initial tension is T_0. When the hanger breaks at time t_1, its tension drops from T_0 to zero during time interval Δt.

Here, Δt is called breakage time of hanger, which takes a value of 0.005 s [5] except when studying the influences of the breakage time on the dynamic responses.

The following analysis processes are adopted for analysis, as shown in Figure 6.

(a) Using nonlinear static analysis method and adjusting the initial stresses of main cables, hangers, girder, and towers, a reasonable static state of the undamaged bridge under dead loads is reached. In this state, the tensions of all hangers are nearly equal, and the

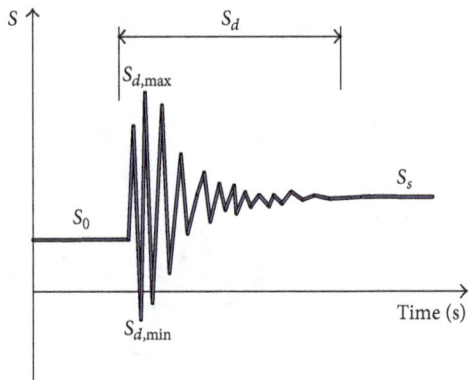

FIGURE 6: Time history of structural response.

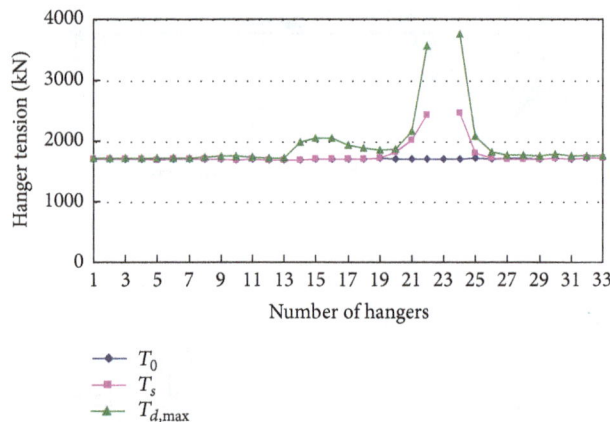

FIGURE 7: Hanger tensions caused by breakage of hanger 23.

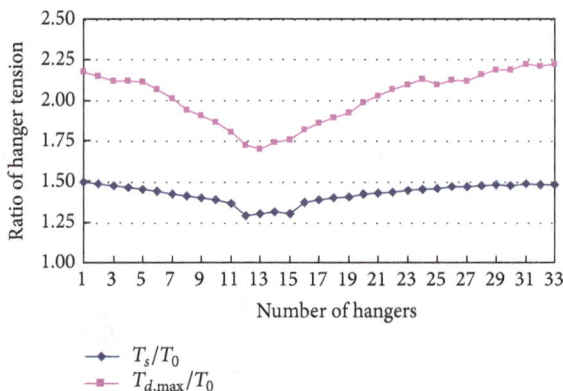

FIGURE 8: Ratios of hanger tensions.

bending moments of the girder and towers are very small.

(b) A hanger element is removed, and its tension is unloaded in time interval Δt.

(c) The structure vibrates strongly due to the sudden breakage of the hanger, and the dynamic analysis is carried out to calculate the vibration of the structure. Because of damping, the vibration attenuates with time increasing. Until the maximum node displacement amplitude of the structure is less than 0.1 mm, the state of bridge is taken as the final equilibrium state under dead loads after breakage of a hanger.

The static state before breakage of the hanger is identified by S_0. After the sudden breakage of a hanger, the structure vibrates strongly, and the dynamic state of the bridge is identified by S_d. It is especially mentioned that the final static state obtained by dynamic analysis is just the same as the static states of bridge without the broken hanger obtained by static analysis. The final static state after breakage is identified by S_s.

After a preliminary numerical study on the dynamic responses of the bridges to pulse loads, the time step of 0.001 s is used to calculate the dynamic process during 10 seconds after the sudden breakage of hanger, and the time step of 0.005 s is used during the other time. The time steps allow us to achieve an important reduction in computing time and to maintain adequate accuracy in the results. A Rayleigh damping of 2% is used in the dynamic analysis.

3. Reponses due to Sudden Breakage of a Single Hanger

Considering that two hangers are not possible to break at the same time, the dynamic responses caused by breakage of a single hanger is studied in this paper. From the dynamic analysis results, it can be found that the sudden breakage of a hanger produces very strong vibration of the structure; some internal forces of the structure become very large during the vibration. The bending moments of girder, bending moments of towers, and tensions of main cables caused by the sudden breakage of a hanger are not large enough to control the

design. The tensions of hangers and reactions of bearings are so large that they may control the design.

Figure 7 shows the initial value T_0, final value T_s, and maximum value $T_{d,\max}$ of tensions of the other hangers when hanger 23 breaks. It can be seen from the figure that breakage of a hanger has large effects on tensions of the hangers near the broken hanger and has little effects on tensions of the hangers far away from the broken hanger. The maximum tension $T_{d,\max}$ of each hanger during vibration is much larger than the corresponding final value T_s.

To obtain the maximum values of tension ratios $T_{d,\max}/T_0$ and T_s/T_0 of every hanger, maximum tension $T_{d,\max}$ and final tension T_s of each hanger are calculated when every one of the other hangers breaks. Figure 8 shows the maximum values of tension ratios $T_{d,\max}/T_0$ and T_s/T_0 of the hangers from 1 to 33. The tension ratio $T_{d,\max}/T_0$ of hanger 30 is the largest, that is, 2.22. However, the tension ratio T_s/T_0 of hanger 30 is only 1.48, which is only 66.7% of the tension ratio $T_{d,\max}/T_0$. So, it can be concluded that the difference of the results between static analysis and dynamic analysis cannot be neglected. Additionally, the hanger is farther from the tower, the vibration caused by its breakage is stronger. So the tension of hanger increases with the distance between hanger and tower increasing.

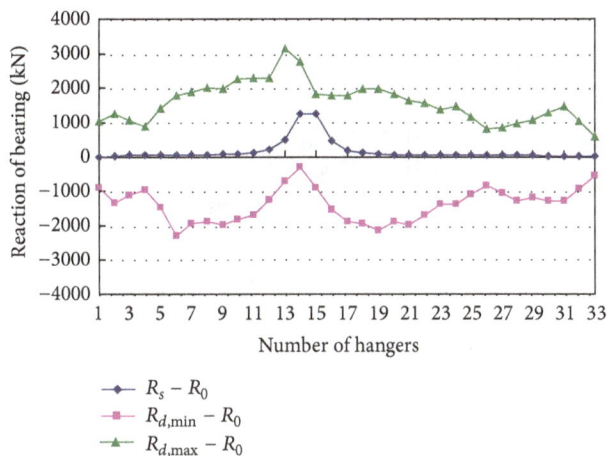

FIGURE 9: Reaction of the left bearing at tower T1.

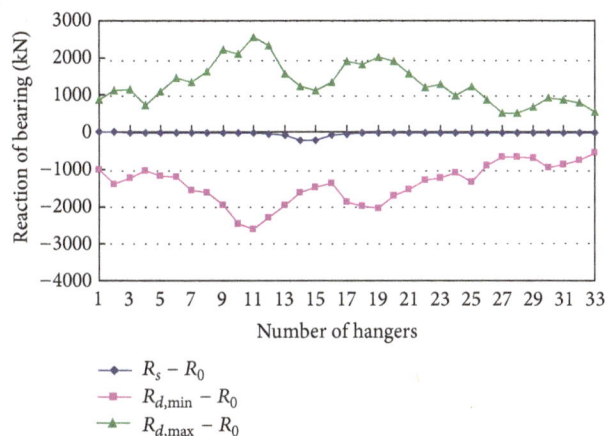

FIGURE 10: Reaction of the right bearing at tower T1.

The dynamic effects on hanger tensions are so marked that the maximum tension reaches 2.2 times of the initial value. If the breakage of a hanger is for the reason of corrosion, the other hangers may be also exposed to corrosion, and their resistance capacities decrease too. The breakage of one hanger may induce one-by-one breakage of the other hangers, and the bridge may collapse after several hangers break. So, the breakage of a hanger can endanger the safety of the bridge seriously.

The reaction forces $R_{d,\max} - R_0$, $R_{d,\min} - R_0$, and $R_s - R_0$ of the two bearings at tower T1 produced by sudden breakage of every one of the hangers from 1 to 33 on the left side are shown in Figures 9 and 10. Because breakage of any one hanger can induce strong vertical and torsional vibrations, vibrations further induce large changes of reaction forces of bearings. The maximum increment of reaction force of the left bearing is produced by breakage of hanger 13, and it is 1.86 times the tension of the broken hanger. The maximum decrement of reaction force of the right bearing is produced by breakage of hanger 6, and it is 1.35 times the tension of the broken hanger. Both the maximum increment and the maximum decrement of reaction force of the right bearing are produced

by breakage of hanger 11, and they are 1.51 and 1.54 times the tension of the broken hanger, respectively. So, the reaction forces of bearings produced by breakage of a hanger cannot be ignored. Additionally, the reaction forces of bearings due to torsion of girder are related to the distance between the two bearings, and the reaction forces increase with the distance decreasing. A relatively large distance is adopted in the bridge analyzed in this paper.

4. Parametric Study on Influencing Factors

In order to find out the relationship between the responses caused by breakage of hanger and the structure, based on the bridge analyzed above, breakage of the left hanger H0 at the position of 1/4 of mid-span, as shown in Figure 11, is taken as an example to study the influences of flexural stiffness of girder, torsion stiffness of girder, flexural stiffness of main cable, weight of girder, weight of main cable, span to sag ratio of main cable, distance of hangers, span length, and breakage time of hanger on the dynamic responses. The mid-span length is 200 m for all above factors except span length. For general suspension bridge, the initial tensions of all hangers under dead load are designed to be a same value. In the suspension bridge analyzed in this paper, the initial tension of each hanger is T_0. To demonstrate the extent of the influences on hanger tensions, ratio of the maximum tension $T_{d,\max}$ to initial tension T_0 is adopted in the following analysis. Because the breakage of hanger H0 has marked influence only on the adjacent hangers HL and HR, only the tensions of the two adjacent hangers are presented when the influencing factors are studied parametrically. Additionally, considering that the bearing reactions change largely during the dynamic responses, they are presented for a part of the influencing factors.

4.1. Flexural Stiffness of Girder. Figure 12 shows relationship between tension ratio $T_{d,\max}/T_0$ of hangers HL and HR and flexural stiffness of girder when hanger H0 breaks suddenly. In the figure, λ_i is ratio of the changed flexural stiffness to the actual flexural stiffness of the designed girder. When the flexural stiffness of girder increases from 0.1 to 100 times of actual value, the tension ratio of hanger HL changes in the range of 2.085~2.099, and it changes only by 0.67%. The tension ratio of hanger HR changes in the range of 2.093~2.100, and it changes only by 0.33%. It can be concluded that flexural stiffness of girder has small effects on hanger tension. Figure 13 shows reactions of the four bearings at the two towers change with the flexural stiffness ratio increasing. It can be seen from the figure that the bearing reactions increase with the flexural stiffness of girder increasing. When the flexural stiffness ratio is between 0.01 and 3.0, the reactions caused by breakage of hanger H0 are nearly same except for the reaction of the left bearing on tower T2. But when the flexural stiffness ratio is larger than 3.0, the bearing reactions change very markedly. Particularly, when the flexural stiffness ratio is about 50.0, the peak values occur in all curves at the same time. The maximum reaction is 3946.5 kN, which is 2.31 times the initial tension of the broken hanger ($T_0 =$

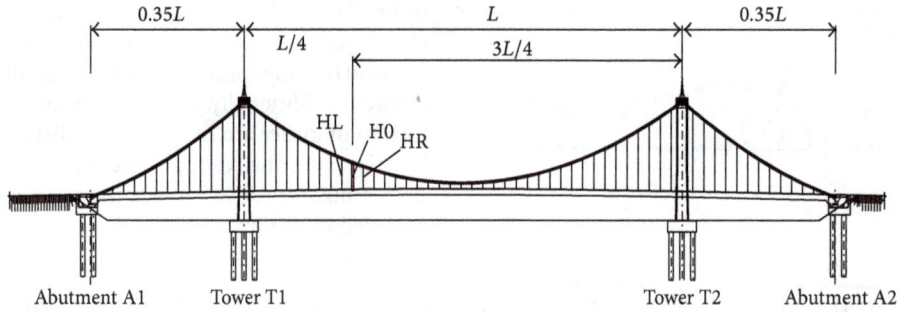

FIGURE 11: Positions of hangers H0, HL, and HR.

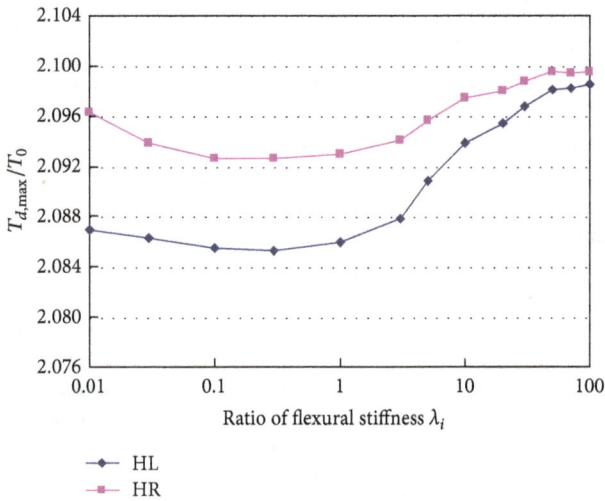

FIGURE 12: Effects of flexural stiffness of girder on hanger tensions.

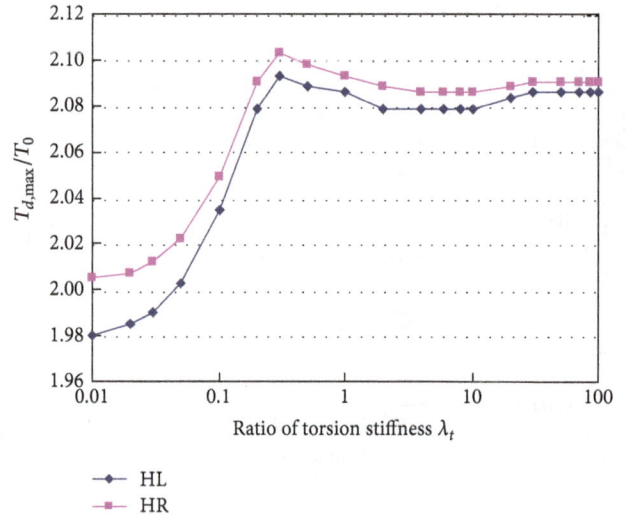

FIGURE 14: Effects of torsion stiffness of girder on hanger tensions.

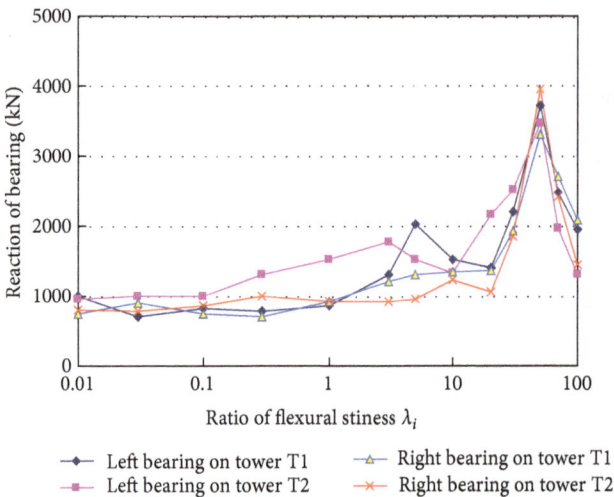

FIGURE 13: Effects of flexural stiffness of girder on bearing reactions.

1705.2 kN). From the vertical vibration of the structure, it can be seen that local vibration of main cable induces resonance of the global bridge when flexural stiffness of girder takes a certain value.

4.2. Torsion Stiffness of Girder. Figure 14 shows relationship between tension ratio $T_{d,\max}/T_0$ of hangers HL and HR and torsion stiffness of girder when hanger H0 breaks suddenly. In the figure, λ_t is ratio of the changed torsion stiffness to the actual torsion stiffness of the designed girder. When the torsion stiffness ratio increases from 0.1 to 0.3, the tension ratio of hanger HL increases from 1.980 to 2.093, the tension ratio of hanger HR increases from 2.005 to 2.103, and they increase 5.7% and 4.9%, respectively. When the torsion stiffness ratio increases from 0.3 to 4.0, the hanger tension ratios decrease a bit. When the torsion stiffness ratio is larger than 4.0, the hanger tension ratios remain nearly unchanged. Figure 15 shows the trend that the bearing reactions increase with the increment of torsion stiffness of girder. When the torsion stiffness ratio increases from 0.01 to about 50, the reactions are nearly in the range of 750~2000 kN. When the torsion stiffness ratio is larger than 50, the reactions increase markedly, and the maximum value is 4814.0 kN at the stiffness ratio of 100.

4.3. Flexural Stiffness of Main Cable. Generally, the flexural stiffness of main cable is neglected and the main cables are modeled using cable elements. The above assumption

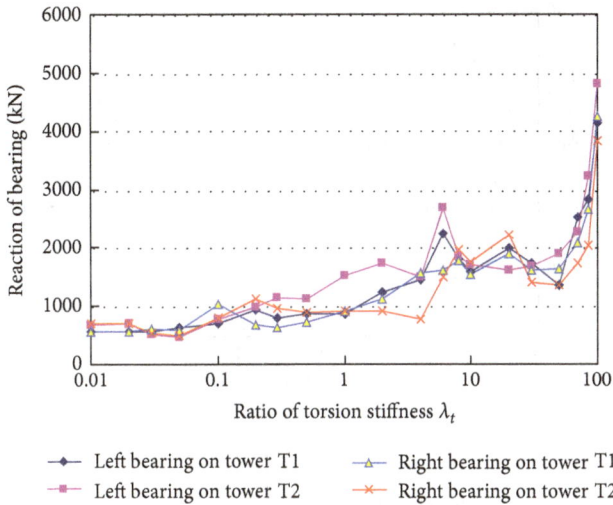

FIGURE 15: Effects of torsion stiffness of girder on bearing reactions.

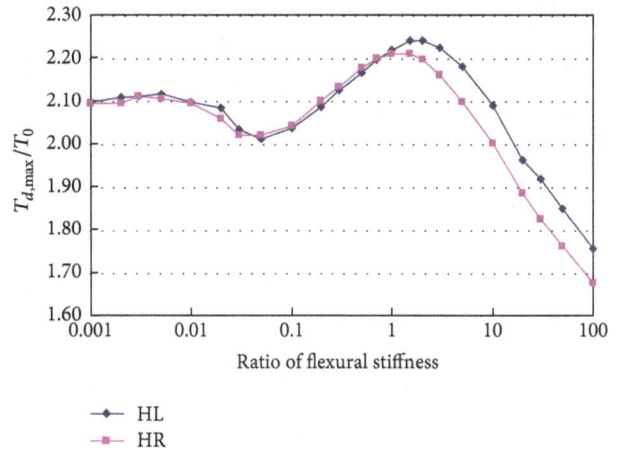

FIGURE 16: Effects of flexural stiffness of main cable on hanger tension.

is reasonable when the local vibration of main cable is not considered in structural analysis of suspension bridge. Because the flexural stiffness has large effects on the dynamic characteristics of the cable, it cannot be ignored in analysis of the violent local vibration of main cable caused by sudden breakage of hanger. In this paper, the main cable is assumed as a solid cylinder with curved geometry, which has the same sectional area as the actual main cable. By changing the flexural stiffness of main cable from 0.001 to 100 times the flexural stiffness of the solid cylinder, the effects of the flexural stiffness of main cable on the dynamic responses caused by breakage of hanger H0 are studied. Figure 16 shows the relationships between tension ratios of hangers HL and HR and the flexural stiffness ratio of main cable. It can be seen that the flexural stiffness of main cable has very marked influences on hanger tensions. When the flexural stiffness ratio of main cable increases from 0.001 to 0.01, the tension ratios nearly remain a constant value of about 2.10. When it increases from 0.01 to 0.05, the tension ratio decreases to be a value of about 2.01. When it increases from 0.05 to 1.5, the tension ratio increases to be a maximum value of 2.240. When it increases from 1.5, the tension ratio decreases rapidly. The tension ratio of hanger HR decreases to be a value of 1.675 when the flexural stiffness ratio is 100.

4.4. Weight of Structure. Weight of structure is one of the primary factors that affect the dynamic characteristics of the bridge. By changing weight of girder and main cable, the effects of weight on the dynamic responses caused by breakage of hanger H0 are studied. Figure 17 shows the changes of tension ratios of hangers HL and HR when the weight of girder increases from 0.8 to 1.2 times the actual weight of the designed girder. The tension ratios change less than 0.2%, so the effects of weight of girder can be ignored.

Figure 18 shows the relationship between tension ratios of hanger and the weight of main cable. It can be seen from the figure that the weight of main cable has larger effects than the weight of girder. When the weight of main cable

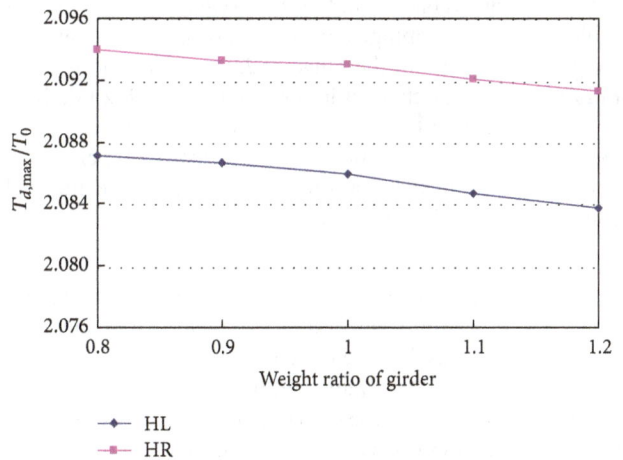

FIGURE 17: Effects of weight of girder on hanger tension.

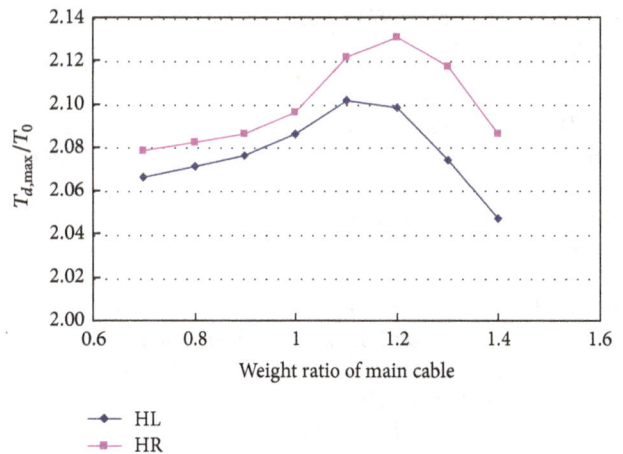

FIGURE 18: Effects of weight of main cable on hanger tension.

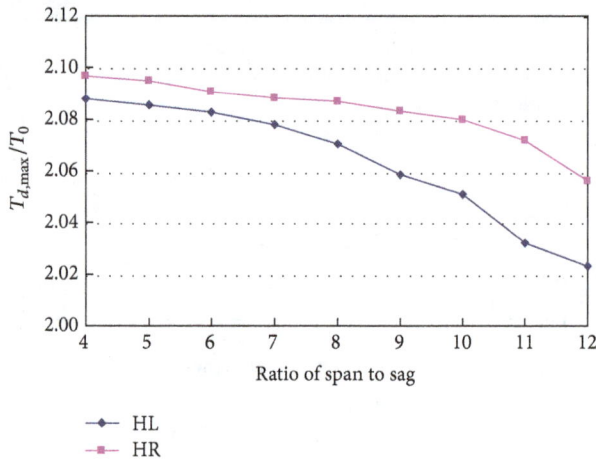

FIGURE 19: Effects of span to sag of main cable on hanger tension.

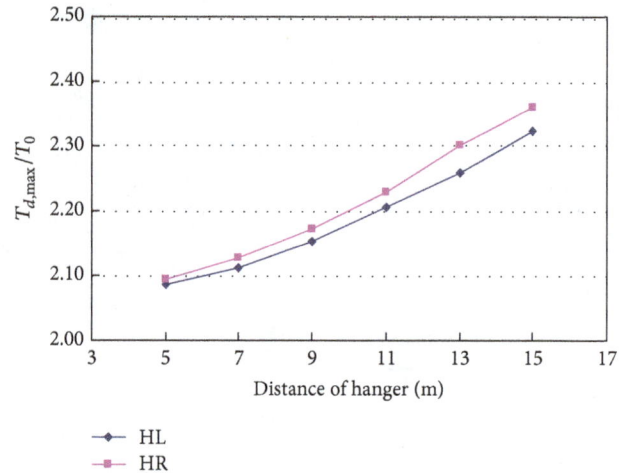

FIGURE 20: Effects of distance of hanger on hanger tension.

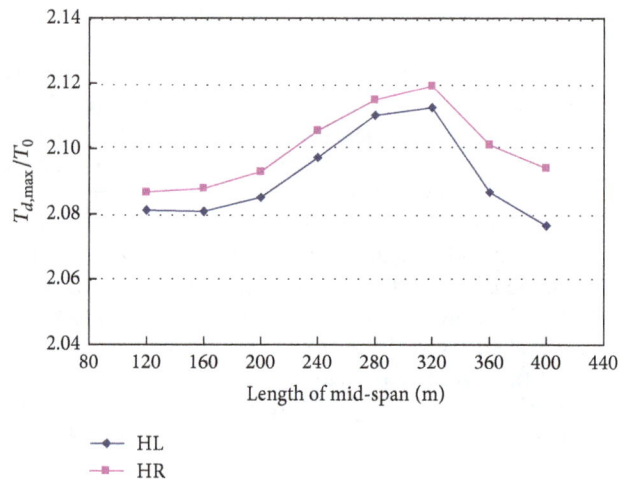

FIGURE 21: Effects of mid-span length on hanger tension.

increases from 0.7 to 1.4 times the actual weight of the designed main cable (including the weight of steel wires of cable, clamps, wrapping, and decoration), tension ratio of hanger HL changes in the range of 2.047~2.101, and tension ratio of hanger HR changes in the range of 2.079~2.131. They change by 2.6% and 2.5%, respectively. From Figures 16 and 18, it can be seen that dynamic characteristics of main cable determined by flexural stiffness and weight have marked effects on the tensions of the hangers.

4.5. Span to Sag Ratio of Main Cable. Span to sag ratio of main cable is a very important design parameter for suspension bridge, and it has large effects on the mechanic state of the bridge. When the span length of the bridge remains unchanged, the effects of span to sag ratio of main cable on the dynamic responses caused by sudden breakage of hanger H0 are studied. Figure 19 shows the relationship between tension ratios of hanger and the ratio of span to sag of main cable. It can be seen from the figure that the tension ratio of hanger decreases with the ratio of span to sag increasing. When the span to sag ratio increases from 4 to 12, tension ratio of hanger HL decreases from 2.088 to 2.023, and tension ratio of hanger HR decreases from 2.097 to 2.056. They decrease by 3.1% and 2.0%, respectively.

4.6. Distance of Hanger. When the span length of the bridge remains unchanged, the effects of distance of hanger on the dynamic responses caused by sudden breakage of hanger H0 are studied. Figure 20 shows the relationship between tension ratios of hanger and the distance of hanger. The tension ratio of hanger increases with the distance of hanger increasing. When the distance of hanger increases from 5 m to 15 m, tension ratio of hanger HL increases from 2.086 to 2.324, and tension ratio of hanger HR increases from 2.093 to 2.360. The tension ratios of the two hangers increase by 11.4% and 12.8%, respectively. So, the distance of hanger has significant effect on the dynamic responses caused by sudden breakage of a hanger.

4.7. Span Length. The mid-span length of the actual bridge is 200 m. In order to study the effects of span length on the dynamic responses caused by sudden breakage of hanger, the mid-span length is changed from 120 m to 400 m. In the parametric study, the distances of hangers remain to be 5 m. Figure 21 shows the relationship between tension ratios of hanger and mid-span length. With the mid-span length increasing, the tension ratio of hanger increases firstly and then decreases. The minimum value of tension ratio of hanger HL is 2.077, and the maximum value is 2.113, which is larger than the minimum value by 1.7%. The minimum value of tension ratio of hanger HR is 2.087, and the maximum value is 2.119, which is larger than the minimum value by 1.5%.

4.8. Breakage Time of Hanger. Figure 22 shows effects of breakage time Δt of hanger on hanger tension ratios. The breakage time of hanger affects the tension ratio of hanger very markedly. The tension ratio of hanger decreases rapidly with the breakage time increasing. When the breakage time increases from 2 ms to 70 ms, tension ratio of hanger HL

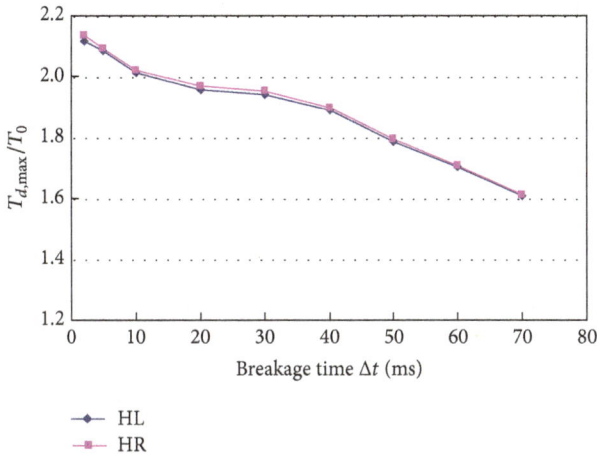

FIGURE 22: Effects of breakage time of hanger on hanger tension.

decreases from 2.118 to 1.607, and tension ratio of hanger HR decreases from 2.135 to 1.612. The tension ratios of the two hangers increase by 24.1% and 24.5%, respectively. So, the breakage time of hanger should be studied in detail through experimental studies to make the numerical study be more accurate.

5. Conclusions

For self-anchored suspension bridge, its main cables are anchored directly to two ends of its stiffening girder, and the stiffening girder subjected to a very large compression force is supported by hangers. So the hangers are very important for the safety of the whole bridge. With the time increasing, the resistance capacities of the hangers decrease due to corrosion, and the hangers may break suddenly. The study on the responses of a self-anchored suspension bridge to breakage of hangers reaches the following conclusions.

(1) The sudden breakage of a hanger produces a very strong vibration and large changes of internal forces of the bridge. During the vibration, the maximum tension of hanger reaches 2.22 times the initial value. The maximum increment of reaction force of bearing is 1.86 times the tension of the broken hanger, and the maximum decrement of reaction force of bearing is 1.54 times of tension of the broken hanger.

(2) Within the studied influencing factors, the flexural stiffness of main cable, distance of hangers, and breakage time of hanger have very significant effects on the dynamic responses caused by sudden breakage of a hanger. Their effects on hanger tension ratio are larger than 10%. Torsion stiffness of girder, weight of main cable, span to sag ratio of main cable, and span length have significant effects on the dynamic responses caused by sudden breakage of a hanger, and their effects on hanger tension ratio are in the range of 2~5%. Flexural stiffness and weight of girder have very small effects on the dynamic responses caused by sudden breakage of hanger, and their effects

on hanger tension ratio are less than 1%. Some of the above influencing factors have marked effects on reactions of the bearings. Particularly, when the factors take some values, the torsion vibrations of girder are very strong, and the bearing reactions occur to be very large values, which are several times of the tension of broken hanger.

Conflict of Interests

The authors declare that there is no conflict of interests regarding the publication of this paper.

Acknowledgment

This work was supported by the National Natural Science Foundation of China (NSFC-51178080).

References

[1] L. L. Wang and W. J. Yi, "Cases analysis on cable corrosion of cable-stayed bridges," Central South Highway Engineering, vol. 32, no. 1, pp. 94–132, 2007 (Chinese).

[2] A. M. Ruiz-Teran and A. C. Aparicio, "Dynamic amplification factors in cable-stayed structures," Journal of Sound and Vibration, vol. 300, no. 1-2, pp. 197–216, 2007.

[3] M. Wolff and U. Starossek, "Robustness assessment of a cable-stayed bridge," in Proceedings of the 4th International Conference on Bridge Maintenance, Safety and Management (IABMAS '08), pp. 690–696, July 2008.

[4] M. Wolff and U. Starossek, "Cable-loss analyses and collapse behavior of cable-stayed bridges," in Proceedings of the 5th International Conference on Bridge Maintenance, Safety and Management (IABMAS '10), pp. 2171–2178, July 2010.

[5] C. M. Mozos and A. C. Aparicio, "Parametric study on the dynamic response of cable stayed bridges to the sudden failure of a stay, Part I: bending moment acting on the deck," Engineering Structures, vol. 32, no. 10, pp. 3288–3300, 2010.

[6] C. M. Mozos and A. C. Aparicio, "Parametric study on the dynamic response of cable stayed bridges to the sudden failure of a stay, Part II: bending moment acting on the pylons and stress on the stays," Engineering Structures, vol. 32, no. 10, pp. 3301–3312, 2010.

[7] C. S. Kao and C. H. Kou, "The influence of broken cables on the structural behavior of long-span cable-stayed bridges," Journal of Marine Science and Technology, vol. 18, no. 3, pp. 395–404, 2010.

[8] Y. Zhou and S. Chen, "Time-progressive dynamic assessment of abrupt cable-breakage events on cable-stayed bridges," Journal of Bridge Engineering, vol. 19, no. 2, pp. 159–171, 2014.

[9] Post-Tensioning Institute (PTI), Recommendations for Stay Cable Design, Post-Tensioning Institute, Farmington Hills, Mich, USA, 2007.

[10] C. M. Mozos and A. C. Aparicio, "Numerical and experimental study on the interaction cable structure during the failure of a stay in a cable stayed bridge," Engineering Structures, vol. 33, no. 8, pp. 2330–2341, 2011.

[11] J. G. Cai, Y. X. Xu, L. P. Zhuang, J. Feng, and J. Zhang, "Comparison of various procedures for progressive collapse analysis of cable-stayed bridges," Journal of Zhejiang University: Science A, vol. 13, no. 5, pp. 323–334, 2012.

[12] Z. L. Qu, X. F. Shi, X. X. Li, and X. Ruan, "Research on dynamic simulation methodology for cable loss of cable stayed bridges," *Structural Engineers*, vol. 25, no. 6, pp. 89–92, 2009 (Chinese).

[13] A. M. Ruiz-Teran and A. C. Aparicio, "Response of under-deck cable-stayed bridges to the accidental breakage of stay cables," *Engineering Structures*, vol. 31, no. 7, pp. 1425–1434, 2009.

[14] W. L. Qiu, M. Jiang, and Z. Zhang, "Influencing factors of ultimate load carrying capacity of self-anchored concrete suspension bridge," *Journal of Harbin Institute of Technology*, vol. 41, no. 8, pp. 128–141, 2009 (Chinese).

[15] C. S. Kao, C. H. Kou, W. L. Qiu, and J. L. Tsai, "Ultimate load-bearing capacity of self-anchored suspension bridges," *Journal of Marine Science and Technology*, vol. 20, no. 1, pp. 18–25, 2012.

[16] Z. Zhang, Q. J. Teng, and W. L. Qiu, "Recent concrete, self-anchored suspension bridges in China," *Proceedings of the Institution of Civil Engineers: Bridge Engineering*, vol. 159, no. 4, pp. 169–177, 2006.

[17] Ministry of Transport of the People's Republic of China, *Design Specifications for Highway Suspension Bridge*, Ministry of Transport of the People's Republic of China, Beijing, China, 2002.

22

Numerical Simulation of Monitoring Corrosion in Reinforced Concrete Based on Ultrasonic Guided Waves

Zhupeng Zheng, Ying Lei, and Xin Xue

Department of Civil Engineering, Xiamen University, Xiamen 361005, China

Correspondence should be addressed to Ying Lei; ylei@xmu.edu.cn

Academic Editor: Ting-Hua Yi

Numerical simulation based on finite element method is conducted to predict the location of pitting corrosion in reinforced concrete. Simulation results show that it is feasible to predict corrosion monitoring based on ultrasonic guided wave in reinforced concrete, and wavelet analysis can be used for the extremely weak signal of guided waves due to energy leaking into concrete. The characteristic of time-frequency localization of wavelet transform is adopted in the corrosion monitoring of reinforced concrete. Guided waves can be successfully used to identify corrosion defects in reinforced concrete with the analysis of suitable wavelet-based function and its scale.

1. Introduction

Civil infrastructures frequently cause failure due to corrosion of steel and rebar in concrete structures. Due to corrosion billions of US dollars should be spent annually in repair, rehabilitation, and reconstruction efforts of reinforced concrete structures. The fact makes it arguably the single largest infrastructural problem facing the industrialized countries [1]. Thus, it is very important to develop effective corrosion monitoring technologies. A wide range of techniques have been reported in the paper that can be employed for the monitoring of corrosion of steel in concrete structures for the purpose of diagnosing the cause and extent of the reinforcement corrosion [2]. Most of the current techniques are based on electrochemical methods such as half-cell potential mapping linear polarization. These techniques relate corrosion rate and extent through assessment on surrounding concrete medium. While many electrochemical techniques have been well established, none of these techniques concentrate on monitoring through direct condition assessment or measurements on embedded steel. As alternative tools for monitoring steel corrosion, some physical based techniques have been proposed [3, 4]. Compared with the electrochemistry based approaches, these physical approaches can not only provide supplemented tools for monitoring steel corrosion, but also conduct more accurate condition assessment of steel

corrosion. Recently, the authors presented a review of some physical based monitoring techniques for condition assessment of corrosion in reinforced concrete in the past decades [5].

Among the current available physical monitoring techniques, the technique based on ultrasonic guided wave (UGW) is popular due to the advantages for monitoring corrosion related damage in reinforcing bars, so it has gained popularities in the recent years [5]. However, one difficulty of guided wave based technique for monitoring corrosion in reinforced concrete is the limitation of monitoring range for certain modes and frequencies [5, 6]. Unlike guided wave propagation in other multilayered systems, such as a metal pipeline in air, wave energy in steel bars embedded in mortar or concrete is lost (i.e., attenuated) at high rates due to leakage into the surrounding concrete. For the defects test of steel bar embedded in concrete, the reflected signals will be very weak, so the general time-frequency methods have difficulty in extracting the weak reflection signals of the defects in the detection signals. Meanwhile, there are many interference factors in the process of the experiment, for example, noise and the ideal boundary conditions which are difficult to achieve, and so forth, so it is hard to extract effective information of damage or defect from the received signals using guided wave methods. It is necessary to first investigate the problem by numerical simulation.

There are two methods widely used for the numerical simulation [7]. One is finite element method (FEM) and the other is boundary element method (BEM). BEM has been used in the wave guide of slab; for example, Cho and Rose [8] analyzed the mode conversion of Lamb wave at the reflection of boundary by BEM; Zhao and Rose [9] explored the guided waves on the identification of the size of defect by simulating various size and depth of defects on the slab using BEM. FEM has been mostly used in the wave guide of tube; for example, Demma [10] showed that a series of models of pipe with defects was calculated using FEM and the results were consistent with the experimental results. Moser et al. [11] simulated the propagation of elastic wave in the sheet and tubular structure using FEM. The results are fully consistent with those from experiment, which further proves the validity of the simulation in wave propagation using FEM. Cheng [12] used shell element to simulate the defect monitoring by longitudinal guided wave and get the relation curves between reflection coefficient and circumferential length or axial length of the defect in pipe. He et al. [13] studied the propagation of guided waves in bending pipe using FEM.

However, most of the previous analyses concentrated on thin wall pipe or slab using shell element in the simulation of guided waves. The shell element can only be used to simulate the pipes with very thin walls and is not suitable for the modes of guided waves with larger radial displacements [8]. There are few simulations for the rod structures. Therefore, it is proposed to explore the numerical simulation of corrosion monitoring of steel bar in reinforced concrete in this paper.

2. Numerical Simulation of Corrosion Monitoring in Reinforcement

When the ultrasonic guided waves are excited at the end of the steel bar and propagate along the axial direction of it, the mutation on the section dimensions or material properties of the steel bar will cause the strong discontinuity of the propagation of guided waves [14], resulting in reflection signals of guide waves from defects in the damage location. By the time process features of guided wave signals for the incidence, reflections from defects, and the end using sensors, the damage identification and location can be achieved.

The corrosion productions of steel are $Fe(OH)_2$, $Fe(OH)_3$, or Fe_3O_4, which are floc and have no strength. The pitting corrosion will cause the reduction in the section area of the steel bar, which is like notches formed in the surface of steel bar. Though there is difference in the geometry between the artificial defect (i.e., notch) and real corrosion in the steel bar, the research results [15, 16] show that the reflections for the mode L(0, 1) are almost the same for the artificial defect and real corrosion defect under the conditions of the same depth and circular length because the axial dimensions of these two defects are much shorter than the wavelength of mode L(0, 1). Therefore, the pitting corrosion is simulated by notch fabricated in the surface of the steel bar in this paper.

2.1. Modeling and Meshing. When the numerical simulation of guided waves is conducted using finite element method,

the first step is to establish finite element model of the simulation component. According to the types of the established models, the finite element models for guided wave detection can be divided into the following three types: plane axial symmetric element model, shell element model, and entity unit model. These three models simplify respectively the actual waveguide in different way, which results in the differences of the rationality and the complexity for the established models [10]. This paper focuses on rod components, so the shell element model is not applicable and only the plane axial symmetric element model and entity unit model can be employed. In order to save the amount of computation, the plane axial symmetric element model is employed when the calculated results are better for the model with axial symmetry properties; otherwise, the entity unit model is used.

When establishing the finite element model, meshing is a very important step, and there is a high requirement on the mesh division for the finite element simulation of guided waves, in which the establishment of defect in the simulation model of the guided wave is one of the important issues. There are two main methods for modeling defect. The first one is establishing directly the component model of rod with defect and then meshing. In this way, defect can be simply established and various types of defects can be done, but the existence of defect destroys the regularity of the model and it is very difficult to mesh the model by mapping because the size of the defect is generally small. The other method is progressing as follows: firstly, establish a zero-defect model and map mesh and, then, remove the unit at the defect position. In this way, the regular grids of model can be achieved but the geometry of defect will be restricted by the element shape. In order to get regular girds, the latter method is adopted in this paper.

For the processing of boundary conditions, rigid boundaries are set at both ends of the cylinder steel bar [17]. At the same time, the axial boundary of concrete is set as no reflection boundary to make the result more reasonable in which uncertainty factors in simulation are eliminated, because the steel bars are distributed along the longitudinal direction of the concrete components in the actual on-line detection and there is no reflection on the border for the propagation of stress wave in concrete.

2.2. Signal Loading and Postprocessing. Research shows that guided waves can be excited in the waveguide when instantaneous displacement is loaded in all nodes of certain section in the model [18]. The incentive signal with the center frequency of 75 KHz is used in this paper for numerical simulation. The time-domain waveform of the signal is shown in Figure 1.

Numerical simulation of guided waves in this paper is the transient dynamic analysis. The excitation of guided waves is completed when axial displacement in the time history curve as shown in Figure 1 is applied in the left side of steel model as "signal-loading." In postprocessing, time-displacement curve will be extracted and analyzed from the nodes at the receiving location and the data of corresponding

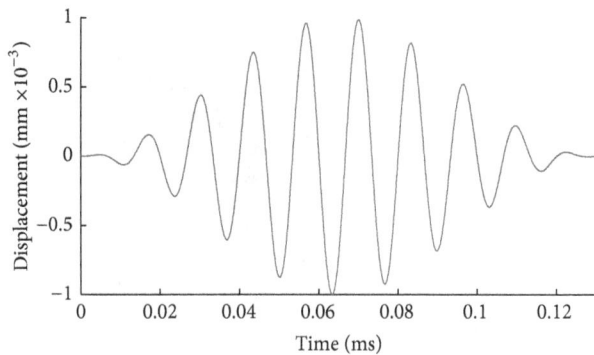

FIGURE 1: Excitation signal (center frequency: 75 KHz).

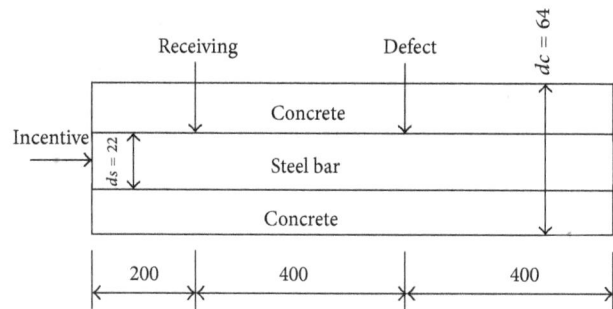

FIGURE 2: Layout of incentive and receiving (unit: mm).

TABLE 1: Material parameters of steel bar and concrete.

Parameter	Steel bar	Concrete
Poisson's ratio	0.2865	0.27
Density (kg/m^3)	7932	2200
Modulus (MPa)	215000	22000

TABLE 2: Discrete parameters.

Parameter	Value
Unit length Δx	$\Delta x \leq \lambda/20$
Time step Δt	$\Delta t \leq 0.8 \times \Delta x/v$

results will be outputted to a text file for later analysis and processing.

2.3. Parameter Selection. It is the subject of transient elastic dynamic problems for the propagation of guided waves which is discrete by the finite element method in time and space domain. There is important significance for the selection of related parameters on the correctness of the results, especially the length of space discrete unit and time step of the discrete time. The main parameters involved in this paper are material, geometric, discrete, and incentive parameters, respectively. The material parameters are elastic modulus, Poisson's ratio, and density which are related to the waveguide shown in Table 1.

Geometric parameters depend on the model, which are mainly including the diameter and the length of the component as well as locations of incentive and receiving, as shown in Figure 2.

Discrete parameter is one of the most important parameters in finite element simulation of guided wave and determines the precision of the finite element model. There are unit length and time step included in the parameters, which should be suitable for the small calculation error and enough precision. Moser et al. [11] proposed that there are at least 20 units needed in each wavelength of guided wave in the propagating direction and time step should be less than 0.8 times of the required time that guided wave propagates through a unit. The expressions are shown in Table 2, where v is the speed of the guided wave propagating in the waveguide. According to Table 2, the axial length unit takes 2 mm and time step takes 0.1 μs in the model in this paper.

Incentive parameters mainly involve the mode and the frequency of excitation. Axial symmetric longitudinal modes are commonly used in the actual detection.

2.4. Excitation and Receiving of the Longitudinal Mode of Guided Wave in Reinforced Concrete. From the basic theory of guided wave, the axial order number for axial symmetric mode of guided wave is zero; consequently, the excitation and receiving are also axially symmetric. There are only the axial and radial displacements, no circumferential displacements for axial symmetric longitudinal mode. Research results show

that the longitudinal mode of guided waves can be excited in the waveguide when instantaneous axial displacement is loaded in all nodes of certain section in the model [19]. In order to eliminate the influence of flexural modes, axial displacements of all the nodes at the receiving location are added up and the better axial symmetric longitudinal mode is obtained. According to this method, the axial symmetric longitudinal mode is excited and received in the reinforcement, and the layout is shown as Figure 2.

According to the dispersion curve of group velocity for the steel bar, the sine signal modulated by Hanning window with ten cycles at the center frequency of 75 KHz is applied to all the nodes at the incentive location, which is along the axial direction of the bar. The model is shown as Figure 3.

After solving, add up the axial displacement of all nodes at the receiving location. The results are shown in Figure 4. From the curve of displacement versus time in Figure 4, it can be seen that there are two groups of sine signals with ten cycles which are the passing signal and end echo signal, respectively. The propagation distance of guided wave from passing signal to echo signal is 1600 mm, and the time interval is 0.3326 ms. Therefore, the velocity of the wave is calculated as 4810 m/s. From the dispersion curve of group velocity of the steel bar, it is known that there are only the modes L(0, 1) and F(1, 1) at the frequency of 75 KHz and the velocity of mode L(0, 1) is 4745 m/s, which is almost the same as the velocity from the above calculation. The results show that the axial symmetric longitudinal mode L(0, 1) of guided waves can be excited in the steel bar by applying instantaneous displacement on the end of it and taking

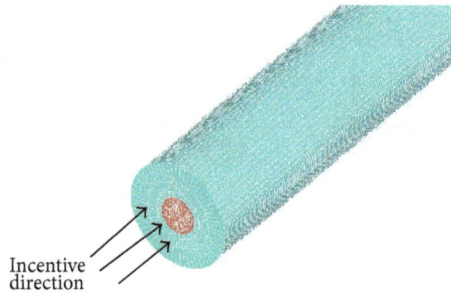

FIGURE 3: Scheme of axial symmetric longitudinal modal incentive.

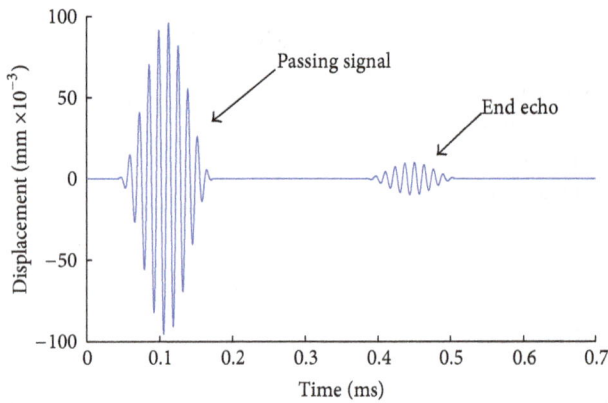

FIGURE 4: Waveform at receiving place after summation of all nodes axial displacement.

the axial displacement of nodes at the receiving location. It is also seen from Figure 4 that there is no other mode of waves besides the passing and echo signals, which show that the flexural mode F(1, 1) can be eliminated to get the pure mode L(0, 1) from the summation of displacement of nodes at the receiving location.

2.5. Pitting Corrosion Simulation. As mentioned earlier, the reflection signal of notch is the same as that of actual pitting corrosion, so the steel pitting corrosion is simulated by removing the unit at the corrosion location in the numerical simulation. Based on the established model, the member of reinforced concrete and the locations of defect and receiving position are designed as shown in Figure 2, which can avoid the overlay signals between the reflection from defect and the end.

The dimensions of defects are 2 mm in width and 1/8~1/2 times the perimeter in circumference length. The circumferential direction of the component is divided into 16 units and 4 units in radial direction. In this model the total number of units and nodes is 214292 and 315556, respectively. It takes 5 hours to work in the laboratory computer. The extraction model of steel bar from concrete is shown in Figure 5.

2.6. Simulation Results. The method is used to simulate the propagation of guided waves in the model with defect. Varying degrees of pitting are simulated by removing the steel

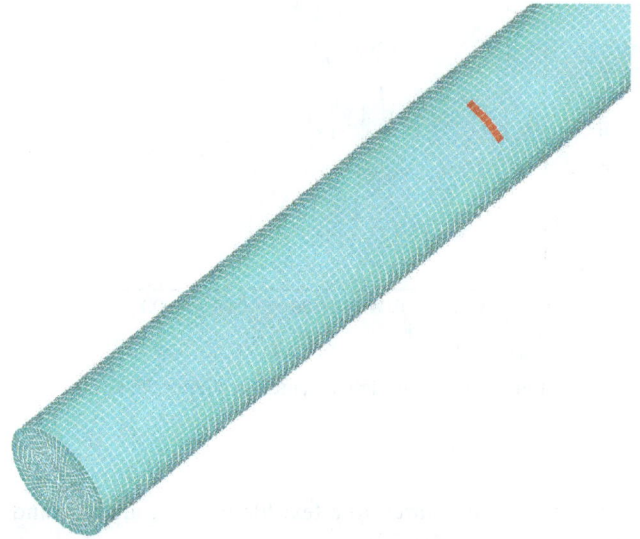

FIGURE 5: Finite element model of steel bar with defect.

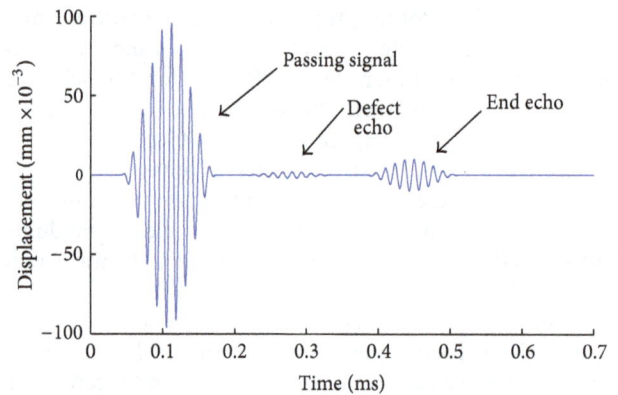

FIGURE 6: Detection signal with length of defect = 1/2 circumference.

units in the finite element model. Results are shown as Figures 6, 7, and 8. From Figure 6, the defect echo and end echo can be obviously identified. The damage location can be determined by the velocity of mode L(0, 1) at the frequency of 75 KHz and the propagating time of echo, which is the time interval between passing signal and the defect echo, that is, 0.206 ms in this case. Thus, the total propagating distance of defect signal is as follows:

$$S = C \times t = 4745 \times 2.06 \times 10^{-4} = 0.977 \, \text{m} = 977 \, \text{mm}. \quad (1)$$

The receiving location is 200 mm far from the left end of steel bar, so the location of defect x is known as

$$x = \frac{S + 200}{2} = 588 \, \text{mm}. \quad (2)$$

Compared with the exact location of the defect in the model, the calculation results are consistent with those of the preset defect (error is within 2%). Results show that it is feasible to simulate the detection of guided waves in reinforced concrete by FEM and the built model is correct.

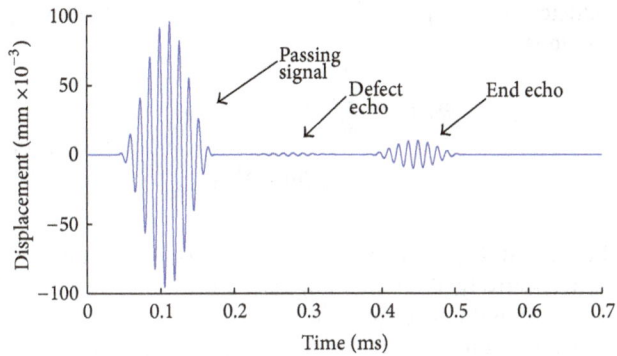

FIGURE 7: Detection signal with length of defect = 1/4 circumference.

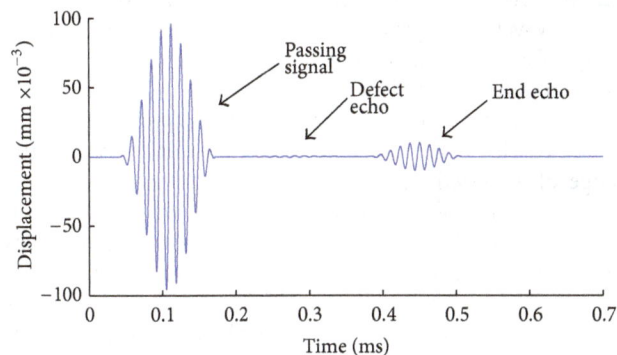

FIGURE 8: Detection signal with length of defect = 1/8 circumference.

However, it is difficult to identify the reflection signal of defect from Figure 8, because the signal is extremely weak and intertwined with other forms of faint waves. As shown in Figure 9, by amplifying directly the signals between 0.17 ms and 0.38 ms in the time domain, it can be seen that the refection wave is mixed together with a large number of narrow pulse type of clutter waves and the waveform is extremely too complex to distinguish the real reflection signal. Therefore, more detailed requirement in time domain or frequency domain is needed in adjusting the size of window of the function. Therefore the following method of wavelet analysis is to be used.

2.7. Identification of Weak Damage Signal by Wavelet Transform. When the guided-wave based technique is used to monitor corrosion defect in reinforced concrete, it is difficult to extract the weak reflection signal of corrosion defect from the detection signals by the general time-frequency method. At the same time, the background of collecting signals is influenced by various noises which is objective existence caused by the experiment environment. As a result it is difficult to intuit the effective identification of damage information from the detected signals. Wavelet transform is a kind of method for analyzing signal by 2 scales on time, which has the characteristics of multiresolution analysis and the very good characterization of local signal in time domain and frequency domain [20, 21]. Therefore special

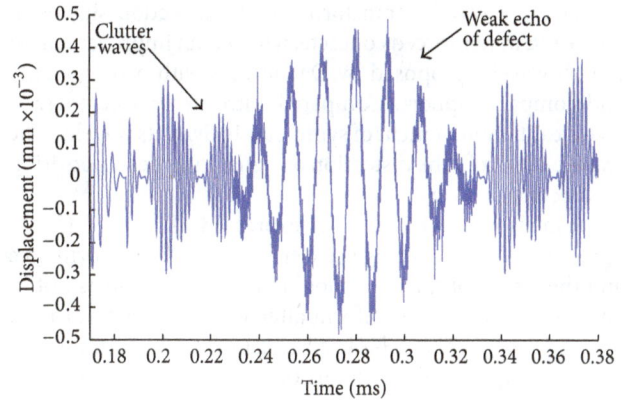

FIGURE 9: Amplification of the original signal.

signals with certain characteristics (such as specific frequency or waveform) can be carried out by this method to be amplified and the extraction of signal detection and damage identification can be achieved.

In fact, wavelet coefficient obtained by wavelet transform is the correlation coefficient between the original signal with wavelet basis function in the corresponding scale and position which reflects the similarity between them [22]. Thus, it is a primary issue to select the appropriate wavelet basis function before wavelet transform because in essence it is the projection from the original signal to a set of wavelet bases which have influence on the results of wavelet transform. According to the characteristics of the wavelet basis function [23], during the selection of wavelet bases, the following factors are usually considered: the computing speed, attenuation of wavelet base, precise reconstructing of signal, presence of phase distortion, characteristics of filtering, time-frequency resolution, orthogonality and how to reduce the edge effect, and so forth. In the analysis of high order singular signal, the disappearance of the wavelet moment must be also considered. The wavelet is used in this paper for the analysis of signals of corrosion monitoring in the reinforced concrete, so the aspects of reasonable selection of wavelet basis function mainly include compact support, vanishing moment, regularity, orthogonality, and waveform.

There are a lot of wavelet functions available, such as Gaussian, Mexican Hat, Daubechies, and Biorthogonal. Because every wavelet function has its own structure and characteristics, the analysis results are also different and it is unadvisable to analyze all the different issues by the same wavelet function. The properties of wavelet function should be paid attention to in the application, which are directly related to the time-frequency localization, singularity detection, decomposition, and precision of reconstruction. Therefore, when wavelet analysis technique is engaged in damage identification for structural health monitoring, the wavelet function should be fit for detection of local mutation signal based on the features and purposes of analysis signals [24].

Based on the above principles and the similarity and power spectral matching between wavelet function and detection signals into account, this paper adopts Symlets wavelet to make the wavelet transform for the detection signals of corrosion in reinforced concrete, which is an improvement of the db wavelet proposed by Daubechies with orthogonality and compact support. Compared with db wavelet, there is significant improvement of symmetry in Symlets wavelet. The feature can avoid the distortion of signal in its decomposition and reconstruction.

Symlets wavelet is usually expressed as sym N (N = 2, 3 . . . 8). When choosing wavelet function, the performance and the length of the filter should be considered so as not to affect the quality of waveform after wavelet transform [25]. Sym 1 wavelet cannot be used because its discontinuity of filter. Though the filter lengths of sym 2 and sym 3 wavelet are short, they are easily influenced by outside interference. The waveforms of scale function and wavelet function of sym 4 and sym 8 are shown in Figures 10(a)–10(d). It is found from the figures that not only the frequency characteristics of wavelet function and scale function of sym 8 are better than those of sym 4, but also the time domain resolution of sym 8 meet the requirements and its shape is more close to the actual curve. Therefore, the sym 8 wavelet with approximate symmetry, compact support, and biorthogonality is selected as the wavelet base to decompose the detected signal of corrosion monitoring in reinforced concrete.

One of the parameters to evaluate the characteristics of wavelet function is the scale, which is in an inverse relationship to the decomposition frequency of the transforming signal as shown in the following equation [26]:

$$F_a \times a \times \Delta = F_c, \tag{3}$$

where a is the wavelet scale, Δ is the sampling period, F_a is the frequency of the transforming signal, and F_c is the center frequency of wavelet function and decided by the choice of wavelet function. Because the center of frequency window is related to the wavelet function, the relationship between the scale and the center frequency depends on the wavelet function and the sampling period.

It can be seen from (3) that the greater the scale, the lower the frequency resolution of the analysis signal, and the more characteristics gained in low frequency and vice-versa. The mother wavelet will turn into a set of wavelet basis functions after translation and scaling and then does the dot product with the signals to be analyzed under different scales. After that, the continuous wavelet transform will be achieved as shown in the following equation:

$$\mathrm{CWT}_x(a, b) = \frac{1}{\sqrt{a}} \int_{-\infty}^{+\infty} x(t) \psi^* \left(\frac{t - b}{a} \right) dt, \tag{4}$$

where $\mathrm{CWT}_x(a, b)$ is the continuous wavelet transform, $\psi(t)$ is the basic wavelet or named mother wavelet, ψ^* is its conjugation, $x(t)$ is the signal to be analyzed, t is the time, a is the scale factor, and b is the time shift parameter.

The wavelet transform is usually conducted using fast algorithms by the software of Matlab in the practical application. The remaining coefficient and wavelet coefficient in the space of scale $j + 1$ will be got after making weighted summation of filter coefficients on the basis of the remaining coefficient in the space of scale j, as shown in the following equations:

$$d_{j+1,k} = \sum_m h_0 (m - 2k) c_{j,m},$$
$$c_{j+1,k} = \sum_m h_1 (m - 2k) c_{j,m}, \tag{5}$$

where $h_0(n)$, $h_1(n)$ are the filter coefficients, $d_{j+1,k}$ and $c_{j+1,k}$ are, respectively, the remaining coefficient and the wavelet coefficient in the space of scale $j + 1$.

After decomposition, the signals in the scale j are corresponding to the wavelet coefficients $c_{j,k}$. The information of signals is complete and all components are preserved due to the integrity of wavelet space, which provides the analysis of the characteristics of the signal energy distribution.

In wavelet analysis, it is the first consideration of the range of wavelet scale and its division because it will lead to low recognition rate if the scale of wavelet is chosen blindly [27]. Therefore, in order to analyze the response signals of structure with target in detailed using wavelet transform, the range of analyzed signal frequency should be determined to get the corresponding scale range of wavelet function according to (3). Therefore the number and size of scales in wavelet transform will greatly affect the results of the identification.

First, the waveform of sym 8 wavelet in the time domain can be calculated using the file named centfrq.m in Matlab [27], which determines the center of frequency window. Results are shown in Figure 11, in which sym 8 wavelet is shown in thick line and its center frequency is around 0.667 Hz. The waveform of sine wave in the same frequency is also shown in fine line for comparison.

In the case of unknown range of frequency in received signals, the number of scales of wavelet transform can be obtained by two ways. One is to estimate the range of signal frequency and then choose the step length of the analyzing frequency band. After dividing the range of frequency evenly, the size of scales of each band can be calculated from (3). The other way is to determine the size and number of the analyzing frequency band directly by evenly divided scales of wavelet, which will better reflect the band-pass filter function of wavelet transform. In this way, the energy characteristics of decomposed signal will be more obvious and the identification precision of signals can be significantly improved than the first one. Therefore, the evenly divided scales of wavelet are directly employed to get the number and size of analyzing band of signal frequency in this paper. The analysis range of the wavelet scales is 1 to 199, and the corresponding range of the signal frequency is 33.5 KHz to 6670 KHz which is consistent with estimated range. The step length of scale is 5; that is, the number of scale is 40. After wavelet transform of the signals in Figure 9, the results are shown in Figure 12.

After trial, it is found that it will make no contribution to improve significantly the identification precision of the signal by increasing the range of scales or decreasing the length of step, because it has met the need of signal analysis when the frequency range is corresponding to the scale range of

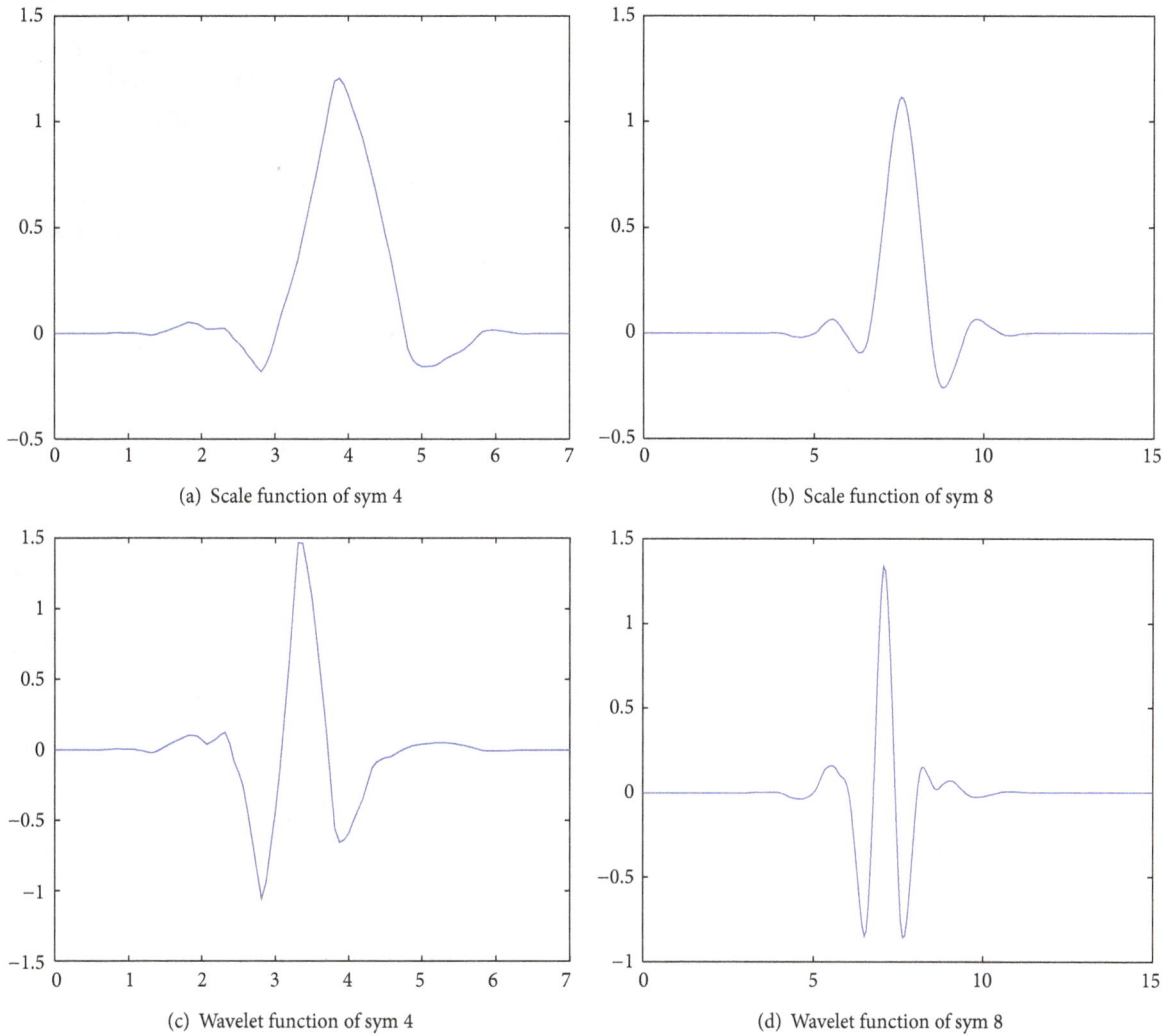

(a) Scale function of sym 4

(b) Scale function of sym 8

(c) Wavelet function of sym 4

(d) Wavelet function of sym 8

FIGURE 10: Wavelet function and scale function of sym 4 & sym 8.

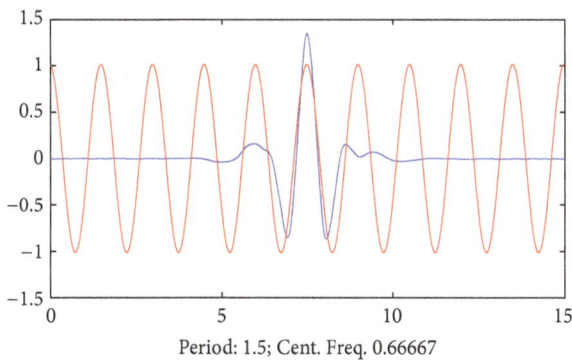

Period: 1.5; Cent. Freq. 0.66667

FIGURE 11: Waveform of sym 8 in time domain (shown in thick line).

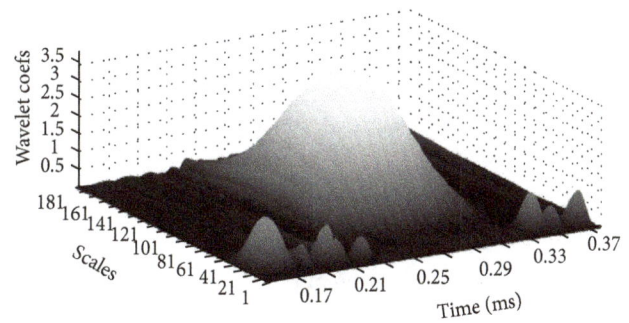

FIGURE 12: Wavelet transform of the detection signals at different scales.

wavelet. It will increase the number and the redundancy of detection signals and be unconducive to monitoring if the range of scales is increased or the length of step is decreased.

There are different time scales for different signals (such as noise, clutter, and defect reflection signal), so the results of wavelet transform will be different due to the chosen scale. In order to reflect the maximum similar degree between wavelet basis function and the defect reflection signal, that is, to get the maximum wavelet coefficient of defect reflection signal

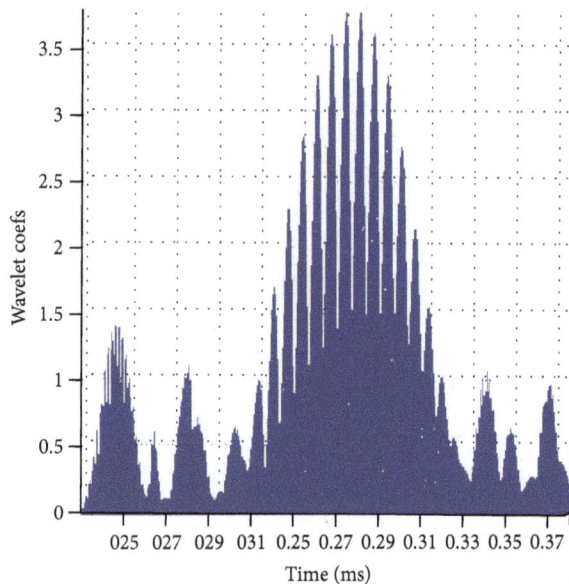

FIGURE 13: Detection signals after wavelet transform at scale of 89.

in the same scale, the corresponding scale can be obtained by the conversion of (3), as shown by

$$a = \frac{F_c}{(F_a \times \Delta)}, \tag{6}$$

where, sampling period Δ refers to the time step in numerical calculation, which is 0.1 μs in this case. F_c and F_a are the center frequency of wavelet function and telltale signal (i.e., defect echo), respectively, which are 0.667 Hz and 75 KHz, respectively, in this case. Thus the corresponding scale is 88.93 after calculation. Therefore, the scale of 89 is chosen and the waveform of wavelet transform to reflect the defect reflection signal is shown in Figure 13.

From the comparison of the results in Figure 13 with those in Figure 9, the defect reflection signal at the time of 0.28 ms from the detection signals which is obtained using appropriate scale by wavelet transform method can be clearly seen. The time of 0.28 ms is the same as the flight time that the echo of preset defect is needed, which proves that it is feasible and effective of the proposed method.

3. Conclusions

In this paper, the application of guided waves in monitoring corrosion of reinforced concrete is numerically simulated by finite element method. The methods for model establishment and critical parameters are analyzed by comparing the characteristic of each model. The approach for exciting and receiving guided wave with the longitudinal axial symmetric mode in finite element simulation is provided. Based on the preliminary numerical testing results, the pitting corrosion in the steel bar is simulated. Wavelet analysis, which has the time-frequency localization characteristic, is applied for the detection of weak damage signal resulting from energy leaking into concrete. The corrosion defects in reinforced

concrete can be identified by selecting suitable wavelet-based function and scale. The numerical simulation is validated by the satisfactory agreement between the simulation and actual results.

Conflict of Interests

The authors declare that there is no conflict of interests regarding the publication of this paper.

Acknowledgments

This research is supported by Key Science and Technology Project from Fujian Province, China (no. 2013Y0079) and the research fund (Grant no. SLDRCE10-MB-01) from the State Key Laboratory for Disaster Reduction in Civil Engineering at Tongji University, China. The authors also appreciate the kind assistance of Dr. Quan Gu at Xiamen University.

References

[1] S. Sharma and A. Mukherjee, "Nondestructive evaluation of corrosion in varying environments using guided waves," *Research in Nondestructive Evaluation*, vol. 24, no. 2, pp. 63–88, 2013.

[2] J. P. Broomfield, *Corrosion of Steel in Concrete: Understanding, Investigation and Repair*, Taylor & Francis, New York, NY, USA, 2nd edition, 2007.

[3] A. E. Aktan and K. A. Grimmelsman, "The role of NDE in bridge health monitoring," in *Nondestructive Evaluation of Bridges and Highways III*, vol. 3587 of *Proceedings of SPIE*, pp. 2–15, Denver, Colo, USA, July 1999.

[4] P. Matt, "Non-destructive evaluation and monitoring of post-tensioning tendons," in *Durability of Post-Tensioning Tendon*, vol. 15 of *FIB Bulletin*, pp. 100–108, 2001.

[5] Y. Lei and Z. P. Zheng, "Review of physical based techniques for monitoring corrosion in reinforced concrete," *Journal of Mathematical Problems in Engineering*, vol. 2013, Article ID 953930, 14 pages, 2013.

[6] B. L. Ervin and H. Reis, "Longitudinal guided waves for monitoring corrosion in reinforced mortar," *Measurement Science and Technology*, vol. 19, no. 1, Article ID 055702, 19 pages, 2008.

[7] W. D. Chen, *Finite element modeling of rod-shaped components detection with guided waves [M.S. thesis]*, Huazhong University of Science and Technology, 2009.

[8] Y. Cho and J. L. Rose, "A boundary element solution for a mode conversion study on the edge reflection of Lamb waves," *Journal of the Acoustical Society of America*, vol. 99, no. 4, pp. 2097–2109, 1996.

[9] X. Zhao and J. L. Rose, "Boundary element modeling for defect characterization potential in a wave guide," *International Journal of Solids and Structures*, vol. 40, no. 11, pp. 2645–2658, 2003.

[10] A. Demma, *The interaction of guided waves with discontinuities in structures [Ph.D. thesis]*, University of London, Imperial College of Science, Technology and Medicine, 2003.

[11] F. Moser, L. J. Jacobs, and J. Qu, "Modeling elastic wave propagation in waveguides with the finite element method," *Journal of NDT and E International*, vol. 32, no. 4, pp. 225–234, 1998.

[12] Z. B. Cheng, *Numerical simulation and experimental investigation on crack detection in pipes using ultrasonic guided waves [M.S. thesis]*, Taiyuan University of Technology, 2004.

[13] C. He, Y. Sun, Z. Liu, X. Wang, and B. Wu, "Finite element analysis of defect detection in curved pipes using ultrasonic guided waves," *Journal of Beijing University of Technology*, vol. 32, no. 4, pp. 289–294, 2006 (Chinese).

[14] W. G. Guo, Y. L. Li, and T. Suo, *Introduction of Basic Stress Wave*, Northwestern Polytechnical University Press, 2007.

[15] C. H. he, B. Wu, and J. W. Fan, "Advances in ultrasonic cylindrical guided waves techniques and their applications," *Journal of Advances in Mechanics*, vol. 3, no. 2, pp. 203–214, 2011.

[16] D. N. Alleyne and P. Cawley, "The interaction of Lamb waves with defects," *IEEE Transactions on Ultrasonics, Ferroelectrics, and Frequency Control*, vol. 39, no. 3, pp. 381–397, 1992.

[17] M. Redwood, *Mechanical Waveguides: The Propagation of Acoustic and Ultrasonic Waves in Fluids and Solids with Boundaries*, pp. 66–71, Pergamon Press, London, UK, 1960.

[18] D. N. Alleyne, M. J. S. Lowe, and P. Cawley, "The reflection of guided waves from circumferential notches in pipes," *Journal of Applied Mechanics*, vol. 65, no. 3, pp. 635–641, 1998.

[19] J. L. Rose, J. J. Ditri, A. Pilarski, K. Rajana, and F. Carr, "A guided wave inspection technique for nuclear steam generator tubing," *Journal of NDT and E International*, vol. 27, no. 6, pp. 307–310, 1994.

[20] H. Li, T. Yi, M. Gu, and L. Huo, "Evaluation of earthquake-induced structural damages by wavelet transform," *Progress in Natural Science*, vol. 19, no. 4, pp. 461–470, 2009.

[21] T. Yi, H. Li, and M. Gu, "Wavelet based multi-step filtering method for bridge health monitoring using GPS and accelerometer," *International Journal of Smart Structures and Systems*, vol. 11, no. 4, pp. 331–348, 2013.

[22] W. G. Wang and R. D. Zhang, "Multi-time-scale analysis of ground water level series with wavelet transform," *Engineering Journal of Wuhan University*, vol. 41, no. 2, pp. 1–4, 2008.

[23] Y. Z. Ji, *Numerical research of wharf pile foundation non-destructive test method based on 3D guided-wave theory [Ph.D. thesis]*, Tianjin University, 2010.

[24] W. X. Ren, J. G. Han, and Z. S. Sun, *The Application of Wavelet Analysis in Civil Engineering Structure*, China railway Publishing House, 2006.

[25] Z. Li, L. Cheng, and W. Tong, "Study of Symlets wavelet amplitude algorithm," *Journal of Electric Power Automation Equipment*, vol. 29, no. 3, pp. 65–68, 2009.

[26] Y. Wang, J. Q. Yang, and Q. C. Zhao, "Influence on steel structural damage identification with the selection of wavelet scales," *Journal of Shanxi Architecture*, vol. 33, no. 30, pp. 99–100, 2007 (Chinese).

[27] Research and Development Center of Science and Technology Product of Feisi, *Wavelet Theory and Matlab 7*, Publishing House of Electronics Industry, 2005.

Permissions

The contributors of this book come from diverse backgrounds, making this book a truly international effort. This book will bring forth new frontiers with its revolutionizing research information and detailed analysis of the nascent developments around the world.

We would like to thank all the contributing authors for lending their expertise to make the book truly unique. They have played a crucial role in the development of this book. Without their invaluable contributions this book wouldn't have been possible. They have made vital efforts to compile up to date information on the varied aspects of this subject to make this book a valuable addition to the collection of many professionals and students.

This book was conceptualized with the vision of imparting up-to-date information and advanced data in this field. To ensure the same, a matchless editorial board was set up. Every individual on the board went through rigorous rounds of assessment to prove their worth. After which they invested a large part of their time researching and compiling the most relevant data for our readers. Conferences and sessions were held from time to time between the editorial board and the contributing authors to present the data in the most comprehensible form. The editorial team has worked tirelessly to provide valuable and valid information to help people across the globe.

Every chapter published in this book has been scrutinized by our experts. Their significance has been extensively debated. The topics covered herein carry significant findings which will fuel the growth of the discipline. They may even be implemented as practical applications or may be referred to as a beginning point for another development. Chapters in this book were first published by Hindawi Publishing Corporation; hereby published with permission under the Creative Commons Attribution License or equivalent.

The editorial board has been involved in producing this book since its inception. They have spent rigorous hours researching and exploring the diverse topics which have resulted in the successful publishing of this book. They have passed on their knowledge of decades through this book. To expedite this challenging task, the publisher supported the team at every step. A small team of assistant editors was also appointed to further simplify the editing procedure and attain best results for the readers.

Our editorial team has been hand-picked from every corner of the world. Their multi-ethnicity adds dynamic inputs to the discussions which result in innovative outcomes. These outcomes are then further discussed with the researchers and contributors who give their valuable feedback and opinion regarding the same. The feedback is then collaborated with the researches and they are edited in a comprehensive manner to aid the understanding of the subject.

Apart from the editorial board, the designing team has also invested a significant amount of their time in understanding the subject and creating the most relevant covers. They scrutinized every image to scout for the most suitable representation of the subject and create an appropriate cover for the book.

The publishing team has been involved in this book since its early stages. They were actively engaged in every process, be it collecting the data, connecting with the contributors or procuring relevant information. The team has been an ardent support to the editorial, designing and production team. Their endless efforts to recruit the best for this project, has resulted in the accomplishment of this book. They are a veteran in the field of academics and their pool of knowledge is as vast as their experience in printing. Their expertise and guidance has proved useful at every step. Their uncompromising quality standards have made this book an exceptional effort. Their encouragement from time to time has been an inspiration for everyone.

The publisher and the editorial board hope that this book will prove to be a valuable piece of knowledge for researchers, students, practitioners and scholars across the globe.

List of Contributors

Jaan Hui Pu
School of Engineering, Nazarbayev University, 53 Kabanbay Batyr Avenue, Astana 010000, Kazakhstan

Songdong Shao
School of Engineering, Design and Technology, University of Bradford, West Yorkshire BD7 1DP, UK

Zhichao Zhang and Xiaohui Cheng
Department of Civil Engineering, Tsinghua University, Beijing 100084, China

Jianxiu Wang, Tianrong Huang and Dongchang Sui
Key Laboratory of Geotechnical and Underground Engineering of Ministry of Education, Tongji University, Shanghai 200092, China
College of Civil Engineering, Tongji University, Shanghai 200092, China

Chien-Ho Ko
Department of Civil Engineering, National Pingtung University of Science and Technology, 1 Shuefu Road, Neipu, Pingtung 912, Taiwan

Hao Lu, Mingyang Wang and Xiaoli Rong
College of Defense Engineering, PLA University of Science and Technology, Nanjing 210007, China

Baohuai Yang
Nanjing Kun Tuo Civil Engineering Technology Co. Ltd., Nanjing 210007, China

Hoai-Nam Ho, Young-Soo Park and Jong-Jae Lee
Department of Civil and Environmental Engineering, Sejong University, Seoul 143-747, Republic of Korea

Jong-Han Lee
Research and Engineering Division, POSCO E&C, Incheon 406-732, Republic of Korea

Jin Qin and Feng Shi
School of Traffic and Transportation Engineering, Central South University, Changsha 410075, China

Ling-lin Ni
Dongfang College, Zhejiang University of Finance and Economics, Hangzhou 310012, China

Xiao-ping Bai and Xi-wei Zhang
School of Management, Xian University of Architecture and Technology, Xian Shanxi 710055, China

Di Hou and Chuang-bing Zhou
State Key Laboratory of Water Resources and Hydropower Engineering Science, Wuhan University, Wuhan, Hubei 430072, China

Guan Rong
State Key Laboratory of Water Resources and Hydropower Engineering Science, Wuhan University, Wuhan, Hubei 430072, China
Earth Sciences Division, Lawrence Berkeley National Laboratory, Berkeley, CA 94720, USA

Guang Liu
State Key Laboratory of Water Resources and Hydropower Engineering Science, Wuhan University, Wuhan, Hubei 430072, China
Key Laboratory of Rock Mechanics in Hydraulic Structural Engineering, Ministry of Education, Wuhan University, Wuhan 430072, China

Tsai-Lung Weng
Physics Division, Tatung University, 40 Zhongshan North Road, 3rd Section, Taipei 104, Taiwan

An Cheng
Department of Civil Engineering, National Ilan University, 1 Shen-Lung Road, Ilan 260, Taiwan

Wei-Ting Lin
Department of Civil Engineering, National Ilan University, 1 Shen-Lung Road, Ilan 260, Taiwan
Institute of Nuclear Energy Research, Atomic Energy Council, Executive Yuan, Taoyuan 325, Taiwan

Huai-Shuai Shang
School of Civil Engineering, Qingdao Technological University, Qingdao 266033, China
State Key Laboratory of Structural Analysis for Industrial Equipment, Dalian University of Technology, Dalian 116024, China

Ting-Hua Yi
School of Civil Engineering, Dalian University of Technology, Dalian 116024, China

Lukasz Sadowski
Institute of Building Engineering, Wroclaw University of Technology, Plac Grunwaldzki 11, 50 377Wroclaw, Poland

Jianxiu Wang, Tianrong Huang and Xiaotian Liu
College of Civil Engineering, Tongji University, Shanghai 200092, China

Zhiying Guo
College of Marine Environment and Engineering, Shanghai Maritime University, Shanghai 200135, China

Pengcheng Wu
College of Marine Environment and Engineering, Shanghai Maritime University, Shanghai 200135, China
Shanghai International Shipping Service Center Development Co., Shanghai 200120, China

Chien-Ho Ko
Department of Civil Engineering, National Pingtung University of Science and Technology, 1 Shuefu Road, Neipu, Pingtung 912, Taiwan

Qun Chen
School of Traffic and Transportation Engineering, Central South University, Railway Campus, Changsha 410075, China

Yuzhi Li
Changsha University of Science and Technology, Changsha 410076, China

Hong-Qiang Fan, Bin Jia, Xin-Gang Li, Jun-Fang Tian and Xue-Dong Yan
MOE Key Laboratory for Urban Transportation Complex Systems Theory and Technology, Beijing Jiaotong University, Beijing 100044, China

Xiaoyan Zhang, Zheming Zhu and Hongjie Liu
Department of Engineering Mechanics, Sichuan University, Chengdu 610065, China

Chengyin Liu
Harbin Institute of Technology, Shenzhen Graduate School, Shenzhen 518055, China
Key Laboratory of C&PC Structures, Southeast University, Nanjing 211189, China

Ning Wu and Chunyu Liu
Harbin Institute of Technology, Shenzhen Graduate School, Shenzhen 518055, China

Xiang Wu
State Key Laboratory of Robotics and System, Harbin Institute of Technology, Dazhi Street, Nangang District, Harbin 150001, China

Xu Wang
State Key Laboratory Breeding Base of Mountain Bridge and Tunnel Engineering, Chongqing Jiaotong University, Chongqing 400074, China

Bin Chen
College of Civil Engineering and Architecture, Zhejiang University, Hangzhou 310058, China

Dezhang Sun
Institute of Engineering Mechanics, China Earthquake Administration, Harbin 150080, China

Yinqiang Wu
Bureau of Public Works of Shenzhen Municipality, Shenzhen 518006, China

In-Mo Lee, Jong-Sun Kim and Jong-Sub Lee
School of Civil, Environmental and Architectural Engineering, Korea University, Seoul 136-701, Republic of Korea

Hyung-Koo Yoon
Department of Geotechnical Disaster Prevention Engineering, Daejeon University, Daejeon 300-716, Republic of Korea

Wenliang Qiu, Meng Jiang and Cailiang Huang
School of Civil Engineering, Dalian University of Technology, Dalian 116024, China

Zhupeng Zheng, Ying Lei and Xin Xue
Department of Civil Engineering, Xiamen University, Xiamen 361005, China

www.ingramcontent.com/pod-product-compliance
Lightning Source LLC
Chambersburg PA
CBHW080257230326
41458CB00097B/5093

9 781632 402585